Aquatic Environmental Bioengineering

Aquatic Environmental Bioengineering

Monitoring and Remediation of Contamination

Rouf Ahmad Bhat
Sri Pratap College
Jammu and Kashmir, India

Mohammad Yaseen Mir
Dept. of School Education
Govt of Jammu and Kashmir, India

Gowhar Hamid Dar
Sri Pratap College
Jammu and Kashmir, India

Moonisa Aslam Dervash
Sri Pratap College
Jammu and Kashmir, India

WILEY

This edition first published 2022

Registered Office(s)
John Wiley & Sons, Inc., 111 River Street, Hoboken, NJ 07030, USA
John Wiley & Sons Ltd, The Atrium, Southern Gate, Chichester, West Sussex, PO19 8SQ, UK

Editorial Office
The Atrium, Southern Gate, Chichester, West Sussex, PO19 8SQ, UK

For details of our global editorial offices, customer services, and more information about Wiley products visit us at www.wiley.com.

Wiley also publishes its books in a variety of electronic formats and by print-on-demand. Some content that appears in standard print versions of this book may not be available in other formats.

Library of Congress Cataloging-in-Publication Data
Names: Bhat, Rouf Ahmad, 1981- author. | Yaseen Mir, Mohammad, author. | Dar, Gowhar Hamid, author. | Dervash, Moonisa Aslam, author.
Title: Aquatic environmental bioengineering : monitoring and remediation of contamination / Rouf Ahmad Bhat, Mohammad Yaseen Mir, Gowhar Hamid Dar, Moonisa Aslam Dervash.
Description: Hoboken, NJ : John Wiley & Sons, [2022] | Includes bibliographical references and index.
Identifiers: LCCN 2021035587 (print) | LCCN 2021035588 (ebook) | ISBN 9781119760948 (hardback) | ISBN 9781119760962 (pdf) | ISBN 9781119760955 (epub) | ISBN 9781119760979 (ebook)
Subjects: LCSH: Water--Purification. | Water--Pollution.
Classification: LCC TD430 .B47 2022 (print) | LCC TD430 (ebook) | DDC 628.1/62--dc23
LC record available at https://lccn.loc.gov/2021035587
LC ebook record available at https://lccn.loc.gov/2021035588

Cover image: © Suriyapong Thongsawang/Getty Images
Cover design by Wiley

Set in 9.5/12.5pt STIXTwo Text by Integra Software Services Pvt. Ltd, Pondicherry, India.

10 9 8 7 6 5 4 3 2 1

Contents

Preface

The remediation of contaminations of aquatic ecosystems is one of the important topics of environmental restoration that is evolving more rapidly. The contamination of water bodies stems primarily from past and present anthropogenic activities and it presently constitutes one of the greatest environmental liabilities for future generations to bear. The significant categories of contaminants usually found in the contaminated sites are halogenated organic compounds, petroleum hydrocarbons, radionuclides, metals and metalloids, pharmaceuticals drugs, microbial toxins and flame retardants.

This book provides outstanding descriptions about the environmentally safe and economically feasible technologies for treatment of contaminated aquatic ecosystems. The process of remediation of containments from aquatic ecosystems involves a phased approach comprising site characterization, risk assessment and remediation technology selection and application. Taking advantage of omic technologies and recent advancements in biotechnology, it has become possible to develop novel transgenic plants which are highly efficient in phytoremediation.

Furthermore, the advancements achieved in the field of remote sensing and geographic information (GIS) has made it possible to conduct large-scale water remediation studies. Using spatial and spectral resolution sensors and geospatial modeling techniques, the water quality of aquatic ecosystems can be assessed at very low cost and with greater accuracy. Microbial fuel cells have also emerged as a promising technology to mitigate environmental pollution, providing a solution to wastewater treatment and the removal and/or recovery of heavy metals. This book will supplement data for budding researchers by providing extensive knowledge regarding remediation techniques of aquatic ecosystems. The book will also provide a simple anthology for undergraduate and postgraduate students to understand environmental engineering with regards to wastewater treatment.

The fourteen chapters address themes such as emerging pollutants remediation water systems, genetic engineering for metal tolerance and accumulation, transgenics approaches for field testing and risk assessment, GIS and remote sensing, photocatalysis, bioelectrochemical systems, environmentally sustainable solutions for wastewater treatment, bioenergy plants for phytoremediation in water systems, biofilms in remediation, graphene-based adsorbents and sewage sludge.

We are extremely grateful to Wiley for their generous cooperation in publishing this book.

Dr. Rouf Ahmad Bhat
Dr. Mohammad Yaseen Mir
Dr. Gowhar Hamid Dar
Dr. Moonisa Aslam Dervash

About the Authors

Dr. Rouf Ahmad Bhat (PhD) pursued his doctorate at Sher-e-Kashmir University of Agricultural Sciences and Technology Kashmir (Division of Environmental Science) and presently works in the Department of School Education, Government of Jammu and Kashmir. Dr. Bhat has been teaching graduate and postgraduate students of Environmental Sciences for the past three years. He is an author of more than 50 research articles and 40 book chapters, and has published over 28 books with international publishers (Springer, Elsevier, CRC Press Taylor & Francis, Apple Academic Press, John Wiley & Sons, IGI Global). He specializes in limnology, toxicology, phytochemistry and phytoremediation. Dr. Bhat has presented and participated in numerous state, national and international conferences, seminars, workshops and symposia. He has worked as an associate environmental expert for the World Bank-funded Flood Recovery Project, and also with the environmental support staff in the Asian Development Bank (ADB)-funded development projects. He has received many awards, appreciations and recognition for his services to the science of water testing, air and noise analysis. He has served as editorial board member and reviewer on reputed international journals. Dr. Bhat continues to write, and to experiment with diverse capacities of plants for use in aquatic pollution remediation.

Dr. Mohammad Yaseen Mir (PhD) is Senior Researcher at the Center of Research for Development, University of Kashmir Srinagar, Jammu and Kashmir India. Dr. Mir has worked as a senior researcher in research projects funded by the Ministry of Environment, Forest and Climate Change, Government of India, along with G.B. Pant National Institute of Himalayan Environment & Sustainable Development under National Mission on Himalayan Studies (NMHS), and has recently joined the Department of Education J&K as a teacher. He gained his MPhil and PhD qualifications from the University of Kashmir. He has published many research articles in reputed, refereed international and national journals including Springer/Elsevier/Hindwaii. He has also authored many books on themes such as sustainable agriculture, biotechniques in plant biology and nanotechnological intervention in agricultural productivity.

Dr. Gowhar Hamid Dar (PhD) is currently working as an Assistant Professor in Environmental Science, Sri Pratap College, Cluster University Srinagar, Department of Higher Education (J&K). He has a PhD in Environmental Science with specialization in Environmental Microbiology (fish microbiology, fish pathology, industrial microbiology, taxonomy and limnology). Over the past few years he has taught graduate and postgraduate students at the Postgraduate Department of Environmental Science, Sri Pratap College, Cluster University Srinagar. Dr. Dar has over 45 research articles published in international and national journals of repute; also several books with international publishers to his credit. Moreover, he is supervising a number of students for the completion of their degrees. Previously, he finished his PhD programme entitled "Profiling and effect of pathogenic bacteria in fishes of Jammu and Kashmir State and correlation of disease incidence to the trophic status of water bodies" from the Department of the Environmental Science University of Kashmir, India, and his MPhil in Environmental Science on "Isolation, identification and characterization of some *Penicillium* isolates for alpha-amylase activity" from the same department. For the past several years he has been working on the isolation, identification and characterization of microbes, their pathogenic behavior, and also the impact of pollution on development of diseases in fish fauna. He has received many awards and appreciations for his services towards science and development. He also acts as a member of various research and academic committees.

Dr. Moonisa Aslam Dervash (PhD) is presently working as an Assistant Professor in the Department of Environment and Water Management and Department of Environmental Science, S.P. College, Cluster University Srinagar (J&K). She is the author of several books, and her scientific research articles and book chapters appear in journals of national and international repute. Dr. Dervash has participated and presented her papers in national and international scientific conferences. She is the recipient of many prestigious awards including the Young Woman Achiever's Felicitation 2021 (J&K Govt), Bharat Excellence Award (Friendship Forum of India, FFI), Young Educationist Award (FFI) and honorary DLitt (*Vidya Sagar*) from Vikramasheela Vidyapeeth, Bhagalpur, India for her literary works. Her specializations include environmental monitoring, climatology and carbon sequestration, aquatic ecology, terrestrial ecology, sustainable agriculture and environmental education. Her focus has remained on many facets of society especially on environmental conservation and women's empowerment. She has been felicitated by the state government (J&K Department of Ecology, Environment and Remote Sensing) for her outstanding contribution to environmental awareness and conservation through electronic media.

1

Emerging Pollutants Remediation Water Systems

Biomass-Based Technologies

CHAPTER MENU

1.1 Introduction

Emerging Pollutants (EPs) or pollutants of emerging concern are labels for a class of chemicals that are intrinsically detrimental to the environment but not essentially lethal [1]. This assemblage of chemicals incorporates an extensive assortment of substances including human and veterinary drugs, personal care products (PCPs), disinfection by-products (DBPs), polycyclic aromatic hydrocarbons (PAHs), perfluorinated compounds (PFCs), heavy metals, surfactants, pesticides, flame retardants, algal contaminants and hormones besides numerous other groups of chemicals. Throughout the preceding few decades, several investigative programmes have been going on to explore and examine the existence and fate besides ecotoxicology, of a lot of emerging pollutants (EPs). The EPs are sampled out from soils [2], sediments [3], wastewaters plus other aquatic

Aquatic Environmental Bioengineering: Monitoring and Remediation of Contamination, First Edition.
Rouf Ahmad Bhat, Mohammad Yaseen Mir, Gowhar Hamid Dar and Moonisa Aslam Dervash.

environs [4]. As far as distribution of EPs is concerned traces of these have been reported from practically all the domains of our environment [5]. These EPs have also been found in the bodies of newborn infants proving their grave peril to the well-being of humans and other organisms alike. Normal treatment procedures are incompetent to eliminate these EPs, resulting in their discharge into receiving aquatic environments. However, a thorough and all-encompassing scientific research concentrated on the investigation of EPs source, presence, effects; and its treatment methods are not well established [6].

EPs are highly reactive and mobile due to which its elimination from aquatic settings is often an uphill task. It is pertinent to mention that minute quantities of EPs can pollute a significant bulk of aquatic ecosystems. Furthermore, EP-laden wastewaters can't be efficiently treated by already existing treatment operations which results in an augmented jeopardy due to their persistence in the environs [7, 8]. Consequently, the existence of EPs in wastewater management facilities is gradually developing into an unintended birthplace of severe contamination hazard to the waterways. Therefore, developing state-of-the-art procedures for management of waters tainted with EPs is an unrelenting and everlasting requisite. In this concern, a number of techniques have been explored and developed for eliminating and mitigating the deleterious effects of EPs on aquatic ecosystems through various physical, chemical and biological approaches. Certain methods are working well, however are costly, while other methods are equally capable and economical but detrimental to the environment. In this backdrop, water management techniques using biomass and related derivatives as a chief element are prospective of remediating the predicament of a number of EPs (Table 1.1). The biomass-based technologies are more efficacious, economical as well as eco-friendly as compared to existing treatment methods. This chapter delivers an overall gist of diverse pioneering biological methods for eliminating an array of EPs as well as outlining forthcoming exploration trends.

Table 1.1 Biosorption of varied biomasses to eliminate certain EPs from contaminated environments.

Emerging pollutant	Biosorbent	Reference
Pharmaceuticals		
• Paracetamol	Sugarcane bagasse	[9]
• Metronidazoles	Siris seed pods	[10]
Endocrine-disrupting chemicals		
• Nonylphenol	*Rhizopus arrhizus*	[11]
Radionuclides		
• Uranium VI	Wheat straw	[12]
• Cesium	*Citrobacter freudii* (bacteria)	[13]
	Fusarium Sp. (fungi)	[14]
	Basil seeds	[15]
	Walnut shells	[16]
Algal toxins		
• Microcystin-LR	Peat	[17]

1.2 Adsorption-Based Remediation

This water treatment method characterizes a vital part of EP remediation systems. Four major processes are discussed here, viz. sorption, adsorption, biosorption and bioadsorption. The terms adsorption and adsorbent refer to the sorbing substances that aren't biomass-based. This involves activated carbons and biopolymers (biomass derivatives), as well as silica, zeolites, clays and synthetic polymers. The terms biosorption and biosorbent (raw or pretreated) encompass the application of dead ("bioadsorption") and standing biomass ("passive" biosorption and/or "active" bioaccumulation).

1.2.1 Biomass

Several types of biosorbents "plants, algae, bacteria, fungi, yeasts" have been quantified in the sources over many years due to their potential for eliminating diverse toxic pollutants. Although EPs are discovered in diverse environments and pose a number of severe undesirable effects on our planet, they aren't presently counted in standard environmental assessment options. There exists a knowledge gap concerning the level of research conducted on the already existing pollutants and the newly emerging pollutants. A scientific supposition stating that conventional contaminants hold greater and confirmed toxicity threat in contrast to the EPs isn't correct at all times. In fact, during the last few years research investigations vis-à-vis EPs have increased exponentially [18].

1.2.2 Terrestrial and Marine Bioresources

Azoimide (hydrazoic acid) is an extremely toxic substance polluting hospital wastes [19]. In order to find a viable way out to deal with this toxic substance, its removal by powdered almond integument from contaminated waters was carried out. It was found that 1 g of almond biomass removed 45 mg of the azoimide toxin [20]. The release of hazardous wastes from nuclear power plants has been a serious cause of concern. To find a solution, investigators explored the adsorption efficacy of cactus fibers (*Opuntia ficus*) to get rid of uranium VI from discharged nuclear wastes and they were quite successful in it [21]. Similarly, the potential of *Posidonia oceanica* for removing surfactants from aqueous solutions was evaluated. The marine algae *Padina pavonia* was used as a biosorbent for the remediation of uranium (VI) from wastewaters. It was discovered that the marine biomass was effective in eliminating uranium with 98% effectiveness [22].

1.2.3 Agro-Industrial Wastes

In Turkey, the biosorption of Gallium III from polluted waters was accomplished via tea waste. It was discovered that this tea waste was effective in removing 77.4% of waste present in the solution [23]. Rice straw was utilized as a biosorbent to eliminate pharmaceuticals [24] which are often identified in wastewaters [25]. The findings confirmed that rice straw was able to remove the pharmaceuticals by means of adsorption. Similarly, a research group from Brazil was successful in removing Diclofenac from aqueous solutions via Isabel grape bagasse [26].

1.2.4 Activated Carbons (ACs)

1.2.5 Bioresources

Various biological precursors are used for manufacturing extremely absorbent activated carbons which can further be used for tackling EPs. An innovative AC was manufactured using Artemisia vulgaris for removing the widely used drug Ibuprofen from aqueous solutions. It was established that A. vulgaris-derived AC removed 17 mg/g of the drug [27]. ACs have also been produced from the leaves of the date palm (*Phoenix dactylifera*). The ACs formed from it were used for elimination of an antibiotic, ciprofloxacin, from contaminated waters [28].

1.2.6 Agro-Industrial Wastes

Chemically activated biochar formed from loblolly pine chips was used for the adsorption of several EPs from contaminated waters. The adsorption findings revealed an enhanced capacity for removing EPs [29]. In another investigation researchers examined the potential of ACs made from sesame stalk derived for removing Phenanthrene. The adsorbent displayed a remarkable efficacy to remove this EP [30].

1.2.7 Activated Sludge (AS)

Activated sludge is often useful in wastewater management plants. Huge quantities of AS are generated regularly, therefore its utilization for removing EPs from aqueous environs has been deliberated upon. Perfluorocarbons are incessant EPs and for tackling this crisis, AS was used in aqueous solution. The results showed that AS was capable of eliminating the EPs with a higher efficacy as compared to other techniques [30].

1.3 Bioremediation

Bioremediation utilizes the intrinsic abilities of particular plants and microbes to biologically eliminate or stop the action of lethal substances from polluting our environment. Bioremediation is also useful to remove EPs from aquatic environs through plants (i.e. phytoremediation) and microbes (i.e. microbial remediation).

1.3.1 Phytoremediation

Callus cultures of the poplar tree (*Populus nigra*) have been inspected for removal of a commonly used drug, Ibuprofen. It has been stated that Ibuprofen was eliminated entirely by poplar cells [31]. In another research vetiver grass was reported to be able to successfully remove cesium and strontium from artificially polluted waters and also from nuclear wastes [32]. Similarly, researchers were successful in using *Eleocharis acicularis* for removal of toxic metals from polluted waters in Japan [33]. Water hyacinth biomass was used for the removal of radioactive waste consisting of cesium and cobalt residues. The chief findings indicated that the Cs uptake by this macrophyte increased with increased biomass amount and sun contact [34].

1.3.2 Constructed Wetlands (CWs)

CWs are fundamentally advanced treatment structures to aid in the natural process of bioremediation. Significant research and development efforts have been carried out using CWs to get rid of EPs in a natural and eco-friendly way. CWs have been used for bioremediation of various EPs such as pharmaceuticals, personal care products, pesticides etc. The removal efficacy varies with the type of EP. Also, the bioremediation processes in CWs include coordinated biodegradation, biosorption and photodegradation [35].

1.3.3 Microbial Remediation

Microbes have been equally useful in bioremediation of EPs as with any other organism. Fungus T. versicolor was used to get rid of many classes of EPs including pharmaceuticals, brominated flame retardants (BFRs) and UV filters. Dioxins and dioxin-like compounds were removed from wastewaters using three brown-rot fungi "*Aspergillus aculeatus, A. flavus, A. fumigatus*" and one white-rot fungus "*Phanerochaete chrysosporium*" [36]. In the same way, a metal-resistant bacteria, Bacillus cereus, was successful in bioremediation of waters polluted by "decabromo diphenyl ether", which is a poisonous polybrominated flame retardant [37]. Thankfully, in our environment, the degradation of contaminants is typically by reason of mutual or synchronized enzymatic activities of numerous microbes, mainly bacteria, microfungi and microalgae. These microbes play a vital role in cleaning up environments polluted by humans.

1.3.4 Biocoagulants and Bioflocculants

Coagulation and flocculation are one of the central steps in wastewater treatment plants (WWTP). As mentioned earlier, WWTP are also developing into an additional source of EPs in water bodies. Numerous investigations have deliberated upon the generation of biomass-based flocculants and coagulants to remove EPs. As a result, many impending bioresources have been investigated to generate recyclable and economical flocculating/coagulating agents to deal with EPs [38]. Figure 1.1 represents a common three-phase extraction and purification arrangement for biocoagulants, as shown by Yin [39]. Moringa tree (*Moringa oleifera*) seeds have demonstrated a remarkable coagulation activity to remove surfactants from contaminated waters with 80% efficiency predominantly via the biocoagulation/bioflocculation method [40]. A biopolymeric flocculant, "MBF-5" made from the *Klebsiella pneumoniae* strain was tested to tackle pathogenic, disinfection-resistant Acanthamoeba cysts from wastewaters [41]. A whopping rate of 79.4% flocculation of this pathogen was attained using the *K. pneumoniae*-based bioflocculant [42]. Sulfamethoxazole, a widely used antibiotic, is one of the most widely found drug products in wastewaters [43]. This EP was removed with 53.27% efficiency using bioflocculant MFX, produced from Klebsiella sp [44].

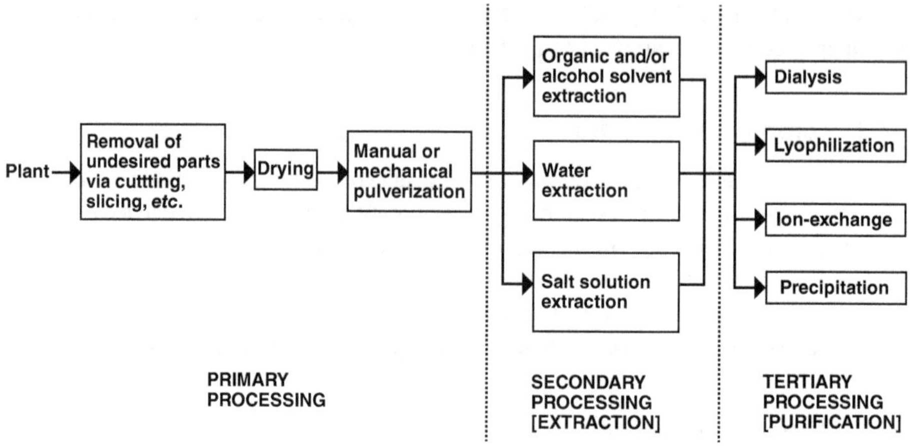

Figure 1.1 Steps involved in preparation of plant-based coagulants [39].

1.4 Multi-Element Water Treatment Process

In this section, multi-element water management methods concerning the application of biomass-based techniques for EPs remediation will be elaborated. The chief objective is to demonstrate the merging of decontamination techniques as a central approach to maximize the benefits.

1.4.1 Membrane Bioreactors (MBRs): Biodegradation and Membrane Filtration

The MBR technique for wastewater management is a precisely guaranteed method. This technique combines biodegradation with membrane filtration of wastes and can be used effectively to get rid of EPs from aquatic environments [45]. The chief leads of MBRs are improved treated wastewater quality and a reduced quantity of sludge output besides advanced functioning to treat EPs with increased elimination efficacy [45-46]. In Japan exclusion of EPs (pharmaceuticals) from wastewaters was achieved through MBRs. The tested MBRs displayed superior EP removal levels for the scrutinized drugs [47]. In a different research study, the removal of two perfluorinated compounds was achieved using activated carbon and the MBR system with an efficiency of 90% [48].

The biological combination is one of the highly noteworthy methods in the field of EP treatment. This association is highly versatile, environmentally sustainable, besides being economical. In every wastewater treatment plant, both adsorption and biodegradation play a major role in removal of contaminants. Adsorption aids in biodegradation which helps in total elimination of EPs. Biodegradation helps in adsorption of pollutants in addition to removing contaminants from the aqueous solution which are capable of constraining microbial growth or action. In a study, biosorption and biodegradation were helpful in removing Triphenyltin, an extremely poisonous EP with the aid of the bacteria *Brevibacillus brevis*. This bacterium adsorbed 97% of the toxin; while as for biodegradation, an elimination efficacy of 60% was recorded [49].

1.4.2 Activated Carbon and Ozone

Water treatment techniques developed from the linking of adsorption and ozonation have successfully demonstrated the removal of numerous kinds of contaminants [50]. This combined method of enhancing removal efficiency demands to midpoint on a particular contaminant and check its elimination levels by means of specific as well as collective approaches. It has been proven that the combination of ozonation and adsorption onto activated carbon is an efficient method to decrease the concentration of micro-pollutants in contaminated waters [51,52].

1.5 Views and Recommendations

By and large, the research endeavours in the domain of water and wastewater management are based on the development of competent, economical and environmentally friendly approaches. With respect to biomass and its eminent benefits (i.e. renewability, accessibility, incredible generation rates, plus biochemical abundance), any wastewater treatment technique using biomass or its derived products as a primary constituent in the treatment method needs to be adaptable and to come with additional improvements. Accordingly, biomass as the nucleus of any decontamination approach will facilitate investigators to pursue the accurate orderliness of importance in their approaches, which are: being environmentally friendly, efficient and economical. Certainly, in formulating a sustainable EPs elimination technique, we can't rescue the environs by substituting one type of contamination with a new one. In this scenario the usage of biomass is highly valuable due to its varied decontamination processes. As soon as the ecological facet of the biomass-centered method is assured, the spotlight will be on optimizing the water treatment procedure itself. Numerous choices are available to deal with EPs because of the accomplishments of worldwide research groups. However, the costs need to be appropriated to adjust the decontamination processes realistically. In this regard, the utilization of biomass is exceedingly profitable as a renewable and economical alternative.

1.6 Conclusion

In facing the whole host of organic and inorganic contaminants of emerging concern that are hazardous to the aquatic environs, multiple research studies were conducted to alleviate this grave issue. Biomass-centered cleansing approaches are among the most remarkable water treatment methodologies. An assortment of resources was explored globally for the expansion of competent, environmentally friendly methods to fight the menace of EPs. In this context, biomass-based remediation techniques hold an important collection of techniques for EP elimination from aquatic environs. Nevertheless, it is necessary to mention that very few of the reviewed research findings incorporated a commercial application. Bioadsorption onto bioresources, organic wastes, and derived activated carbons and biopolymers were utilized for the management of wastewaters containing radioactive wastes, pharmaceuticals, polyaromatic hydrocarbons etc. with encouraging outcomes. Further methods to deal with the subject of water pollution by EPs comprise bioremediation. This is centered on the ability of plants and microbes to disintegrate, amass, diminish, precipitate or restrain noxious pollutants. In addition, numerous research activities projected multicomponent water treatment approaches founded on the management of biomass entirely or when merged with other nonbiological substances. The associated results disclosed that this mutual approach augmented the elimination of EPs from wastewaters. This was the case for membrane bioreactors (membrane filtration/microbial degradation) or collective adsorption/biodegradation, which illustrated fascinating cleansing potentials towards pharmaceuticals, pesticides, toxic polychlorinated compounds and varying contaminants of mounting concern. Generally, this technique is in the budding stage as major detailed research studies are executed in research facilities only and chiefly bounded to distinct constituent pollution in aqueous solutions. Scarcely any major investigations have been conducted with actual wastewater or other multifaceted real-case set-ups via these resources to minimize contaminants in wastewaters. Therefore, these biomass-based remediation techniques for EPs need to be attested in outsized pilot-scale systems.

References

1 Daughton, C.G. (2004). Non-regulated water contaminants: Emerging research Environ. *Impact Assess. Rev.* 24: 711–732.

2 Caliman, F.A. and Gavrilescu, M. (2009). Pharmaceuticals, personal care products and endocrine disrupting agents in the environment: A review. *CLEAN – Soil Air Water* 37: 277–303.

3 Bigus, P., Tobiszewski, M., and Namiesnik, J. (2014). Historical records of organic pollutants in sediment cores. *Mar. Pollut. Bull.* 78: 26–42.

4 Stuart, M., Lapworth, D., Crane, E., and Hart, A. (2012). Review of risk from potential emerging contaminants in UK groundwater. *Sci. Total Environ.* 416: 1–21.

5 Naidu, R. and Wong, M.H. (2013). Contaminants of emerging concern. *Sci. Total Environ.* 463–464: 1077–1078.

6 Chaalal, O., Zekri, A.Y., and Soliman, A.M. (2015). A novel technique for the removal of strontium from water using thermophilic bacteria in a membrane reactor. *J. Ind. Eng. Chem.* 21: 822–827.

7 Stackelberg, P.E., Gibs, J., Furlong, E.T., et al. (2007). Efficiency of conventional drinking-water-treatment processes in removal of pharmaceuticals and other organic compounds. *Sci. Total Environ.* 377: 255–272.

8 Westerhoff, P., Yoon, Y., Snyder, S., and Wert, E. (2005). Fate of endocrine-disruptor, pharmaceutical, and personal care product chemicals during simulated drinking water treatment processes. *Environ. Sci. Technol.* 39: 6649–6663.

9 Ribeiro, A.V.F.N., Belisario, M., Moretto Galazzi, R., Cazoni Balthazar, D., Pereira, M.D.G., and Ribeiro, J.N. (2011). Evaluation of two bioadsorbents for removing paracetamol from aqueous media. *Electron. J. Biotechnol.* 14: 1–10.

10 Ahmed, M.J. and Theydan, S.K. (2013). Microwave assisted preparation of microporous activated carbon from siris seed pods for adsorption of metronidazole antibiotic. *Chem. Eng. J.* 214: 310–318.

11 Lang, W., Dejma, C., Sirisansaneeyakul, S., and Sakairi, N. (2009). Biosorption of nonylphenol on dead biomass of rhizopus arrhizus encapsulated in Chitosan Beads. *Bioresour. Technol.* 100: 5616–5623.

12 Wang, X., Xia, L., Tan, K., and Zheng, W. (2012). Studies on adsorption of Uranium (VI) from aqueous solution by Wheat Straw. *Environ. Prog. Sustain. Energy* 31: 566–576.

13 Xie, S., Yang, J., Chen, C., Zhang, X., Wang, Q., and Zhang, C. (2008). Study on biosorption kinetics and thermodynamics of Uranium by Citrobacter freudii. *J. Environ. Radioact.* 99: 126–133.

14 Chen, F., Tan, N., and Long, W. (2014). Enhancement of Uranium (VI) biosorption by chemically modified marine-derived mangrove endophytic fungus fusarium sp. #ZZF51. *J. Radioanal. Nucl. Chem.* 299: 193–201.

15 Chakraborty, D., Maji, S., Bandyopadhyay, A., and Basu, S. (2007). Biosorption of Cesium-137 and Strontium-90 by mucilaginous seeds of ocimum basilicum. *Bioresour. Technol.* 98: 2949–2952.

16 Ding, D., Zhao, Y., Yang, S., Shi, W., Zhang, Z., Lei, Z., and Yang, Y. (2013). Adsorption of cesium from aqueous solution using agricultural residue walnut shell: Equilibrium, kinetic and thermodynamic modeling studies. *Water Res.* 47: 2563–2571.

17 Sathishkumar, M., Pavagadhi, S., Vijayaraghavan, K., Balasubramanian, R., and Ong, S.L. (2010). Experimental studies on removal of microcystin-LR by Peat. *J. Hazard. Mater.* 84: 417–424.

18 Tabb, M.M. and Blumberg, B. (2006). New modes of action for endocrine- disrupting chemicals. *Mol. Endocrinol.* 20: 475–482.

19 Graham, J.D.P. (1949). Actions of sodium azide. *Br. J. Pharmacol.* 4: 1–6.

20 Hosseini, S., Malekbala, M.R., and Hosseini, S. (2013). Biosorption of Azoimide on Almond Integument: Kinetics, Isotherm and Thermodynamics Studies. *J. Environ. Chem. Eng.* 1: 696–702.

21 Prodromou, M. and Pashalidis, I. (2013). Uranium adsorption by non-treated and chemically modified cactus fibres in aqueous solutions. *J. Radioanal. Nucl. Chem.* 298: 1587–1595.

22 Aytas, S., Gunduz, E., and Gok, C. (2014). Biosorption of Uranium Ions by Marine Macroalga Padina pavonia. *CLEAN Soil Air Water* 42: 498–506.

23 Eroglu, H., Varoglu, E., Yapici, S., and Sahin, A. (2010). An environmentally friendly batch bioadsorption study of the radionuclides 67Ga from aqueous solutions by Fibrous Tea Waste. *Chem. Eng. J.* 165: 563–572.

24 Salgado, R., Oehmen, A., Carvalho, G., Noronha, J.P., and Reis, M.A. (2012). Biodegradation of clofibric acid and identification of its metabolites. *J. Hazard. Mater.* 241: 182–189.

25 Zhou, Q., Deng, S., Zhang, Q., Fan, Q., Huang, J., and Yu, G. (2010). Sorption of perfluorooctane sulfonate and perfluorooctanoate on activated sludge. *Chemosphere* 81: 453–458.

26 Antunes, M., Esteves, V.I., and Guegan, R. (2012). Removal of diclofenac sodium from aqueous solution by Isabel Grape Bagasse. *Chem. Eng. J.* 192: 114–121.

27 Dubey, S.P., Dwivedi, A.D., Sillanp€a€a, M., and Gopal, K. (2010). Artemisia vulgaris-derived mesoporous honeycomb-shaped activated carbon for ibuprofen adsorption. *Chem. Eng. J.* 165: 537–544.

28 El-Shafey, E.I., Al-Lawati, H., and Al-Sumri, A.S. (2012). Ciprofloxacin adsorption from aqueous solution onto chemically prepared carbon from date palm leaflets. *J. Environ. Sci.* 24: 1579–1586.

29 Jung, C., Park, J., Lim, K.H., Park, S., Heo, J., Her, N., Oh, J., et al. (2013). Adsorption of selected endocrine disrupting compounds and pharmaceuticals on activated biochars. *J. Hazard. Mater.* 263: 702–710.

30 Kong, H., He, J., Wu, H., Wu, H., and Gao, Y. (2012). Phenanthrene removal from aqueous solution on sesame stalk-based carbon. *CLEAN Soil Air Water* 40: 752–759.

31 Iori, V., Pietrini, F., and Zacchini, M. (2012). Assessment of ibuprofen tolerance and removal capability in Populus nigra L. by in vitro culture. *J. Hazard. Mater.* 229: 217–223.

32 Singh, S., Eapen, S., Thorat, V., Kaushik, C.P., Raj, K., and D'Souza, S.F. (2008). Phytoremediation of 137Cesium and 90Strontium from solutions and low-level nuclear waste by Vetiveria zizanoides. *Ecotoxicol. Environ. Saf.* 69: 306–311.

33 Sakakibara, M., Ohmori, Y., Ha, N.T., Sano, S., and Sera, K. (2011). Phytoremediation of Heavy Metal-Contaminated Water and Sediment by Eleocharis acicularis. *CLEAN Soil Air Water* 39: 735–741.

34 Saleh, H.M. (2012). Water hyacinth for phytoremediation of radioactive waste simulate contaminated with cesium and cobalt radionuclides. *Nucl. Eng. Design* 242: 425–432.

35 Matamoros, V. and Salvado, V. (2012). Evaluation of the seasonal performance of a water reclamation pond-constructed wetland system for removing emerging contaminants. *Chemosphere* 86: 111–117.

36 Perlatti, B., Silva, M.F., Fernandes, J.B., and Forim, M.R. (2013). Biodegradation of 1,2,3,4-Tetrachlorodibenzo-p-Dioxin in liquid broth by Brown-Rot Fungi. *Bioresour. Technol* 148: 624–627.

37 Lu, M., Zhang, Z.Z., Wu, X.J., Xu, Y.X., Su, X.L., Zhang, M., and Wang, J.X. (2013). Biodegradation of Decabromodiphenyl Ether (BDE- 209) by a Metal Resistant Strain, Bacillus cereus JP12. *Bioresour. Technol* 149: 8–15.

38 Zhao, Y.X., Gao, B.Y., Shon, H.K., Wang, Y., Kim, J.H., Yue, Q.Y., and Bo, X.W. (2012). Anionic polymer compound bioflocculant as a coagulant aid with aluminum sulfate and titanium tetrachloride. *Bioresour. Technol* 108: 45–54.

39 Yin, C.Y. (2010). Emerging usage of plant-based coagulants for water and wastewater treatment. *Process Biochem* 45: 1437–1444.

40 Kansal, S.K. and Kumari, A. (2014). Potential of M. oleifera for the treatment of water and wastewater. *Chem. Rev.* 14 (9): 4993–5010.

41 Coulon, C., Collignon, A., and McDonnell, G. (2010). Resistance of acanthamoeba cysts to disinfection treatments used in health care settings. *J. Clin. Microbiol.* 48: 2689–2697.

42 Zhao, H., Liu, H., and Zhou, J. (2013). Characterization of a bioflocculant MBF-5 by Klebsiella pneumonia and its application in acanthamoeba cysts removal. *Bioresour. Technol* 137: 226–232.

43 Heberer, T. (2002). Occurrence, fate, and removal of pharmaceutical residues in the aquatic environment: A review of recent research data. *Toxicol. Lett.* 131: 5–17.

44 Xing, J., Yang, J.X., Li, A., Ma, F., Liu, K.X., Wu, D., and Wei, W. (2013). Removal efficiency and mechanism of sulfamethoxazole in aqueous solution by bioflocculant MFX. *J. Anal. Methods Chem.* 568614: 1–8.

45 Brindle, K. and Stephenson, T. (1996). Mini review, the application of membrane biological reactors for the treatment of wastewater. *Biotechnol. Bioeng.* 49: 601–610.

46 Basile, T., Petrella, A., Petrella, M., Boghetich, G., Petruzzelli, V., Colasuonno, S., and Petruzzelli, D. (2011). Review of endocrine-disrupting- compound removal technologies in water and wastewater treatment: An EU perspective. *Ind. Eng. Chem. Res.* 50: 389–8401.

47 Kimura, K., Hara, H., and Watanabe, Y. (2007). Elimination of selected acidic pharmaceuticals from municipal wastewater by an activated sludge system and membrane bioreactors. *Environ. Sci. Technol.* 41: 3708–3714.

48 Yu, J., He, C., Liu, X., Wu, J., Hu, Y., and Zhang, Y. (2014). Removal of perfluorinated compounds by membrane bioreactor with powdered activated carbon (PAC): Adsorption onto sludge and PAC. *Desalination* 334: 23–28.

49 Ye, J., Yin, H., Peng, H., Bai, J., Xie, D., and Wang, L. (2013). Biosorption and biodegradation of triphenyltin by Brevibacillus brevis. *Bioresour. Technol* 129: 236–241.

50 Akhtar, J., Amin, N.S., and Aris, A. (2011). Combined adsorption and catalytic ozonation for removal of sulfamethoxazole using Fe2O3/CeO2 loaded activated carbon. *Chem. Eng. J.* 170: 136–144.

51 Joss, A., Siegrist, H., and Ternes, T.A. (2008). Are we about to upgrade wastewater treatment for removing organic micropollutants. *Water Sci. Technol* 57: 251–255.

52 Zhou, X.F., Dai, C.M., Zhang, Y.L., Surampalli, R.Y., Zhang, T.C., and Preliminary, A. (2010). Study on the occurrence and behavior of carbamazepine (CBZ) in aquatic environment of Yangtze River Delta, China. *Environ. Monit. Assess.* 173: 45–53.

2

Genetic Engineering for Metal Tolerance and Accumulation

CHAPTER MENU

2.1 Introduction

Active development due to human activities in the industrial, urban and agriculture fields resulted in the development of chemical pollution of the environment during the twentieth century [1]. Globally, the concentration of metallic elements has risen to such an extent that it has become a threat to the health of human beings [2]. The occurrence of many environmental hazards is due to the release of small particles into the open environment [3, 4]. For the treatment and prevention of such types of harmful pollution, different techniques are available and developed to eliminate these toxic metals from the surroundings. However, the conventional physical and chemical techniques for the remediation are very expensive and harmful for the microorganisms and properties of soil [5, 6]. Currently, adjacent to these conventional techniques that involve

Aquatic Environmental Bioengineering: Monitoring and Remediation of Contamination, First Edition.
Rouf Ahmad Bhat, Mohammad Yaseen Mir, Gowhar Hamid Dar and Moonisa Aslam Dervash.

physico-chemical treatment for the management of polluted areas, scientists have developed a new horizon by using microorganisms, i.e. bioremediation and plant-means phytoremediation to eliminate these hazardous contaminants from the surroundings [1, 7, 8]. “Phytoremediation” is a technology for the soil that is polluted by various heavy metals in which plant genotypes with elevated stabilizing capacity or extracting metal are used. “Phytoremediation” techniques are used to treat the sites that are contaminated with heavy metals such as “phytoextraction”, “phytostabilization”, “rhizofiltration” and “phytovolatilization” [9, 10]. Plants take up these metals and cause translocation and concentration in the plant organs that are above ground level in “phytoextraction”, and on the other hand in “phytostabilization” metal ions are upholded in the polluted substrates by bringing them to halt around the roots. Therefore, stabilization of heavy metals is done in the rhizosphere. Additionally, blowing of dust and water erosion from the ground is prevented by the vegetation cover [6, 11]. In “rhizofiltration”, root systems of phytoremediating plants collect contaminants through root absorption, concentration and precipitation from polluted water or contaminated effluents [12]. “Phytoextraction” and “rhizofiltration” requires collection of plant biomass with reserved harmful metals. After this plant material is burned, the residue formed from it is used for the recycling of metals [9, 13].

A number of wide-ranging, mining and industrial activities are continuously releasing high number of toxic metals into the environment. For the growth of organisms metals like Zn, Cu and Fe are necessary but at high concentrations these can be harmful for the environment. Toxic heavy metals like cadmium and mercury are usually found to be tremendously hazardous in nature because these are highly reactive toward the sulphur and nitrogen atoms present in the side chains of amino acids. In order to make themselves resistant to toxic metals present in the soil, plants have developed different methods or techniques. The uptake and distribution of different metal ions and detoxification of ions are controlled by different networks of the transport of metals such as “chelating”, “trafficking” and “sequestration” [14]. “Hyperaccumulators” is a term which has been coined by Brooks. Plants that collect a variety of essential and nonessential toxic metals to an extremely large extent in the plant parts that are above ground level come under the term hyperaccumulators [15]. Around 400 different species from a wide array of taxa have been characterized as “hyperaccumulators” [16, 17]. For the cleaning of fields that are polluted with heavy metals, “hyperaccumulators” are used widely. However, “hyperaccumulators” are slow growing and have small shoot biomasses so cannot be applied for the extensive decontamination of contaminated soil. To overcome this limitation, the use of selected genes has been suggested in more favorable plant species, resulting in increased efficiency of “phytoremediation” [18, 19]. The properties of “hyperaccumulators” which make them an ideal choice for decontamination are tolerance and particular toxic metal hyperaccumulation. For the study of tolerance of toxic metal and accumulation two species of plants are considered as ideal, such as *Arabidopsis helleri* and *Thlaspi* species. Genes that are concerned about tolerance of heavy metals and their accumulation were extracted from these “hyperaccumulators” [20–25]. Homologs of these genes show over-expression in transgenic plants and elevation in the heavy metals tolerance and accumulation [26–30].

Harvested biomass is prepared from heavy metals-hyperaccumulating plants, and further this biomass is used for the production of "eco-catalysts" for use in the chemical industry [31, 32]. A volatilization process is used by the plants to eradicate inorganic contaminants from the soil. In the process of volatilization, there is transformation of metals biologically into gaseous form which is then eliminated into the surroundings [7, 33]. The major growth in selecting effective phytoremediating plants is due to the selection of genotypes that are growing in naturally polluted sites [34–36]. Some growth also comes from the selection of long-term toxic metals tolerant replica in in-vivo and greenhouse conditions [37–39]. Genetic engineering is used to improve responsive genotypes and helps to makes the process of "phytoremediation" possible for heavy metal contaminated sites. Plant transformation is actually its modification by the use of various genes that perform the function of metal transport and homeostasis of metals and the defense response of plants to oxidative stress or detoxification of xenobiotics.

2.2 Mechanisms of Metal Uptake and Their Transport in Plants

In plant tissues the assimilation of heavy metals is mediated outside a cell or inside the body cells by the use of varied physical reactions, biochemical mechanisms and molecular mechanisms in contaminated soil [40]. Plants carry out distribution of metals in different ways internally. Moreover, heavy metals are either limited to small areas such as the root or stem or collected in various plant parts in nonhazardous form for additional use and distribution. Heavy metals follow symplastic and apoplastic pathways and cortical cells of the plant roots uptake these heavy metals by the competitive uptake with essential elements [41]. The following factors are responsible for the movement of metals to the surface of roots: mass flow is the metal ions absorption by the surface of roots and takes place through a diffusion process and root interception developed by growth of roots [42]. Depending upon the particular metal, the apical region of the root or the entire surface may be responsible for the metal uptake. In addition, root capacity and growth also influence the uptake of metals. Extracellular means apoplastic and intracellular means symplastic and are the entrance routes for heavy metals to plant cells. The most permeable route for solutes is the apoplastic way of the root, where root cells absorb the metals because these have negatively charged sites [43], then metals are translocated into the tissues of roots while suberin lamellae inhibit the solutes from the apoplast to the xylem [44]. For the transfer of metal ions from the endodermis to the xylem of roots they have to transport through the protein [45]. Chelation of several heavy metals is done through the organic acids [46]. But the chelation process and mechanism of transportation into the root xylem is unknown for some metals [45], because their charged metal ions are unable to cross the lipophilic structure of cell walls easily. Therefore, by transporter proteins, ions transport must be intervened by the cell [43] Cysteine-ProlineX-CPx-ATPases, the natural resistance-associated macrophage proteins, also called the "cation diffusion facilitator family" [47], and ZRT-IRT-like proteins family are the different transports present in plants which are responsible for the uptake of metals and homeostasis for the tolerance of metal [48, 49]. Furthermore, through Fe21 transporters, cadmium-like heavy metals take entry to the plant [28]. The extracellular domain is there on the membrane transporter on which a particular metal binds and a transmembrane domain also remains on extracellular and intracellular membranes, and the transportation of metals from outer surface to within the cell is one of its functions [43]. This type of transporters shows a definite type of transport capability (Vmax) and ion affinity (Km) [50]. Specific types of transport mechanisms are there for the heavy metals transport from plant root to plant shoot and take place via the xylem. For example, loading of nickel in the tissues of xylem may be because of the complexation of Ni to free His (histidine) [41]. Metal translocation in the tissues of xylem mostly takes place all the way through the transpiration-driven mass-flow [41]. The cell wall of xylem possesses a "high cation-exchange capacity"; as a result noncationic chelate-metal complexers ought to be transported easily in xylem [51]. Chelation in the form of cadmium (II)-citrate is responsible for the translocation of Cd into the tissues of xylem, as phytochelatins and other ligands that contains the thiol group do not engage in transportation of cadmium directly [41]. From the root cell to the stem cell, the translocation of metals in the xylem mostly takes place through the process of transpiration, i.e. for creating a -ve pressure for the upward transportation of water and solutes [44], and the membrane transport protein transport the metals in the leaf. There are specific membrane transport proteins for the transport of metals and when they are taken up internally and divide into different parts in specific organelles, then these can no longer be harmful to vital cellular processes. Heavy metals are mostly accumulated into the cell wall or vacuole [52, 53], while in tissue heavy metals are collected in the epidermis or hairs.

2.2.1 Heavy Metals Tolerance (Mechanism) in Plants

Plants acquire various types of cellular mechanisms such as extracellular and intracellular, for avoiding toxicity produced by the toxic metals. The entry of metals is checked by root cells at the primary stage, by taking up an avoidance plan, but when metals enter the root cells by some means, some mechanisms such as cell wall binding, organic acids, "chelation", and "sequestration", detoxify these kinds of heavy metals and this is called heavy metals tolerance strategy by the plants. In addition to this, anti-oxidative defense mechanisms and proteins that are related to stress formed in plants are also the mechanism of tolerance in heavy metal toxic plants. As a result, the plants show two types of strategies besides the heavy metals toxicity, "avoidance" and "tolerance", in which avoidance means limitation to metal uptake and "intracellular detoxification" is said to be tolerance [54]. The leaf cells bounded by the chelators sequester these heavy metals. Chelators include the tripeptide glutathione and its oligomer; the phytochelatins and metallothioneins shows its vigorous role in metal "sequestration" or "tolerance" and in essential metal regulation.

2.2.2 Mechanisms of Avoidance in Plants

The heavy metals uptake is limited by the avoidance strategies and these are present to ensure the metal ions entry into the various cells of plants through the root tissues. The first line of defense in plants against the heavy metals toxicity is the extracellular defense. Several strategies come under this extracellular defense system such as halting by mycorrhiza, formation of coordination complexes through root exudates, and change in the rhizospheric pH, metal-binding organic acids secretion or redox barrier development.

The mycorrhizal association is ectomycorrhizas and arbuscular mycorrhiza, which grow on soil that is polluted with heavy metals. The mechanism of absorption, mechanism of adsorption, or chelation mechanisms are the mycorrhizal efficient exclusion barriers; these barriers are there to check the heavy metal influx into the host plant [54]. In addition to this, amino acids or organic acids, water, inorganic ions, and carbohydrates are also released by roots and bicarbonates, protons and carbon dioxide, are excreted by roots; the roots also secretes "mucilage", "siderophores", "allelopathic compounds", etc. These are generally called root exudates and these root exudates form stable ligand complexes due to this metal becoming less toxic which helps the plant to survive in contaminated areas [54]. Another very important adaptation strategy besides the heavy metal toxicity is cellular exclusion. The apoplastic space of roots mostly contains a large number of toxic metals [40].

The possible sites for the tolerance of toxic metals are the cell wall and the plasma membrane because these have accumulation properties like in "Italian ryegrass (*Lolium multiflorum*)"; copper accumulation of about 60% occurs in the cell walls of root and cell membranes [55]. Plant cation exchange capacity depends upon the roots cell wall exchange sites [56]. Tolerant cultivars have more cell wall cation exchange capacity to prevent the entry of heavy metals than sensitive wheat cultivars [40, 57]. It was described that, in order to generate the tolerance for metals by active efflux, plants show some utilization strategy that reduces the intracellular concentration to subtoxic levels [58].

2.2.3 Binding of Metal to the Cell Wall

Extracellular carbohydrates are present in the plant cell wall to examine the heavy metal uptake into the cytosol. Polygalacturonic acid is present in the cell wall pectins, and its role is to exchange the cations and bind the heavy metals into their carboxyl group and avoid heavy metal uptake. It has been reported in many studies that metal uptake influences the heavy metal tolerance adapted by the chemical properties of the cell wall. Moreover, due to the deficiencies of metal absorption sites the plant cell wall is unable to influence the heavy metal tolerance. As a result, there is no complete or well understood mechanism which shows the plant cell wall mechanism against heavy metals toxicity [54]. Plants have the capacity to reduce the adverse impacts of toxic metals. These control the metal allocation and localization within the plant cells. Besides hyperaccumulators, plant roots contain high amounts of metals as compared to the shoots to reduce the transport and accumulation of heavy metals to the plant shoot cells [40]. Consequently, there is a smaller amount of heavy metals in the shoot in the tolerant genotypes than the sensitive genotypes. So, in the leaf of the genotype of "tolerant cowpea (*Vigna unguiculata*)" there is homogenized allocation of manganese achieved thoroughly, as the nontolerant genotypes collect the manganese in the particular area of leaf in the form of dark brown spots of MnO_2 precipitates [40].

2.2.4 Mechanisms of Tolerance in Plants

Heavy metal tolerance mechanisms comprise heavy metal accumulation, heavy metal restoration and heavy metal immobilization because metals form bonds with the "amino acids, proteins, or peptides", thus forming compounds. However, this is considered as the "plant's second line of defense", which primarily speeds up the intracellular detoxification of metals in the plants. Moreover, it was described that on heavy metals toxicity metal ions were bound or modified by the plants mainly to reduce the transport of metal across the cell membrane [59]. The detoxification of metal ions that enter inside the plant cell is done by inactivating them or changing them into less hazardous forms [54]. Once a heavy metal enters inside the cell, plants employ different strategies to manage it, such as transporting these heavy metals outside the cells, and sequestration of ions into the vacuole or other organelles of the cell where there is an occurrence of susceptible metabolic processes [14]. Consequently, the appropriate site for the storage of toxic metals accumulation is the central vacuole. The vacuolar uptake of solutes catalyzed by the channels or transporters is facilitated by the two vacuolar proton pumps such as a "vacuolar proton-ATPase" and a "vacuolar proton pyrophosphatase". Polypeptides are the chelators that bind the sequested heavy metals in the vacuole. Metallothioneins (MTs) and phytochelatins (PCs) are found in plants that are two important metal-binding polypeptides (Figure 2.1). Metallothioneins are named as encoding genes or disulphide-rich peptides with lower molecular weight because of their significant properties [60]. Several metallothionein genes are metallothioneins1, metallothioneins2, metallothioneins3 and metallothioneins4; these all are present in various types of higher plants like *Arabidopsis thaliana*. Phytochelatins also come under class III metallothioneins and these may be cysteine rich. Glutathione forms the phytochelatins with the help of a particular transpeptidase, i.e. "γ-glutamylcysteine dipeptidyl transpeptidase", also named "phytochelatin synthase (PCS)" [61]. The phytochelatin synthase is necessary for post-translational activation by the heavy metals [62]. Cd is thought to be a good activator for phytochelatin synthase. Ag, Pb, Zn are the moderate activators of phytochelatin synthase [63, 64]. The

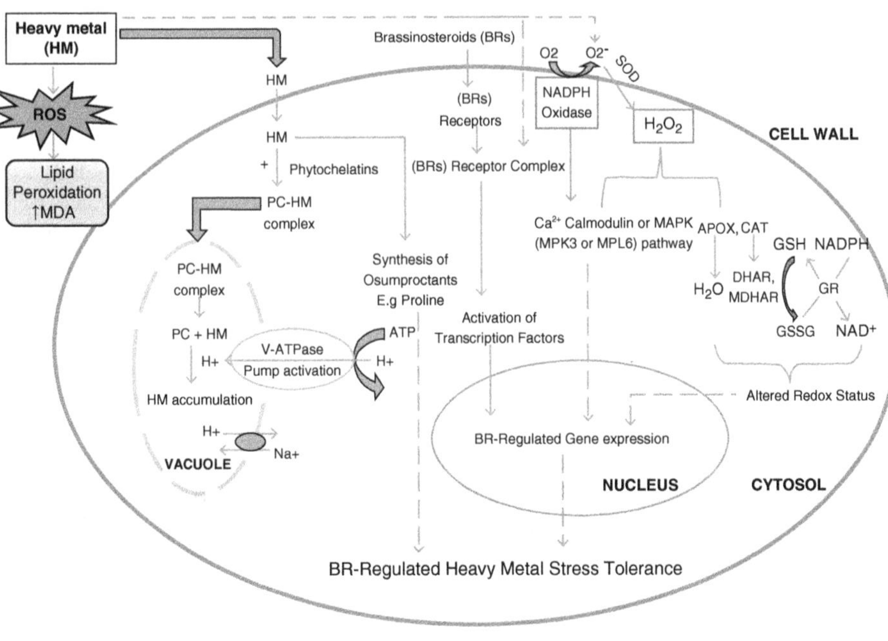

Figure 2.1 Showing mechanism of heavy metal stress tolerance.

heavy metal phytochelatin complexes use the help of the tonoplast membrane through ABC transporters to sequester phytochelatins in the vacuoles of cells [65], and further stabilization of these complexes is carried out by the sulfhydryl groups [53]. The prohibition of Cd and Ar is done by phytochelatins because these have a very important role in their prohibition. On the other hand, these have an insignificant role in the improvement of heavy metals like copper, zinc, nickel and selenium trioxide [63]. Extra mechanisms are shown by hyperaccumulators for metal detoxification. For example, "*Thlaspi goesingense*" (nickel hyperaccumulator) has a higher tolerance of nickel because it forms a complex with His (Histidine) which is further responsible for the metals inactivation [46]. Phytochelatins and metallothioneins show a very important role in the tolerance of metal at cytoplasmic level; these form the complexes with metals, and then complexes are stored into the vacuole without any negativity [49].

2.3 Phytoremediation Using Genetic Engineering Stress-Tolerant Plants

The achievements in "genetic engineering stress-tolerant plants" have been the conclusion of major works in recognizing important genes. These genes on over-expression elevate tolerance to a particular metal or common stress-related tolerance that results in the elevated biomass of the plant. A clear-cut strategy for elevating the plant tolerance to a heavy metal is to elevate its capability to alter the metal to a less toxic form. In this way, such plants become highly capable of accumulating large amounts of the detoxified form. Over-expression of the gene encoding "selenocysteine methyltransferase (SMT)" in Indian mustard is an example of this approach [66]. The selenium hyperaccumulator *Astragalus bisulcatus* is the plant in which the gene that encodes for selenocysteine methyltransferase (SMT) was first identified. And it was found that this gene is an important component of the hyperaccumulation mechanism [67]. The Se toxicity is due in part to its chemical resemblance to sulfur. Because of the misincorporation of seleno-amino acids analogs of the sulfur-amino acids such as cysteine and methionine, into proteins there is change in the structure of enzyme and activity. In hyperaccumulation, selenocysteine methyltransferase performs the function of methylation of selenocysteine to the nonprotein selenoamino acid, i.e. methylselenocysteine. Therefore, the detoxification of Se is done by removal of Se from the selenoamino acids, and then from proteins [68]. Over-expression of "selenocysteine methyltransferase" in transgenic Indian mustard showed elevated tolerance and accumulation of Se [66]. For phytoremediation there is another particular pathway for the detoxification of metals; the bacterial mer operons that were changed into plants [69]. The mer operon genes are merA and merB that encode "mercuric reductase" and "methylmercury lyase", respectively. These enzymes change organic or inorganic mercury to elemental mercury when expressed together. Elemental mercury can escape from the plant in unstable form. There are various species of plants over-expressing merA gene and merB gene which have elevated tolerance and accumulation of both forms of Hg (organic and inorganic forms) [70]. The introduction of MerA genes and merB genes into plants is done by using chloroplast transformation and nuclear transformation [71]. The introduction of particular detoxification mechanisms may be performed at the time when there is increase in gene expression involved in plant stress response. For instance, in the development of *Arabidopsis thaliana* over-expression of ArsC, encoding "arsenate reductase", and glutathione synthetase II, the gene encoding γ-glutamylcysteine synthetase (ECS) for elevated Ar resistance [72]. In the first step in arsenic detoxification is that there is reduction of arsenate to arsenite. However, aresenate is less toxic that arsenite. One advantage is that its detoxification can be done through chelation with the compounds that contain the thiol group including GSH and PCs. Over-expression of an arsenate reductase in Arabidopsis leads to the increase in the proportion of arsenic present as toxic arsenite [72] Because Arabidopsis would unable to detoxify a wide range of arsenite efficiently without associated elevation in compounds that contain the thiol group, the authors over-expressed γ-glutamylcysteine synthetase in the same plants. In the production of phytochelatins γ-glutamylcysteine synthetase (ECS) catalyzes a rate-limiting step, short peptides that contains Cys known to chelate diversity of metals and metalloids [73].

There are many symptoms which show toxicity caused by heavy metals, including short growth, damage in roots, necrosis, etc. However, oxidative stress is also involved in the damage mechanism which further causes the production of reactive oxygen species [74]. Further, genes that ameliorate the "oxidative stress" may also protect against a number of toxic contaminants, which is a plus point for heavy metal contaminated sites. In order to "genetically engineer Indian mustard" for increased heavy metals tolerance and accumulation, there was development of transgenic plants over-expressing "glutathione synthetase (GS)" and "γ-glutamylcysteine synthetase (ECS)". The rate-limiting steps in the generation of "glutathione" and "phytochelatins" (PC) are catalyzed by the enzymes GS and ECS [75, 76]. In chelation and sequestration of Cd the function of phytochelatins was well developed and it was expected that these plants would have increased tolerance of cadmium and its accumulation [75–77]. These plants have since been exposed to tolerate and accumulate a wide diversity of toxic metals [78].

Likewise, for the phytoremediation of selenium Indian mustard, over-expressing the Arabidopsis gene encoding "ATP sulfurylase" (APS) was initially developed [79]. In the sulphur/selenium assimilation pathways in plants a rate-limiting step is catalyzed by APS, and its over-expression resulted in elevated uptake of selenate and ultimately it reduced to selenite [80]. Simultaneously, there is also elevation in the uptake and reduction of sulfate on over-expression of APS. Indeed, wild type plants were less efficient than APS Indian mustard plants in tolerating and isolating a number of toxic metals from polluted soil [81]. Recently, there was a study in the US with Indian mustard lines over-expressing "ATP sulfurylase", "γ-glutamylcysteine synthetase" and "glutathione synthetase" respectively, during the first successful trial of transgenic plants in phytoremediation [82]. High amounts of selenium are accumulated by all three lines in their shoots, up to 430% more in the case of ATP sulfurylase, although there is no increase in selenium by the γ-glutamylcysteine synthetase and glutathione synthetase [77].

GSH production is thought to be the plant's protective response to a number of stresses such as salinity, heat stress, attacks by pathogens, reactive oxygen species, HM-stress, and the enhanced growth of the glutathione synthetase line on contaminated soil could well have been due to an increased concentration of GSH [83–86]. It has been observed that besides glutathione, elevation in the levels of metallothioneins, SOD (superoxide dismutase), ascorbate and organic acids like malate and citrate other reducing agents, also show a response to several stresses like wounds, attacks by pathogens, drought, heat stress, and metal stress [46, 87–89]. The over-expression of genes that reduce oxidative stress could indeed also protect against various stresses faced by the plants such as the primary metal contaminant and secondary environmental conditions used in phytoremediation.

2.3.1 Selenium Accumulation by Plants

The presence of selenium is not required in angiosperms but there is evidence that it stimulates the growth of plants and their survival in certain environmental conditions [90, 91]. Sufficient amounts of selenium reduce oxidative stress in plants induced by several ecological circumstances, while optimal concentrations of selenium help in protection against various pathogens and plant-eater herbivores [92, 93]. However, overload in selenium accretion introduces toxicity into plants by introducing unselective Se-amino acids, selenomethinone (SeMet) and selenocysteine (SeCys) into proteins which obstruct their functions and lead to physiological and biochemical faults [94–96]. Selenium concentration in different plants is heavily dependent on selenium phytoavailability in the soil, although different species of plants have different mechanisms to attain selenium in distinguishable ecological situations [95, 97, 98]. Se concentration in soil varies greatly. Although in most soil, selenium concentration varies from 0.01 to 2.0 mg/Kg, some soils associated to certain environmental developments may have selenium absorptions up to 1200 mg/Kg [99–101]. Organic matter content, redox potential and pH are some factors that affect the bioavailability of selenium in the soil [100–103]. Accumulation of selenium in the plant roots takes place in the form of selenite (SeO_3^{2-}, $HSeO^{3-}$, H_2SeO^3), selenate (SeO_4^{2-}) ions or organo-selenium complexes including SeMet, SeCys, but these are incapable of absorbing selenide species, i.e. (Se^{2-}) [98]. Selenate is the only water soluble form of selenium in toxic soils (Ph + Pe > 15), while in the case of anaerobic soils the most soluble form is selenite (SeO_3^{2-}). Selenides are the ions which are only found in rigorously anaerobic soils and usually soils with acidic pH (pH + Pe < 7.5). Selenite is somewhat free in the top soil elucidation, but it is readily taken up through Fe and Al_2O_3 and to some level, by clays and organic matter [100, 101]. Well-marked plant populations are present in soils with higher selenium availability [94, 104]. Meanwhile, superfluous accumulation of selenium in soil is very toxic to plants, as plants growing on selenium-rich soil are capable of eliminating or removing selenium from their cells and tolerating the concentration of selenium in soils (Figure 2.2). Different plant species show noticeable differences in their capacities to handle selenium inside their tissues. In the meantime, plants growing on seleniferous lands exhibit selenium with increased concentrations. It is believed that the establishment of seleniferous soils asks for minimum capability to endure raised Se concentration in tissues [92, 94, 104]. Depending on the selenium accumulation capabilities of a plant, angiosperm classes have been divided into different types [92, 94, 95, 97, 98, 104]. Furthermore, angiosperm species are allocated as "nonaccumulator" species. These nonaccumulator plant species are not capable of enduring higher than 10–100 μg/g dry matter selenium concentrations and for that reason they cannot established themselves on selenium-rich lands [95, 98, 100, 104].

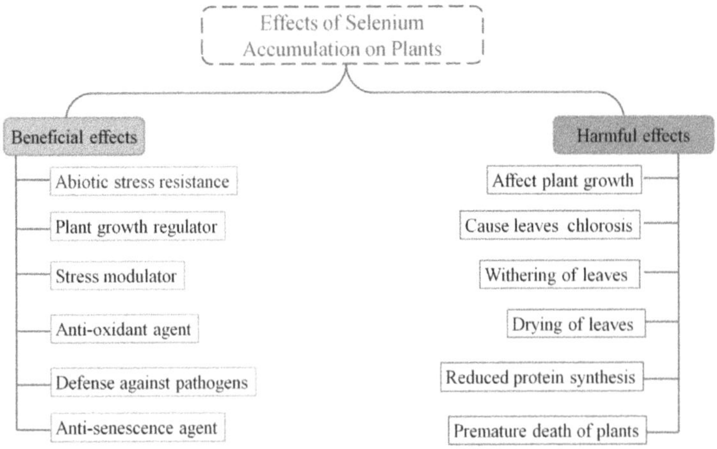

Figure 2.2 Showing effects of accumulation of Selenium on plants.

Several plant species which are capable of tolerating soft tissue Se absorptions reaching up to 1 mg/g dry matter are allocated as "Se-indicator" species. These species of plants can establish themselves in both seleniferous as well as nonseleniferous lands [104]. Selenium concentrations inside tissues of these plants are openly connected to the phytoavailability of selenium in the soil, in this manner, "specifying" the phytoavailability of selenium in soil [105]. A number of plant species which are capable of enduring selenium concentration in tissues higher than 1 mg/g DM are allotted as "Se-accumulator" plant species. Se-accumulator plant species are commonly confined to seleniferos soils [92, 94, 104]. Only those plants get the designation of "Se-hyperaccumulators" that have leaves retaining more than 1 mg/g dry matter when taken from ordinary surroundings [98, 106], even though some professionals believe that the lower limit must reduce to 100 μg/g DM [106]. While there are dissimilarities amid species based on their response toward selenium accumulation, which are exhibited, e.g., in their organic approaches toward soil holding high selenium phytoavailability, there seems to be no systematic phylogenetic effects on shoots selenium concentrations within the angiosperm species [95, 97, 107]. Maximum discrepancy in selenium concentration in shoots of angiosperm plants seems to be present among species in orders and between ecotypes in species [95, 97, 107–110]. Consequently, discrepancy in selenium concentration in shoots of plants is most dominant in plant species having "selenium-accumulators" or "selenium-indicator plants" [95, 97].

2.3.2 Genetics of Plants Selenium Accumulation

The importance of selenium metal can be seen not only in plants but also in animals [91]. Adequate amounts of selenium reduces oxidative stress in plants induced by several ecological circumstances while optimal concentrations of selenium help in protection against various pathogens and plant-eater herbivores [92]. Nevertheless, an excess amount of selenium in plant tissues induces toxicity to various plant species, which are capable of tolerating increased concentration of selenium in tissues and are able to establish themselves on selenium accumulated soil [94, 104]. The potential of selenium hyperaccumulator plant species to endure excessive selenium concentrations in tissues seems to be evolved from convergent evolution in various angiosperm species via related variations in the metabolism of selenium [94, 95, 97, 111].

Just as hyperaccumulator plants are a great source for noble genes, used in phytoremediation of Se, microbes, and especially those which have developed under extreme environments, they may also be used for providing a new library of genes that can help in enhancing tolerance and phytoremediation capacity in certain plants [112]. The upregulation of bacteriological genes has been successfully introduced in plants to enhance their tolerance against substantial toxic metals [75, 76, 113]. The involvement of many pathways and genes to promote plant tolerance against heavy metals is present as microbes have been used first, rather than plants, due to the broad genetic and biochemical study on microbes [114]. In order to survive extreme environments, organisms must develop successful mechanisms of tolerance. Unfavorable environmental surroundings are not capable of harboring plant life and growth periods due to their excessive harmfulness. Alternatively, microbes have been isolated from environments where there is no existence of life. Hence, these extreme ecological conditions such as hotsprings, alkaline lakes, rivers and saline atmospheres are potential sites for numbers of previously undiscovered microbial lineages having a wide range of mechanisms for their survival [115–118]. The biological characterization of microorganisms has led to the discovery of noble classes of proteins when taken from the sites, for instance, the heat shock proteins are present in hyper-thermophiles. By using these kinds of strategies, various microorganisms were collected from hypersaline solar evaporation ponds and then classified [119]. These types of ponds contain high amounts of selenium, especially in the form of selenate; besides that they also contain raised levels of sulphur with salt that inhibits plant growth, except for microbes. At Red Rock Ranch, microorganisms taken from a solar evaporation pond were separated and characterized by the use of 16S ribosomal DNA phylogenetic analysis. This pond contained only bacterial microorganisms and no Archea were established. The major species available was a previous unaffiliated group of uncivilized microbes belong to the order Cytophagales. Cultural γ-Proteobacteria is the second most dominating group from the species Halomonas. These cultural strains of bacteria are well known for their capacity to endure elevated volumes of selenium and salt, and also to collect and volatilize selenium and its different forms. With time, the selenium accumulation power of microbes also increases. The solar evaporation pond from where the microbes were taken is largely aerobic, and hence the microorganisms decrease the concentration of Se through assimilation and not from the dissimilatory pathway, resulting in the production of elemental Se. This is important as reduction of Se in plants also takes place through assimilation paths [120]. Various studies on the biochemical and genetic characterizations of the bacteria resulted in the discovery of new enzymatic activity and genes that might be added in plants [119]. There is as yet no simple experiment to promote the capacity of plant performing phytoremediation with the help of genes of microbial origin [121]. One significant consideration is the probability for dissimilarities used in codon [122]. Perhaps the bacterial gene merA coding for a mercuric ion reductase changes harmful Hg^{2+} into less toxic metallic mercury, Hg^0. Preliminary efforts to demonstrate this protein in the plant were unsuccessful despite using effective plant expression systems. The sequence was rich in CpG dinucleotides, making it vulnerable to methylation and gene slicing in plants. This problem was solved by appropriate modification of approximately 9% of the coding sequence and then introducing this mutagenized sequence to *Arabidopsis* [122]. Even at low expression levels of protein, transgenic seedlings were capable of growing on 100 μM Hg^{2+} and toxic levels of Au^{3+}. Another difficulty with the search for novel genes in tolerant microbes is that the operon encoding for resistance to heavy metal contains more than one gene [121]. Nevertheless, the use of bacterial genes helps in preventing co-suppression that can arise from attempts to over-express a plant gene that is identical to the endogenous version [123].

Up-regulation of enzymes clearly directed the catalyzing of the occasional limiting step in selenium incorporation pathways, and speedily promoted the assimilation and accumulation of selenium [79]. The only restriction to this method is that incapacitating one rate limiting step leads to the formation of another such step. Up-regulation of certain genes in transgenic plants performs the absorption pathways of selenium and sulphur. Although it is complicated and tiresome to evolve a plant by single gene transformations, similar results can be attained by over-expression of a single transcription factor, in charge for the stimulation of many co-regulated genes. Promoter regions bear shared sequences of gene expression and these expressions are controlled by similar transcription factors which bind to the sequenced genes [124] A similar perspective was introduced to grow cold-tolerant Arabidopsis effectively [125]. Over-expression of CBF1, which is a cold-linked transcription factor in Arabidopsis helps to up-regulate the expression of cold-promoting genes considered as enduring freezing in genetically modified plants. Maybe the possibility is there to locate a common transcript factor, to modulate the expression of co-ordinately genes involved in the selenium and sulphur incorporation passageway. Similar modifications with respect to the above pathway can be seen in the regulation of genes including APS reductase, ATP sulfulyrase and sulphate permease [126–132]. The efficiency of these regulatory genes improves during the sulphur starvation circumstances, while during sufficient sulphur conditions it declines in various plant species. The probability of sharing common gene sequences in the promotor regions of cells is very high, which is later on controlled by a single transcription factor. In higher plants, signal transduction pathway modifying co-ordinate regulation of these genes has not yet been known [133]. However, sulphur acclimatization genes have been discovered from the algae *Chlamydomonas* with the help of mutant analysis [134, 135]. *Chlamydomonas* mutants were not capable of increasing the expression of their higher affinity toward sulphate transport systems under the circumstances of sulphate starvation as seen in the wild type. One of the mutants identified as Sacl was unable to elevate its high-affinity sulfate transport system or transcription of ATP sulfurase under the circumstances of sulphur starvation [134, 135]. The "Sacl protein" is thought to be an essential membrane protein, which may act as a transporter for signaling compounds or sensors of low cellular sulfur levels or both ([134, 135]. "Sac3 mutants" are not capable of producing advanced levels of the high-affinity sulfate transporters. The "Sac3 amino acid" sequence show little resemblance to SNF1-related protein kinase 2 (a family of kinases), of which at least 15 have been discovered in angiosperms [135]. Various studies on *Chlamydomonas* have been revealed the presence of sulphur assimilatory genes which might be present in higher plant groups also.

2.3.3 Proteins for Metal Accumulation

By screening a tobacco through the cDNA library and using radiolabeled recombinant calmodulin as a probe, a protein called NtCBP4 was discovered [136], which is capable of controlling the heavy metal enduring capacity of plants. Numerous independent transgenic lines coding for *Nicotiana tabacum* plasma membrane channel protein, had higher values of displaying increased tolerance to nickel and lead hypersensitivity, which is linked to decreased accumulation of nickel and increased accumulation of lead, respectively. NtCBP4 was the first kind of plant protein to be reported which modulates the plant endurance and accretion of lead. These kinds of genes can be used as an effective tool for refining phytoremediation approaches [137]. Partial peptides expressions from the C terminus of the "TcHMA4 (the Thlaspi heavy metal ATPase) protein" comprising several heavy metal bindings to genes histidine and cystein repeats residues, results in improved cadmium tolerance and hyperaccumulation potential in yeast. The TcHMA4 possesses a high capability in remediation of heavy metals studies and is used for augmenting the metal acceptance and phytoremediation prospective of higher plants [23].

Metallothioneins are genetically engineered plants displaying metallothioneins (these are the proteins which are capable of binding metals and are used to increase tolerance and accumulation to heavy metals in plants) have been developed and granting it boosts tolerance against high metal concentrations; it does not show any effect on the uptake of metal. To increase metal sequestration in higher plants, the "yeast metallothionein cup1" was established into tobacco plants, and the "cup1 gene" expression and phytoextraction of cadmium was resolute [138]. Over-expression of Cu inducible metallothioneins cup1 also showed increased copper tolerance in plants [139]. More than fifty metallothioneins were identified by the researchers successfully in various plants divided into four categories of metallothioneins proteins [53]. A wide range of metallothionein genes, isolated from many sources (such as human, mouse, Chinese hamsters and yeast) have been over-expressed in plants [138, 140–142].

"Chloroplast engineering": Some regulatory barriers are there to get transgenic plants in the field, remediating polluted sites. These limitations encouraged the investigators to develop different techniques to transform the chloroplast of particular plants, which ultimately prevents cross genes from breaking off into weed and plants through pollens [143]. The above process can be used to merge the bacterial gene, i.e. merAB, into the genetic material of the choloroplast of tobacco. Hence, the genetically modified plant showed increased resistance toward greatly contaminated organic mercury, present in the phenylmercuric acetate form, as compared to wild plant species [144]. Other advantages of chloroplast modification are codon optimization [145].

2.4 Genetically Modified Plants Against Uptake, Tolerance and Detoxification of Heavy Metals

Plant biotechnology is now focussing more on the genetic and biochemical basis. A basic understanding regards the metal uptake and translocation processes in normal plants and metal hyperaccumulators, and the regulatory control of plant activities and the use of tissue specific promoters [146]. The genetically engineered plants including "*Populus angustifolia*", "*Nicotiana tabacum*" or "*Silene cucubalis*" over-express glutamyl cysteine syntlietase, and therefore, increase in heavy metal accumulation as compared to corresponding wild type plants. The genetically engineered plant *Brassica junca* was specially developed to inspect rate-limiting factors for the formation of glutathione and phytochelatins. "*Escherichia coli* shl gene" was introduced to attain this. The gamma-glutamylcysteine synthetase transgenic seedlings demonstrated enhanced tolerance toward Cd and possess high concentration of PCs, γ-L-Glutamyl-L-cysteine, GSH and total nonprotein thiols in respect to the wild type seedlings [147]. It has been reported in some studies that "c-glutamylcystein" asynthetase inhibitor, "L-buthionine-[S, R]-sulphoximine" (BSO), rapidly increase arsenic concentration in nonadopted and arsenic-hyper-tolerant plants as well, exhibiting that "phytochelations-based sequestration" is necessary for both normal "constitutive tolerance" and "adaptive hypertolerance" to this metalloid [65]. Enzyme "selenocysteine methyltransferase" (SMT) is responsible for the conversion of protein amino acid selenocysteine into a nonprotein amino acid called methylselenocysteine. By introducing genes from "selenocysteine methyltransferase" from the selenium hyperaccumulators "*Astragalus bisulcatus*", it changes the flow of selenium from the selenocysteineamino acids which then would have entered the protein, resulting in modifications in enzyme arrangement, utility and harmfulness. Transgenic plants over-expressing "selenocysteine methyltransferase" displayed increased tolerance toward Se, especially selenite, and manufactured 3- to 7-fold more biomass than the wild type and 3-fold longer root lengths [148]. The selenocysteine methyltransferase plants collected up to 4-fold more selenium than the wild type, with higher proportions present in the form of MetSeCys. In addition, "selenocysteine methyltransferase" Arabidopsis and "selenocysteine methyltransferase" Indian mustard expressed two to three times more volatilization of Se, when used with Se Cys and selenate, correspondingly. In another study, cystathionine-synthetase (CGS) was over-expressed in the Indian mustard plant. Various studies have revealed that with cystathionine-synthetase or CGS the Indian mustard plant displayed increased sufferance to selenite and volatilization of selenium which increased by two to three times as compared to wild type plant species, but at the same time reduced the accumulation of selenium in different plant parts including roots and shoots [149]. A number of scientists at the University of California integrated the bacteriological gene encoding "pentaerythritol tetranitrate reductase" in transgenic tobacco, which then boosted the survival capability of plants on soil contaminated with nitrates and ester [150]. Additional inspection showed an increased deprivation of these toxic compounds by genetically modified tobacco plants with respect to the same plantlets. Different studies have presented that over-expression of bacterial gene merA (expressing enzyme mercuric reductase) in "*A. thaliana*" displayed increased conflict toward $HgCl_2$ along with full of atmosphere volatilization. This modus operandi in the building up of transgenic yellow poplar increases the volatilization rate of elemental mercury by ten times, as compared to untransformed plants [151]. Furthermore, over-expression of the merB gene (coding for organo-mercuric lyase) in *A. thaliana* boosted its conflict to methylmercury and its resistance is enhanced when enzyme targets its endoplasmic reticulum, providing access to its hydrophobic substrate [152]. Incorporating merA and merB genes in the tobacco plant also allows for moderate levels of resistance to methylmercury. "Symmetric somatic hybridizations" and "asymmetric somatic hybridizations" are also being used as gene transformers. It is even now utilized in incorporating contaminated metals resistant traits in *Thalspica erulescens* and *Brassica juncea* and has resulted in enhanced accumulation of heavy metals, metals tolerance and increased biomass yield in crossbreed plants [153]. Biomass accretion in plants may be attained by acquainting the most effective genes that accumulate metal in the plants which are much higher than natural plants, resulting in amplified biomass production [75]. At present, plant bioengineer experiments take place in Beltsville, for the purpose of creating highly zinc-tolerant capacity plants called pennycress (*Thalspi*). The plants are taller than normal ones and are expected to have increased biomass and are therefore more capable of taking large number of contaminated metals. The incorporation of the bacterial gene "1-aminocyclopropane-1-carboxylic acid deaminase" into tomato plants resulted in increased accumulation of metal and tolerance level for a variety of toxic metals like cadmium, copper, nickel, manganese, lead and zinc as compared to unchanged plants [154].

2.5 Cadmium Tolerance and Accumulation Mechanisms in Plants

A number of processes are the result of the absorbed metals neutralization; these processes include holding in the cell wall, complex formation with chelators, the elimination of toxic metals from plant cells, categorization, etc. It should be highlighted that basic tolerance and hypertolerance against cadmium is developed by these mechanisms. However, there are several dissimilarities. Compartmentalizations in cadmium vacuoles that are attached to PCs or other chelators are responsible for the occurrence of basic tolerance. In this case, cadmium is present at higher amounts within the roots and only a small part of it goes to the organs that are above ground level (Figure 2.3).

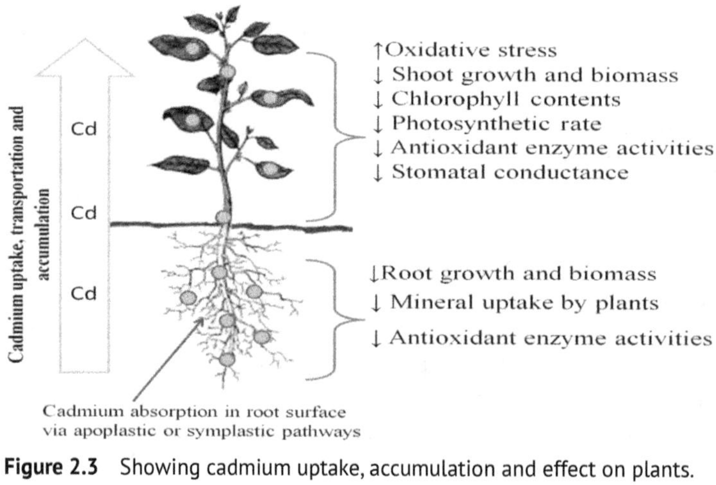

Figure 2.3 Showing cadmium uptake, accumulation and effect on plants.

The phytochelatins synthesis does not have a higher effect on hypertolerance, simultaneously, the transport of cadmium in xylem was elevated efficiently, resulting in enhanced movement of cadmium in the organs that are above ground level where it is compartmentalized [155].

2.5.1 Immobilization

Restriction of ions of cadmium in the plant cell wall takes place due to both the irreversible binding to the restricted sites present on the cell wall surface and the synthesis of complexes with carbohydrates present at outer surface of cell like callose, the "protein histidyl groups" and "carboxyl groups" present on the surface of pectin [156].

2.5.2 Chelation Using Organic Acids and Amino Acids

At high concentrations, the OAs like citrate and malate behave as hyperaccumulators in various plants [157], though their function in hypertension has not been accepted due to their low constant association with toxic metals [158]. Tryptophan plays a major role in the basic plant tolerance against cadmium [158]. The "*A. thaliana* trp51 mutant" is categorized by both excess production of tryptophan and increased tolerance against cadmium, while the "trp21 mutant" auxotrophic for tryptophan showed compact tolerance. It was also noted that the expression of the "AtTSB1 gene" coding for β-sub unit of tryptophan synthetase was encouraged on exposure to cadmium. Over-expression of the "AtTSB1 gene" in transgenic plants resulted in enhanced levels of tryptophan, therefore causing increased tolerance against Cd but did not showed any effect on its accumulation. Tryptophan is supposed to carry out a chelating function with cadmium ions. Increased concentrations of tryptophan in transgenic *A. thaliana* plant prevent the expression of genes coding for metal transporters. The above findings have opened up new viewpoints for breeding crop plants consisting of enhanced tolerance to cadmium, decreased accumulation and increased concentration of tryptophan [158].

2.5.3 Stress Peptide Synthesis

Various studies showed that cadmium in numerous plants bring about the production of various peptides such as "heat shock proteins (HSP)" and chaperons [14]. *A. thaliana* contain "CdI19 metallochaperone protein" whose molecular mass is 42.2 KDa and two CXXC motifs for the binding of metal are present in it, and two Cys residues responsible for attachment to metal are also present in it [159]. Along with cadmium, the above peptide also binds to Hg and Cu. "CdI19" is categorized by the addition of isoprenoid at its C-terminus through post-translation and is responsible for the addition of peptide to the cell membrane. Over-expression of CdI19 in transgenic plants resulted in elevated tolerance toward Cd as compared to normal plants. "Metallochaperones" are believed to play an essential role in conserving the concentrations of metals in cells at the optimum level through binding and transporting them to the chelators, such as PCs [159].

2.5.4 Cd Transporters

This was attainable because of the enlarged application of molecular-biological procedures using *S. cerevisiae* and other eukaryotes such as *A. thaliana* [160]. It has been reported in many studies that plant cells consist of several metal transporter protein families, which helps in sustaining homeostasis of metals and plays a major role in plant metal tolerance. At the present time, transcriptomes of numerous "hyperaccumulators" and other non "hyperaccumulating species" are being inspected, such as *Arabidopsis helleri* and *Arabidopsis thaliana* [161–163]. It has been observed that "hyperaccumulating species" are categorized by the increase in the gene expression that codes for metal transporters. It has been reported in many studies that increased levels of transcription are not linked to deviation of genes of hyperaccumulating and nonhyperaccumulating species but are rather triggered by the variations in the gene level regulation or the number of gene copies [155, 164].

2.5.5 Genetic Analysis of Cadmium Tolerance and Accumulation in Plants

There are two viewpoints which have a better understanding regarding the control of tolerance to metals by genes: assessment of dissimilarities in these traits within natural populations and "experimental mutagenesis". It was indicated in the genetic assessment of population of various higher plants that basic tolerance in several metals like Ar, Cu and Zn is regulated by gene-modifiers and main genes which adjust the level of tolerance [165, 166]. In some metals heavy metal tolerance is controlled by those genes which are dissimilar from the genes which provide tolerance to other metals [167]. Studies conducted on population of naturally growing grain plants like groundnut, pea plant, maize plant, wheat plant, rice plant, soybean plant, barley plant and hyperaccumulators of "*B. Juncea*" and "*T. Caerulescens*" indicated that genome-wide polymorphism is linked to the basic Cd tolerance and accumulation and its accumulation and "hyperaccumulation" [168–174]. For example, in pea, the content of cadmium present in dry biomass varied in different genotypes ranging from 35 mg/kg to 135 mg/kg and the tolerance index varied from 54% to 100% when grown in a cadmium-rich medium [170]. It was demonstrated in durum wheat that a low concentration of cadmium is modulated by the single gene dominant allele [172], which is usually present on the "5B chromosome" [175]. Lines with a low concentration of cadmium trait were marked due to the decreased transport of cadmium into the above-ground part from the roots system, resulting in the limited absorption by grain [176]. It was demonstrated that decreased absorption of cadmium does not influence the absorption of other elements; however some studies have shown that it is related to the reduced accumulation of Zn in Zn efficiency in grain [177]. New selections of wheat having small amount of cadmium were developed and registered in a 10-years program launched in Canada. One such variety was Strong field which was cultivated in 2004, which accounts for 25% of the area cultivated with durum wheat in Canada [175]. Nevertheless, it is still unknown whether "cadmium hypertolerance" and "hyperaccumulation" are interconnected or individual events. To resolve this issue, researchers inspected the outcomes of directed crossbreeding of plants showing variation in toxic metal tolerance and accumulation along with inter-specific hybrids

[178]. They intersected two plant hyperaccumulators, *A. helleri* and nonaccumulators *A. petraea* to acquire the F1 generations, which were later on intersected with the parent species. The F1 generations showed remarkable difference in the accumulation of zinc; analysis of individual plants showed that they either contained as small an amount of zinc as *A. petraea* or as high as *A. helleri* and intermediated variants. Genetic inspection of tolerance traits revealed that tolerance of copper, cadmium and zinc is mediated by single gene [179]. The examination of the F2 generations attained by crossing *A. helleri* and *A. patraea* [178, 180], and generations attained by crossing of numerous "*T. Caerulescens*" populations [181] demonstrated that "hyperaccumulation" of cadmium and tolerance of cadmium were not interconnected. Nevertheless, a study conducted by Lombi and co-workers on four dissimilar "*T. Caerulescens*" populations, gave contrary results [182]. They discovered the dissimilarity between populations in terms of their capability to collect cadmium in shoots and in traits of heavy metals tolerance. Populations of plants accumulating maximum amount of cadmium showed most tolerance to its toxic effects. Studies demonstrated the control of genetics on traits such as hyperaccumulation and hypertolerance of cadmium in plants [182]. Even though "experimental mutagenesis" is an efficient method to understand molecular-genetic mechanisms, very few mutant plants having mutated tolerance and accumulation toward cadmium have been studied. A mutant plant of cad 11 "*A. thaliana*" displayed growth in nutritious containing 0.6 μM cadmium in contrast to the wild type [183]. Chlorosis of the leaves and brown pigment formation in the roots has been observed by the researchers. The formation of metal sulphide complexes is due to the brown pigment formation. These characters of mutant plants were then used to recognize three other cad1 allele mutants. Chemical diepoxybutane mutagenesis was used to develop the cad13 mutant [184]. The cad 14 and cad15 were extracted using ethyl methane sulfonate (EMS). On comparison, the cad11, cad13, cad 14 and cad15 displayed different sensitivity toward cadmium at a concentration of 0.6 μM, 0.3 μM, and 1.5 μM, respectively. It was demonstrated using "high performance gel filtration" that the cad11 and cad13 mutants possess a decreased capability to synthesize complexes that binds Cd. The ability of cad1 mutants to store phytochelatins was unveiled using HPLC. Phytochelatins amounts in each mutant were dependent on its sensitivity toward cadmium. So it was shown that in the cad1 mutant's phytochelatin synthase activity is reduced [185]. Expression of the cad1 gene in circular DNA in *E. coli* cells showed that extracts of these cells have the capacity to catalyze the phytochelatin production that are activated by using heavy metals. This shows that coding of phytochelatins synthase is done by the cad1 gene [186]. Cad2 is the mutant of *A. thaliana* that is susceptible to cadmium and was extracted during the studies. In the absence of cadmium, differences in growth were not observed between the cad11 and cad 2 mutants and the wild type. The wild type showed normal growth at 6 μM concentrations while on the other hand the cad11 mutant displayed a showdown in growth at Cd concentration in 1 μM. The cad2 mutant displayed an intermediate phenotype: growth hinderance was noticed at 3 μM cadmium concentrations and was not spotted at the critical concentrations to the "cad 11 mutant" [187]. The "cad2 mutant" displayed a decreased capability to extract cadmium. The level of cadmium complexes in-vivo was decreased in cad2 plants with respect to the wild type.

2.6 Heavy Metal ATPases (HMA)

Heavy metal ATPases comes under the large ATPases superfamily of the P-type. The P1B subfamily that belongs to ATPases has two categories: carriers for monovalent and bivalent metals [160]. Eight types of proteins come under the P1B subfamily in *A. thaliana* [188]. In the zinc and cadmium transport from the root to the plant parts that are above ground level, an important role is played by heavy metal ATPases2 (HMA2), P1B and heavy metal ATPases4 (HMA4) that are present in the plasma lemma [189]. Over-expression of the HMA4 gene in transgenic plants displayed increased tolerance of cadmium and accumulation of cadmium in leaves [30]. It has been reported in some studies that "HMA3 protein" present in the membrane of vacuoles participates in the transportation of zinc and cadmium in the vacuoles. Plants containing the over-expressed HMA3 gene were categorized by increased tolerance against cadmium [190-193]. Using RNA interference, it has been seen that HMA4 protein is of great importance in zinc hyperaccumulation and zinc and cadmium tolerance in *A. helleri*. A. helleri possess hyperaccumulation and hypertolerance against cadmium because triplication and variation in "cis-regulation" of the *A. helleri* heavy metal ATPase 4 gene, causing its enhanced expression in *Arabidopsis helleri* as compared to *Arabidopsis thaliana* [155].

References

1 Gavrilescu, M., Demnerová, K., Aamand, J., Agathos, S., and Fava, F. (2015). Emerging pollutants in the environment: Present and future challenges in biomonitoring, ecological risks and bioremediation. *N. Biotechnol.* 32: 147–156.

2 Kelishadi, R., Moeini, R., Poursafa, P., Farajian, S., Yousefy, H., and Okhovat-Souraki, A.A. (2014). Independent association between air pollutants and vitamin D deficiency in young children in Isfahan, Iran. *Paediatr. Int. Child Health* 34: 50–55.

3 Serbula, S.M., Miljkovic, D.D., Kovacevic, R.M., and Ilic, A.A. (2012). Assessment of airborne heavy metal pollution using plant parts and topsoil. *Ecotoxicol. Environ. Saf.* 76: 209–214.

4 Kularatne, K.I.A. and De Freitas, C.R. (2013). Epiphytic lichens as biomonitors of airborne heavy metal pollution. *Environ. Exp. Bot.* 88: 24–32.

5 Fesenko, E. and Edwards, R. (2014). Plant synthetic biology: A new platform for industrial biotechnology. *J. Exp. Bot.* 65: 1927–1937.

6 Singh, N.P. and Santal, A.R. (2015). Phytoremediation of heavy metals: the use of green approaches to clean the environment. In: *Phytoremediation* (ed. A.A. Ansari, S.S. Gill, R. Gill, G.R. Lanza and L. Newman), pp. 115–1297. Cham: Springer.

7 Kang, J.W. (2014). Removing environmental organic pollutants with bioremediation and phytoremediation. *Biotechnol. Lett.* 36: 1129–1139.

8 Mani, D. and Kumar, C. (2014). Biotechnological advances in bioremediation of heavy metals contaminated ecosystems: An overview with special reference to phytoremediation. *Int. J. Environ. Sci. Technol.* 11: 843–872.

9 Ali, H., Khan, E., and Sajad, M.A. (2013). Phytoremediation of heavy metals: concepts and applications. *Chemosphere* 91: 869–881.

10 Lee, J.H. (2013). An overview of phytoremediation as a potentially promising technology for environmental pollution control. *Biotechnol. Bioprocess Eng.* 18: 431–439.

11 Pilon-Smits, E. and Pilon, M. (2002). Phytoremediation of metals using transgenic plants. *CRC Crit. Rev. Plant Sci.* 21: 439–456.

12 Raskin, I., Smith, R.D., and Salt, D.E. (1997). Phytoremediation of metals: Using plants to remove pollutants from the environment. *Curr. Opin. Biotechnol.* 8: 221–226.

13 Jabeen, R., Ahmad, A., and Iqbal, M. (2009). Phytoremediation of heavy metals: Physiological and molecular mechanisms. *Bot. Rev.* 75: 339–364.

14 Clemens, S. (2001). Molecular mechanisms of plant metal tolerance and homeostasis. *Planta* 212: 475–486.

15 Brooks, R.R. (1998). Geo-botany and hyper-accumulators. *Plant that Hyper-accumulate Heavy Metals* 55–94.

16 Baker, A.J. and Brooks, R. (1989). Terrestrial higher plants which hyper-accumulate metallic elements. A review of their distribution, ecology and phytochemistry. *Biorecovery* 1: 81–126.

17 Baker, A.J.M. (2000). Metal hyperaccumulator plants: A review of the ecology and physiology of a biological resource for phytoremediation of metal-polluted soils. *Phytoremed. Contaminat. Soil Water.*

18 Cherian, S. and Oliveira, M.M. (2005). Transgenic plants in phytoremediation: Recent advances and new possibilities. *Environ. Sci. Technol.* 39: 9377–9390.

19 Krämer, U. (2005). Phytoremediation: Novel approaches to cleaning up polluted soils. *Curr. Opin. Biotechnol.* 16: 133–141.

20 Assunção, A.G.L., Martins, P.D.C., De Folter, S., Vooijs, R., Schat, H., and Aarts, M.G.M. (2001). Elevated expression of metal transporter genes in three accessions of the metal hyper-accumulator. *Thlaspicaerulescens. Plant Cell Environ.* 24: 217–226.

21 Kim, D., Gustin, J.L., Lahner, B., Persans, M.W., Baek, D., Yun, D.J., and Salt, D.E. (2004). The plant CDF family member TgMTP1 from the Ni/Zn hyperaccumulatorThlaspigoesingense acts to enhance efflux of Zn at the plasma membrane when expressed in Saccharomyces cerevisiae. *Plant J.* 39: 237–251.

22 Lasat, M.M., Pence, N.S., Garvin, D.F., Ebbs, S.D., and Kochian, L.V. (2000). Molecular physiology of zinc transport in the Zn hyperaccumulator Thlaspicaerulescens. *J. Exp. Bot.* 51: 71–79.

23 Papoyan, A. and Kochian, L.V. (2004). Identification of Thlaspicaerulescens genes that may be involved in heavy metal hyperaccumulation and tolerance. Characterization of a novel heavy metal transporting ATPase. *Plant Physiol.* 136: 3814–3823.

24 Pence, N.S., Larsen, P.B., Ebbs, S.D., Letham, D.L., Lasat, M.M., Garvin, D.F., Eide, D., and Kochian, L.V. (2000). The molecular physiology of heavy metal transport in the Zn/Cd hyperaccumulator*Thlaspicaerulescens. Proc. Natl. Acad. Sci.* 97: 4956–4960.

25 Persans, M.W., Nieman, K., and Salt, D.E. (2001). Functional activity and role of cation-efflux family members in Ni hyperaccumulation in *Thlaspigoesingense. Proc. Natl. Acad. Sci.* 98: 9995–10000.

26 Hirschi, K.D., Korenkov, V.D., Wilganowski, N.L., and Wagner, G.J. (2000). Expression of Arabidopsis CAX2 in tobacco. Altered metal accumulation and increased manganese tolerance. *Plant Physiol.* 124: 125–134.

27 Song, W.Y., Sohn, E.J., Martinoia, E., Lee, Y.J., Yang, Y.Y., Jasinski, M., Forestier, C., Hwang, I., and Lee, Y. (2003). Engineering tolerance and accumulation of lead and cadmium in transgenic plants. *Nat. Biotechnol.* 21: 914–919.

28 Thomine, S., Wang, R., Ward, J.M., Crawford, N.M., and Schroeder, J.I. (2000). Cadmium and iron transport by members of a plant metal transporter family in Arabidopsis with homology to Nramp genes. *Proc. Natl. Acad. Sci.* 97: 4991–4996.

29 Van Der Zaal, B.J., Neuteboom, L.W., Pinas, J.E., Chardonnens, A.N., Schat, H., Verkleij, J.A., and Hooykaas, P.J. (1999). Overexpression of a novel Arabidopsis gene related to putative zinc-transporter genes from animals can lead to enhanced zinc resistance and accumulation. *Plant Physiol.* 119: 1047–1056.

30 Verret, F., Gravot, A., Auroy, P., Leonhardt, N., David, P., Nussaume, L., Vavasseur, A., and Richaud, P. (2004). Overexpression of AtHMA4 enhances root-to-shoot translocation of zinc and cadmium and plant metal tolerance. *FEBS Lett.* 576: 306–312.

31 Escande, V., Olszewski, T.K., and Grison, C. (2014). Preparation of ecological catalysts derived from Zn hyperaccumulating plants and their catalytic activity in Diels–Alder reaction. *ComptesRendusChimie* 17: 731–737.

32 Escande, V., Olszewski, T.K., and Grison, C. (2015). From biodiversity to catalytic diversity: How to control the reaction mechanism by the nature of metallophytes. *Environ. Sci. Pollut. Res.* 22: 5653–5666.

33 Ruiz, O.N. and Daniell, H. (2009). Genetic engineering to enhance mercury phytoremediation. *Curr. Opin. Biotechnol.* 20: 213–219.

34 Reeves, R.D. and Brooks, R.R. (1983). Hyperaccumulation of lead and zinc by two metallophytes from mining areas of Central Europe. *Environ. Pollut. Ser. A Ecol. Biol.* 31: 277–285.

35 Pollard, A.J., Reeves, R.D., and Baker, A.J. (2014). Facultative hyperaccumulation of heavy metals and metalloids. *Plant Sci.* 217: 8–17.

36 Muszyńska, E., Hanus-Fajerska, E., and Ciarkowska, K. (2015). Studies on Gypsophila fastigiata parameters verifying its suitability to reclamation of post-flotation Zn-Pb wastes. *Geol. Geophys. Environ.* 41: 17–24.

37 Iori, V., Pietrini, F., Massacci, A., and Zacchini, M. (2012). Induction of metal binding compounds and antioxidativedefence in callus cultures of two black poplar (P. nigra) clones with different tolerance to cadmium. *Plant Cell Tissue Organ Cult.* 108: 17–26.

38 Wiszniewska, A., Hanus-Fajerska, E., Smoleń, S., and Muszyńska, E. (2015). In vitro selection for lead tolerance in shoot culture of Daphne species. *ActaScientiarumPolonorumHortorumCultus* 14: 129–142.

39 Muszyńska, E. and Hanus-Fajerska, E. (2017). In vitro multiplication of Dianthus carthusianorum calamine ecotype with the aim to revegetate and stabilize polluted wastes. *Plant Cell Tissue Organ Cult.* 128: 631–640.

40 Hossain, M.A., Piyatida, P., Da Silva, J.A.T., and Fujita, M. (2012). Molecular mechanism of heavy metal toxicity and tolerance in plants: Central role of glutathione in detoxification of reactive oxygen species and methylglyoxal and in heavy metal chelation. *J. Bot.*

41 Salt, D.E., Prince, R.C., Pickering, I.J., and Raskin, I. (1995). Mechanisms of cadmium mobility and accumulation in Indian mustard. *Plant Physiol.* 109: 1427–1433.

42 Anjum, N.A., Pereira, M.E., Ahmad, I., Duarte, A.C., Umar, S., and Khan, N.A. (eds.) (2012). *Phytotechnologies: Remediation of Environmental Contaminants*. CRC Press.

43 Lasat, M.M. (2002). Phytoextraction of toxic metals: A review of biological mechanisms. *J. Environ. Qual.* 31: 109–120.

44 Taiz, L. and Zeiger, E. (2002). *Plant Physiology*. Sinauer Associates.

45 Pilon-Smits, E. (2005). Phytoremediation. *Annu. Rev. Plant Biol.* 56: 15–39.

46 Krämer, U., Cotter-Howells, J.D., Charnock, J.M., Baker, A.J., and Smith, J.A.C. (1996). Freehistidine as a metal chelator in plants that accumulate nickel. *Nature* 379: 635–638.

47 Williams, L.E., Pittman, J.K., and Hall, J.L. (2000). Emerging mechanisms for heavy metal transport in plants. *Biochim. Biophys. Acta Biomembr.* 1465: 104–126.

48 Guerinot, M.L. (2000). The ZIP family of metal transporters. *Biochim. Biophys. Acta Biomem.* 1465: 190–198.

49 Hall, J.Á. (2002). Cellular mechanisms for heavy metal detoxification and tolerance. *J. Exp. Bot.* 53: 1–11.

50 Axelsen, K.B. and Palmgren, M.G. (2001). Inventory of the superfamily of P-type ion pumps in Arabidopsis. *Plant Physiol.* 126: 696–706.

51 Senden, M.H.M.N. and Wolterbeek, H.T. (1990). Effect of citric acid on the transport of cadmium through xylem vessels of excised tomato stem-leaf systems. *Actabotanicaneerlandica* 39: 297–303.

52 Burken, J.G. (2003). Uptake and metabolism of organic compounds: Green-liver model. *Phytoremediation: Transformation and Control of Contaminants* 59: 59–84.

53 Cobbett, C. and Goldsbrough, P. (2002). Phytochelatins and metallothioneins: Roles in heavy metal detoxification and homeostasis. *Annu. Rev. Plant Biol.* 53: 159–182.

54 Dalvi, A.A. and Bhalerao, S.A. (2013). Response of plants towards heavy metal toxicity: An overview of avoidance, tolerance and uptake mechanism. *Ann. Plant Sci.* 2: 362–368.

55 Iwasaki, K.Z., Sakurai, K., and Takahashi, E. (1990). Copper binding by the root cell walls of Italian ryegrass and red clover. *Soil Sci. Plant Nutr.* 36: 431–439.

56 Horst, W.J. and Marschner, H. (1978). Effect of silicon on manganese tolerance of bean plants (Phaseolus vulgaris L.). *Plant Soil* 50: 287–303.

57 Masion, A. and Bertsch, P.M. (1997). Aluminium speciation in the presence of wheat root cell walls: A wet chemical study. *Plant Cell Environ.* 20: 504–512.

58 Reichman, S.M. (2002). *The Responses of Plants to Metal Toxicity: A Review Forusing on Copper, Manganese & Zinc*, 22–26. Melbourne: Australian Minerals & Energy Environment Foundation.

59 Tong, Y.P., Kneer, R., and Zhu, Y.G. (2004). Vacuolar compartmentalization: A second-generation approach to engineering plants for phytoremediation. *Trends Plant Sci.* 9: 7–9.

60 Robinson, N.J., Tommey, A.M., Kuske, C., and Jackson, P.J. (1993). Plant metallothioneins. *Biochem. J.* 295: 1–10.

61 Vatamaniuk, O.K., Mari, S., Lang, A., Chalasani, S., Demkiv, L.O., and Rea, P.A. (2004). Phytochelatin synthase, a dipeptidyltransferase that undergoes multisite acylation with γ-glutamylcysteine during catalysis: Stoichiometric and site-directed mutagenic analysis of Arabidopsis thaliana PCS1-catalyzedphytochelation?>synthesis. *J. Biol. Chem.* 279: 22449–22460.

62 Klapheck, S., Schlunz, S., and Bergmann, L. (1995). Synthesis of phytochelatins and homo-phytochelatins in Pisum sativum L. *Plant Physiol.* 107: 515–521.

63 Cobbett, C.S. (2000). Phytochelatins and their roles in heavy metal detoxification. *Plant Physiol.* 123: 825–832.

64 Pickering, I.J., Prince, R.C., George, M.J., Smith, R.D., George, G.N., and Salt, D.E. (2000). Reduction and coordination of arsenic in Indian mustard. *Plant Physiol.* 122: 1171–1178.

65 Schat, H., Llugany, M., Vooijs, R., Hartley-Whitaker, J., and Bleeker, P.M. (2002). The role of phytochelatins in constitutive and adaptive heavy metal tolerances in hyperaccumulator and non-hyperaccumulatormetallophytes. *J. Exp. Bot.* 53: 2381–2392.

66 LeDuc, D.L., Tarun, A.S., Montes-Bayon, M., Meija, J., Malit, M.F., Wu, C.P., AbdelSamie, M., Chiang, C.Y., Tagmount, A., deSouza, M., and Neuhierl, B. (2004). Overexpression of selenocysteine methyltransferase in Arabidopsis and Indian mustard increases selenium tolerance and accumulation. *Plant Physiol.* 135: 377–383.

67 Neuhierl, B. and Böck, A. (1996). On the mechanism of selenium tolerance in selenium-accumulating plants: Purification and characterization of a specific selenocysteine methyltransferase from cultured cells of astragalusbisculatus. *Eur. J. Biochem.* 239: 235–238.

68 Brown, T.A. and Shrift, A. (1981). Exclusion of selenium from proteins of selenium-tolerant Astragalus species. *Plant Physiol.* 67: 1051–1053.

69 Summers, A.O. (1986). Organization, expression, and evolution of genes for mercury resistance. *Annu. Rev. Microbiol.* 40: 607–634.

70 Rugh, C.L., Bizily, S.P., and Meagher, R.B. (2000). Phytoreduction of environmental mercury pollution. In: *Phytoremediation of Toxic Metals-Using Plants to Clean up the Environment* (ed. I. Raskin and B.D. Ensley).

71 Ruiz, O.N., Hussein, H.S., Terry, N., and Daniell, H. (2003). Phytoremediation of organomercurial compounds via chloroplast genetic engineering. *Plant Physiol.* 132: 1344–1352.

72 Dhankher, O.P., Li, Y., Rosen, B.P., Shi, J., Salt, D., Senecoff, J.F., Sashti, N.A., and Meagher, R.B. (2002). Engineering tolerance and hyperaccumulation of arsenic in plants by combining arsenate reductase and γ-glutamylcysteine synthetase expression. *Nat. Biotechnol.* 20: 1140–1145.

73 Grill, E., Löffler, S., Winnacker, E.L., and Zenk, M.H. (1989). Phytochelatins, the heavy-metal-binding peptides of plants, are synthesized from glutathione by a specific γ-glutamylcysteine dipeptidyl transpeptidase (phytochelatin synthase). *Proc. Natl. Acad. Sci.* 86: 6838–6842.

74 Gallego, S.M., Benavides, M.P., and Tomaro, M.L. (1996). Effect of heavy metal ion excess on sunflower leaves: Evidence for involvement of oxidative stress. *Plant Sci.* 121: 151–159.

75 Zhu, Y.L., Pilon-Smits, E.A., Jouanin, L., and Terry, N. (1999). Overexpression of glutathione synthetase in Indian mustard enhances cadmium accumulation and tolerance. *Plant Physiol.* 119: 73–80.

76 Zhu, Y.L., Pilon-Smits, E.A., Tarun, A.S., Weber, S.U., Jouanin, L., and Terry, N. (1999). Cadmium tolerance and accumulation in Indian mustard is enhanced by overexpressing γ-glutamylcysteinesynthetase. *Plant Physiol.* 121: 1169–1177.

77 Zenk, M.H. (1996). Heavy metal detoxification in higher plants-a review. *Gene* 179: 21–30.

78 Bennett, L.E., Burkhead, J.L., Hale, K.L., Terry, N., Pilon, M., and Pilon-Smits, E.A. (2003). Analysis of transgenic Indian mustard plants for phytoremediation of metal-contaminated mine tailings. *J. Environ. Qual.* 32: 432–440.

79 Pilon-Smits, E.A., Hwang, S., Lytle, C.M., Zhu, Y., Tai, J.C., Bravo, R.C., Chen, Y., Leustek, T., and Terry, N. (1999). Overexpression of ATP sulfurylase in Indian mustard leads to increased selenate uptake, reduction, and tolerance. *Plant Physiol.* 119: 123–132.

80 De Souza, M.P., Pilon-Smits, E.A.H., Lytle, C.M., Hwang, S., Tai, J.C., Honma, T.S.U., Yeh, L., and Terry, N. (1998). Rate-limiting steps in selenium volatilization by Brassica juncea. *Plant Physiol.* 117: 1487–1494.

81 Huysen, T., Terry, N., and Pilon-Smits, E.A.H. (2004). Exploring the selenium phytoremediation potential of transgenic Indian mustard overexpressing ATP sulfurylase or cystathionine-γ-synthase. *Int. J. Phytoremed.* 6: 111–118.

82 Bañuelos, G., Terry, N., LeDuc, D.L., Pilon-Smits, E.A., and Mackey, B. (2005). Field trial of transgenic Indian mustard plants shows enhanced phytoremediation of selenium-contaminated sediment. *Environ. Sci. Technol.* 39: 1771–1777.

83 May, M.J., Vernoux, T., Leaver, C., Montagu, M.V., and Inzé, D. (1998). Glutathione homeostasis in plants: Implications for environmental sensing and plant development. *J. Exp. Bot.* 49: 649–667.

84 Noctor, G., Arisi, A.C.M., Jouanin, L., Kunert, K.J., Rennenberg, H., and Foyer, C.H. (1998). Glutathione: Biosynthesis, metabolism and relationship to stress tolerance explored in transformed plants. *J. Exp. Bot.* 49: 623–647.

85 Kocsy, G., Galiba, G., and Brunold, C. (2001). Role of glutathione in adaptation and signalling during chilling and cold acclimation in plants. *Physiol. Plant.* 113: 158–164.

86 Ruiz, J. and Blumwald, E. (2002). Salinity-induced glutathione synthesis in Brassica napus. *Planta* 214: 965–969.

87 Farmer, E.E. and Ryan, C.A. (1992). Octadecanoid precursors of jasmonic acid activate the synthesis of wound-inducible proteinase inhibitors. *Plant Cell* 4: 129–134.

88 Buchanan-Wollaston, V. (1994). Isolation of cDNA clones for genes that are expressed during leaf senescence in Brassica napus (identification of a gene encoding a senescence-specific metallothionein-like protein). *Plant Physiol.* 105: 839–846.

89 Choi, D., Kim, H.M., Yun, H.K., Park, J.A., Kim, W.T., and Bok, S.H. (1996). Molecular cloning of a metallothionein-like gene from Nicotianaglutinosa L. and its induction by wounding and tobacco mosaic virus infection. *Plant Physiol.* 112: 353–359.

90 Pilon-Smits, E.A. and LeDuc, D.L. (2009). Phytoremediation of selenium using transgenic plants. *Curr. Opin. Biotechnol.* 20: 207–212.

91 White, P.J. and Brown, P.H. (2010). Plant nutrition for sustainable development and global health. *Ann. Bot.* 105: 1073–1080.

92 El Mehdawi, A.F. and Pilon-Smits, E.A.H. (2012). Ecological aspects of plant selenium hyperaccumulation. *Plant Biol.* 14: 1–10.

93 Feng, R., Wei, C., and Tu, S. (2013). The roles of selenium in protecting plants against abiotic stresses. *Environ. Exp. Bot.* 87: 58–68.

94 Brown, T.A. and Shrift, A. (1982). Selenium: Toxicity and tolerance in higher plants. *Biol. Rev.* 57: 59–84.

95 White, P.J., Bowen, H.C., Parmaguru, P., Fritz, M., Spracklen, W.P., Spiby, R.E., Meacham, M.C., Mead, A., Harriman, M., Trueman, L.J., and Smith, B.M. (2004). Interactions between selenium and sulphur nutrition in Arabidopsis thaliana. *J. Exp. Bot.* 55: 1927–1937.

96 Van Hoewyk, D. (2013). A tale of two toxicities: Malformed selenoproteins and oxidative stress both contribute to selenium stress in plants. *Ann. Bot.* 112: 965–972.

97 White, P.J., Bowen, H.C., Marshall, B., and Broadley, M.R. (2007). Extraordinarily high leaf selenium to sulfur ratios define 'Se-accumulator' plants. *Ann. Bot.* 100: 111–118.

98 White, P.J. (2016). Selenium accumulation by plants. *Ann. Bot.* 117: 217–235.

99 Dhillon, K.S. and Dhillon, S.K. (2003). Distribution and management of seleniferous soils. *Adv. Agron.* 79: 119–184.

100 Fordyce, F.M. (2013). Selenium deficiency and toxicity in the environment. In: *Essentials of Medical Geology*, 375–416. Dordrecht: Springer.

101 Barker, A.V. and Pilbeam, D.J. (eds.) (2015). *Handbook of Plant Nutrition*. CRC Press.

102 Mikkelsen, R.L., Page, A.L., and Bingham, F.T. (1989). Factors affecting selenium accumulation by agricultural crops. *Selenium Agric. Environ.* 23: 65–94.

103 White, P.J., Broadley, M.R., Bowen, H.C., and Johnson, S.E. (2007). Selenium and its relationship with sulfur. In: *Sulfur in Plants an Ecological Perspective*, 225–252. Dordrecht: Springer.

104 Rosenfeld, I. and Beath, O.A. (1964). *Selenium: Geobotany, Biochemistry, Toxicity, and Nutrition*. Academic Press.

105 Baker, A.J. (1981). Accumulators and excluders-strategies in the response of plants to heavy metals. *J. Plant Nutr.* 3: 643–654.

106 Reeves, R.D. (2000). Metal-accumulating plants. Phytoremediation of Toxic Metals: Using Plants to Clean up the Environment.

107 Watanabe, T., Broadley, M.R., Jansen, S., White, P.J., Takada, J., Satake, K., Takamatsu, T., Tuah, S.J., and Osaki, M. (2007). Evolutionary control of leaf element composition in plants. *New Phytol.* 174: 516–523.

108 Feist, L.J. and Parker, D.R. (2001). Ecotypic variation in selenium accumulation among populations of Stanleyapinnata. *New Phytol.* 149: 61–69.

109 Cappa, J.J., Cappa, P.J., El Mehdawi, A.F., McAleer, J.M., Simmons, M.P., and Pilon-Smits, E.A. (2014). Characterization of selenium and sulfur accumulation across the genus Stanleya (Brassicaceae): A field survey and common-garden experiment. *Am. J. Bot.* 101: 830–839.

110 El Mehdawi, A.F., Paschke, M.W., and Pilon-Smits, E.A. (2015). S ymphyotrichumericoides populations from seleniferous and nonseleniferous soil display striking variation in selenium accumulation. *New Phytol.* 206: 231–242.

111 Cappa, J.J. and Pilon-Smits, E.A. (2014). Evolutionary aspects of elemental hyperaccumulation. *Planta* 239: 267–275.

112 Krämer, U. and Chardonnens, A. (2001). The use of transgenic plants in the bioremediation of soils contaminated with trace elements. *Appl. Microbiol. Biotechnol.* 55: 661–672.

113 Rugh, C.L., Wilde, H.D., Stack, N.M., Thompson, D.M., Summers, A.O., and Meagher, R.B. (1996). Mercuric ion reduction and resistance in transgenic Arabidopsis thaliana plants expressing a modified bacterial merA gene. *Proc. Natl. Acad. Sci.* 93: 3182–3187.

114 Ehrlich, H.L. (1997). Microbes and metals. *Appl. Microbiol. Biotechnol.* 48: 687–692.

115 Colwell, R.R. (1997). Microbial diversity: The importance of exploration and conservation. *J. Ind. Microbiol. Biotechnol.* 18: 302–307.

116 Huber, H. and Stetter, K.O. (1998). Hyperthermophiles and their possible potential in biotechnology. *J. Biotechnol.* 64: 39–52.

117 Jones, B.E., Grant, W.D., Duckworth, A.W., and Owenson, G.G. (1998). Microbial diversity of soda lakes. *Extremophiles* 2: 191–200.

118 Srinivasan, M.C. and Ingale, S. (1999). Production and application of enzymes stable to and active under extreme environments: An overview. *PINSA B* 65: 143–162.

119 De Souza, M.P., Amini, A., Dojka, M.A., Pickering, I.J., Dawson, S.C., Pace, N.R., and Terry, N. (2001). Identification and characterization of bacteria in a selenium-contaminated hypersaline evaporation pond. *Appl. Environ. Microbiol.* 67: 3785–3794.

120 Terry, N., Zayed, A.M., De Souza, M.P., and Tarun, A.S. (2000). Selenium in higher plants. *Annu. Rev. Plant Biol.* 51: 401–432.

121 Kärenlampi, S., Schat, H., Vangronsveld, J., Verkleij, J.A.C., Van Der Lelie, D., Mergeay, M., and Tervahauta, A.I. (2000). Genetic engineering in the improvement of plants for phytoremediation of metal polluted soils. *Environ. Pollut.* 107: 225–231.

122 Rugh, C.L., Senecoff, J.F., Meagher, R.B., and Merkle, S.A. (1998). Development of transgenic yellow poplar for mercury phytoremediation. *Nat. Biotechnol.* 16: 925–928.

123 Jorgensen, R. (1990). Altered gene expression in plants due to trans interactions between homologous genes. *Trends Biotechnol.* 8: 340–344.

124 Pearce, R.S. (1999). Molecular analysis of acclimation to cold. *Plant Growth Regul.* 29: 47–76.

125 Jaglo-Ottosen, K.R., Gilmour, S.J., Zarka, D.G., Schabenberger, O., and Thomashow, M.F. (1998). Arabidopsis CBF1 overexpression induces COR genes and enhances freezing tolerance. *Science* 280: 104–106.

126 Schmidt, A. (1986). Regulation of sulfur metabolism in plants. In: *Progress in Botany*, 133–150. Berlin, Heidelberg: Springer.

127 Clarkson, D.T., Hawkesford, M.J., and Davidian, J.C. (1993). Membrane and long-distance transport of sulfate. *Sulfur Nutrition and Sulfur Assimilation in Higher Plants: Fundamental, Environmental and Agricultural Aspects* 3: 19.

128 Lappartient, A.G. and Touraine, B. (1996). Demand-driven control of root ATP sulfurylase activity and SO42-uptake in intact canola (the role of phloem-translocated glutathione). *Plant Physiol.* 111: 147–157.

129 Setya, A., Murillo, M., and Leustek, T. (1996). Sulfate reduction in higher plants: Molecular evidence for a novel 5′-adenylylsulfate reductase. *Proc. Natl. Acad. Sci.* 93: 13383–13388.

130 Hell, R. (1997). Molecular physiology of plant sulfur metabolism. *Planta* 202: 138–148.

131 Takahashi, S., Yip, W.C., and Tamura, G. (1997). Purification and characterization of ferredoxin–sulfite reductase from Turnip (Brassica rapa) leaves and comparison of properties with ferredoxin–sulfite reductase from Turnip roots. *Biosci. Biotechnol. Biochem.* 61: 1486–1490.

132 Bolchi, A., Petrucco, S., Tenca, P.L., Foroni, C., and Ottonello, S. (1999). Coordinate modulation of maize sulfate permease and ATP sulfurylase mRNAs in response to variations in sulfur nutritional status: Stereospecific down-regulation by L-cysteine. *Plant Mol. Biol.* 39: 527–537.

133 Leustek, T., Martin, M.N., Bick, J.A., and Davies, J.P. (2000). Pathways and regulation of sulfur metabolism revealed through molecular and genetic studies. *Annu. Rev. Plant Biol.* 51: 141–165.

134 Davies, J.P., Yildiz, F.H., and Grossman, A. (1996). Sac1, a putative regulator that is critical for survival of Chlamydomonas reinhardtii during sulfur deprivation. *EMBO J.* 15: 2150–2159.

135 Yildiz, F.H., Davies, J.P., and Grossman, A. (1996). Sulfur availability and the SAC1 gene control adenosine triphosphate sulfurylase gene expression in Chlamydomonas reinhardtii. *Plant Physiol.* 112: 669–675.

136 Arazi, T., Sunkar, R., Kaplan, B., and Fromm, H. (1999). A tobacco plasma membrane calmodulin-binding transporter confers Ni2+ tolerance and Pb2+ hypersensitivity in transgenic plants. *Plant J.* 20: 171–182.

137 Alkorta, I., Hernández-Allica, J., Becerril, J.M., Amezaga, I., Albizu, I., and Garbisu, C. (2004). Recent findings on the phytoremediation of soils contaminated with environmentally toxic heavy metals and metalloids such as zinc, cadmium, lead, and arsenic. *Rev. Environ. Sci. Biotechnol.* 3: 71–90.

138 Thomas, J.C., Davies, E.C., Malick, F.K., Endreszl, C., Williams, C.R., Abbas, M., Petrella, S., Swisher, K., Perron, M., Edwards, R., and Ostenkowski, P. (2003). Yeast metallothionein in transgenic tobacco promotes copper uptake from contaminated soils. *Biotechnol. Prog.* 19: 273–280.

139 Hamer, D.H. (1986). Metallothionein. *Annu. Rev. Biochem.* 55: 913–951.

140 Misra, S. and Gedamu, L. (1989). Heavy metal tolerant transgenic Brassica napus L. and Nicotiana tabacum L. plants. *Theor. Appl. Genet.* 78: 161–168.

141 Pan, A., Yang, M., Tie, F., Li, L., Chen, Z., and Ru, B. (1994). Expression of mouse metallothionein-I gene confers cadmium resistance in transgenic tobacco plants. *Plant Mol. Biol.* 24: 341–351.

142 Hattori, J., Labbé, H., and Miki, B.L. (1994). Construction and expression of a metallothionein–β-glucuronidase gene fusion. *Genome* 37: 508–512.

143 Bizily, S.P., Rugh, C.L., and Meagher, R.B. (2000). Phytodetoxification of hazardous organomercurials by genetically engineered plants. *Nat. Biotechnol.* 18: 213–217.

144 Heaton, A.C., Rugh, C.L., Wang, N.J., and Meagher, R.B. (2005). Physiological responses of transgenic merA-tobacco (Nicotianatabacum) to foliar and root mercury exposure. *Water Air Soil Pollut.* 161: 137–155.

145 Bücking, H. and Heyser, W. (2003). Uptake and transfer of nutrients in ectomycorrhizal associations: Interactions between photosynthesis and phosphate nutrition. *Mycorrhiza* 13: 59–68.

146 Chaney, R.L., Malik, M., Li, Y.M., Brown, S.L., Brewer, E.P., Angle, J.S., and Baker, A.J. (1997). Phytoremediation of soil metals. *Curr. Opin. Biotechnol.* 8: 279–284.

147 Ow, D.W. (1996). Heavy metal tolerance genes: Prospective tools for bioremediation. *Resour. Conserv. Recycl.* 18: 135–149.

148 Lee, J., Bae, H., Jeong, J., Lee, J.Y., Yang, Y.Y., Hwang, I., Martinoia, E., and Lee, Y. (2003). Functional expression of a bacterial heavy metal transporter in Arabidopsis enhances resistance to and decreases uptake of heavy metals. *Plant Physiol.* 133: 589–596.

149 Van Huysen, T., Abdel-Ghany, S., Hale, K.L., LeDuc, D., Terry, N., and Pilon-Smits, E.A. (2003). Overexpression of cystathionine-γ-synthase enhances selenium volatilization in Brassica juncea. *Planta* 218: 71–78.

150 French, C.E., Rosser, S.J., Davies, G.J., Nicklin, S., and Bruce, N.C. (1999). Biodegradation of explosives by transgenic plants expressing pentaerythritoltetranitrate reductase. *Nat. Biotechnol.* 17: 491–494.

151 Meagher, R.B. (2000). Phytoremediation of toxic elemntal and organic pollutants. *Curr. Opin. Plant Biol.* 3: 153–162.

152 Bizily, S.P., Kim, T., Kandasamy, M.K., and Meagher, R.B. (2003). Subcellular targeting of methylmercury lyase enhances its specific activity for organic mercury detoxification in plants. *Plant Physiol.* 131: 463–471.

153 Dushenkov, S., Skarzhinskaya, M., Glimelius, K., Gleba, D., and Raskin, I. (2002). Bioengineering of a phytoremediation plant by means of somatic hybridization. *Int. J. Phytoremed.* 4: 117–126.

154 Grichko, V.P., Filby, B., and Glick, B.R. (2000). Increased ability of transgenic plants expressing the bacterial enzyme ACC deaminase to accumulate Cd, Co, Cu, Ni, Pb, and Zn. *J. Biotechnol.* 81: 45–53.

155 Hanikenne, M., Talke, I.N., Haydon, M.J., Lanz, C., Nolte, A., Motte, P., Kroymann, J., Weigel, D., and Krämer, U. (2008). Evolution of metal hyperaccumulation required cis-regulatory changes and triplication of HMA4. *Nature* 453: 391–395.

156 Zornoza, P., Vázquez, S., Esteban, E., Fernández-Pascual, M., and Carpena, R. (2002). Cadmium-stress in nodulated white lupin: Strategies to avoid toxicity. *Plant Physiol. Biochem.* 40: 1003–1009.

157 Ueno, D., Ma, J.F., Iwashita, T., Zhao, F.J., and McGrath, S.P. (2005). Identification of the form of Cd in the leaves of a superior Cd-accumulating ecotype of Thlaspicaerulescens using 113 Cd-NMR. *Planta* 221: 928–936.

158 Callahan, D.L., Baker, A.J., Kolev, S.D., and Wedd, A.G. (2006). Metal ion ligands in hyperaccumulating plants. *JBIC J. Biol. Inorganic Chem.* 11: 2–12.

159 Suzuki, N., Yamaguchi, Y., Koizumi, N., and Sano, H. (2002). Functional characterization of a heavy metal binding protein CdI19 from Arabidopsis. *Plant J.* 32: 165–173.

160 Hall, J.Á. and Williams, L.E. (2003). Transition metal transporters in plants. *J. Exp. Bot.* 54: 2601–2613.

161 Becher, M., Talke, I.N., Krall, L., and Krämer, U. (2004). Cross-species microarray transcript profiling reveals high constitutive expression of metal homeostasis genes in shoots of the zinc hyperaccumulator Arabidopsis halleri. *Plant J.* 37: 251–268.

162 Weber, M., Harada, E., Vess, C., Roepenack-Lahaye, E.V., and Clemens, S. (2004). Comparative microarray analysis of Arabidopsis thaliana and Arabidopsis halleri roots identifies nicotianamine synthase, a ZIP transporter and other genes as potential metal hyperaccumulation factors. *Plant J.* 37: 269–281.

163 Talke, I.N., Hanikenne, M., and Krämer, U. (2006). Zinc-dependent global transcriptional control, transcriptional deregulation, and higher gene copy number for genes in metal homeostasis of the hyperaccumulator Arabidopsis halleri. *Plant Physiol.* 142: 148–167.

164 Oomen, R.J., Wu, J., Lelièvre, F., Blanchet, S., Richaud, P., Barbier-Brygoo, H., Aarts, M.G., and Thomine, S. (2009). Functional characterization of NRAMP3 and NRAMP4 from the metal hyperaccumulatorThlaspicaerulescens. *New Phytol.* 181: 637–650.

165 Schat, H., Kuiper, E., Ten Bookum, W.M., and Vooijs, R. (1993). A general model for the genetic control of copper tolerance in Silene vulgaris: Evidence from crosses between plants from different tolerant populations. *Heredity* 70: 142–147.

166 Smith, S.E. and Macnair, M.R. (1998). Hypostatic modifiers cause variation in degree of copper tolerance in Mimulusguttatus. *Heredity* 80: 760–768.

167 Tilstone, G.H., Macnair, M.R., and Smith, S.E. (1997). Does copper tolerance give cadmium tolerance in Mimulusguttatus. *Heredity* 79: 445–452.

168 Arao, T. and Ae, N. (2003). Genotypic variations in cadmium levels of rice grain. *Soil Sci. Plant Nutr.* 49: 473–479.

169 McLaughlin, M.J., Bell, M.J., Wright, G.C., and Cruickshank, A. (1997). Inter-and intra-specific variation in accumulation of cadmium by peanut, soybean, and navybean. *Aust. J. Agric. Res.* 48: 1151–1160.

170 Belimov, A.A., Safronova, V.I., Tsyganov, V.E., Borisov, A.Y., Kozhemyakov, A.P., Stepanok, V.V., Martenson, A.M., Gianinazzi-Pearson, V., and Tikhonovich, I.A. (2003). Genetic variability in tolerance to cadmium and accumulation of heavy metals in pea (Pisum sativum L.). *Euphytica* 131: 25–35.

171 Belimov, A.A., Safronova, V.I., Demchinskaya, S.V., and Dzyuba, O.O. (2007). Intraspecific variability of cadmium tolerance in hydroponically grown Indian mustard (Brassica juncea (L.) Czern.) seedlings. *ActaPhysiologiaePlantarum* 29: 473–478.

172 Clarke, J.M., Leisle, D., and Kopytko, G.L. (1997). Inheritance of cadmium concentration in five durum wheat crosses. *Crop Sci.* 37: 1722–1726.

173 Liu, J., Zhu, Q., Zhang, Z., Xu, J., Yang, J., and Wong, M.H. (2005). Variations in cadmium accumulation among rice cultivars and types and the selection of cultivars for reducing cadmium in the diet. *J. Sci. Food Agric.* 85: 147–153.

174 Roosens, N., Verbruggen, N., Meerts, P., Ximenez-Embun, P., and Smith, J.A.C. (2003). Natural variation in cadmium tolerance and its relationship to metal hyperaccumulation for seven populations of Thlaspicaerulescens from western Europe. *Plant Cell Environ.* 26: 1657–1672.

175 Grant, C.A., Clarke, J.M., Duguid, S., and Chaney, R.L. (2008). Selection and breeding of plant cultivars to minimize cadmium accumulation. *Sci. Total Environ.* 390: 301–310.

176 Chan, D.Y. and Hale, B.A. (2004). Differential accumulation of Cd in durum wheat cultivars: Uptake and retranslocation as sources of variation. *J. Exp. Bot.* 55: 2571–2579.

177 Hart, J.J., Welch, R.M., Norvell, W.A., Clarke, J.M., and Kochian, L.V. (2005). Zinc effects on cadmium accumulation and partitioning in near-isogenic lines of durum wheat that differ in grain cadmium concentration. *New Phytol.* 167: 391–401.

178 Macnair, M.R., Bert, V., Huitson, S.B., Saumitou-Laprade, P., and Petit, D. (1999). Zinc tolerance and hyperaccumulation are genetically independent characters. *Proc. R. Soc. Lond. B Biol. Sci.* 266: 2175–2179.

179 Schat, H., Vooijs, R., and Kuiper, E. (1996). Identical major gene loci for heavy metal tolerances that have independently evolved in different local populations and subspecies of Silene vulgaris. *Evolution* 50: 1888–1895.

180 Bert, V., Meerts, P., Saumitou-Laprade, P., Salis, P., Gruber, W., and Verbruggen, N. (2003). Genetic basis of Cd tolerance and hyperaccumulation in Arabidopsis halleri. *Plant Soil* 249: 9–18.

181 Zha, H.G., Jiang, R.F., Zhao, F.J., Vooijs, R., Schat, H., Barker, J.H.A., and McGrath, S.P. (2004). Co-segregation analysis of cadmium and zinc accumulation in Thlaspicaerulescensinterecotypic crosses. *New Phytol.* 163: 299–312.

182 Lombi, E., Zhao, F.J., Dunham, S.J., and McGrath, S.P. (2000). Cadmium accumulation in populations of Thlaspicaerulescens and Thlaspigoesingense. *New Phytol.* 145: 11–20.
183 Howden, R. and Cobbett, C.S. (1992). Cadmium-sensitive mutants of Arabidopsis thaliana. *Plant Physiol.* 100: 100–107.
184 Dolezal, O. and Cobbett, C.S. (1991). Arabinose kinase-deficient mutant of Arabidopsis thaliana. *Plant Physiol.* 96: 1255–1260.
185 Howden, R., Andersen, C.R., Goldsbrough, P.B., and Cobbett, C.S. (1995). A cadmium-sensitive, glutathione-deficient mutant of Arabidopsis thaliana. *Plant Physiol.* 107: 1067–1073.
186 Ha, S.B., Smith, A.P., Howden, R., Dietrich, W.M., Bugg, S., O'Connell, M.J., Goldsbrough, P.B., and Cobbett, C.S. (1999). Phytochelatin synthase genes from Arabidopsis and the yeast Schizosaccharomycespombe. *Plant Cell* 11: 1153–1163.
187 Howden, R., Goldsbrough, P.B., Andersen, C.R., and Cobbett, C.S. (1995). Cadmium-sensitive, cad1 mutants of Arabidopsis thaliana are phytochelatin deficient. *Plant Physiol.* 107: 1059–1066.
188 Mills, R.F., Krijger, G.C., Baccarini, P.J., Hall, J.L., and Williams, L.E. (2003). Functional expression of AtHMA4, a P1B-type ATPase of the Zn/Co/Cd/Pb subclass. *Plant J.* 35: 164–176.
189 Wong, C.K.E. and Cobbett, C.S. (2009). HMA P-type ATPases are the major mechanism for root-to-shoot Cd translocation in Arabidopsis thaliana. *New Phytol.* 181: 71–78.
190 Morel, M., Crouzet, J., Gravot, A., Auroy, P., Leonhardt, N., Vavasseur, A., and Richaud, P. (2009). AtHMA3, a P1B-ATPase allowing Cd/Zn/co/Pb vacuolar storage in Arabidopsis. *Plant Physiol.* 149: 894–904.
191 Sanjaya, Hsiao, P.Y., Su, R.C., Ko, S.S., Tong, C.G., Yang, R.Y., and Chan, M.T. (2008). Overexpression of Arabidopsis thaliana tryptophan synthase beta 1 (AtTSB1) in Arabidopsis and tomato confers tolerance to cadmium stress. *Plant Cell Environ.* 31: 1074–1085.
192 Davies, J.P., Yildiz, F.H., and Grossman, A.R. (1999). Sac3, an Snf1-like serine/threonine kinase that positively and negatively regulates the responses of Chlamydomonas to sulfur limitation. *Plant Cell* 11: 1179–1190.

3

Transgenic Approaches for Field Testing and Risk Assessment

CHAPTER MENU

3.1 Introduction

A major environmental concern has been the pollution of the water systems with diverse contaminants. While a variety of technologies have been used as remediation techniques, including chemical/thermal treatments, they have numerous inadequacies due to high cost, poor performance, secondary pollutant generation, etc. [1]. The usage of plant systems for the remediation of water and soil by the extraction, sequestration or detoxification of pollutants has also been recognized as one of the environmentally sustainable techniques [2, 3]. Plants can accumulate various heavy metals, inorganic and organic contaminants which later on get converted into other nontoxic or less toxic forms via plant metabolomics [4]. He and Yang [5] pointed out that phytovolatilization, rhizofiltration, phytoextraction and phytostabilization are types of phytoremediation, where bioaccumulation of pollutants from contaminated water and soils will be taken up into shoots from roots which is known as the process of phytoextraction [6]. As per Yadav et al. [7], adsorption, concentration, and precipitation are different stages of rhizofiltration in which plant roots decontaminate water, while mobility of pollutants decreases in phytostabilization by plant adsorption or precipitation, while induction and release into the

Aquatic Environmental Bioengineering: Monitoring and Remediation of Contamination, First Edition.
Rouf Ahmad Bhat, Mohammad Yaseen Mir, Gowhar Hamid Dar and Moonisa Aslam Dervash.

environment following transition into a volatile state is carried out by phytovolatilization [8]. In phytoremediation, approximately 400 species of plants are hyperaccumulators which can absorb large amounts of pollutants by their roots with 50–100 times more affinity than other plants [9]. McGrath and Zhao [10] observed that competent root-to-shoot transport and a factor of bioconcentration are the parameters responsible for high detoxification capacity of hyperaccumulators. Watanabe [11] highlighted that ultimate hyperaccumulators grow faster by having extensive root systems, producing high biomass, and also by having pest or disease resistance. This was shown in a trial to boost the efficiency of contaminant remediation plants. In this respect, an approach was implemented to genetically modified plants with better remediation ability through transgenic modification and technology for plant transformation [12]. Genetically modified plants are the plants that are genetically engineered involving expression of genes that are not native to the plant using recombinant DNA technology [13]. Relevant hyperaccumulator genes are then established and eventually transmitted to plant species with accelerated growth rates, deep rooting mechanisms, resistance to diseases, pests, various climatic events and so on [14, 15]. Consequently, through tissue culture methods, transformed plants may be regenerated. This has so far been described as a highly productive approach to enhancing the potential for phytoremediation. This chapter is therefore intended to provide a description of the advantages and uses of transgenic plants in plant remediation. It will also address existing limits and future threats involved with the use of transgenic plants.

3.2 Transgenic Plants for Environmental Remediation

In the 1940s Sanderman [16] also showed that plants have the capability to degrade xenobiotic compounds like pesticides. Afterwards, genomics, proteomics and metabolomics played a great role in optimizing or modifying the metabolism of many xenobiotic toxins in plants [17–19]. While phytoremediation was initially used to extract inorganic pollutants from the soil, it has eventually proved to be useful in treating organic pollutants, like polyaromatic hydrocarbons, chlorinated solvents and explosives [20, 21]. In order to attenuate the loss of crop yield caused by insect damage while decreasing the quantity of pesticides needed, the primary generation of commercially viable transgenic plants are produced (e.g. Bt toxin expressing plants). Genetically modified plants for remediation were primarily developed to remediate heavy-metal-contaminated soils like *Arabidopsis thaliana* which over-expressed the mercury tolerant mercuric ion reductase gene, or *Nicotiana tabaccum* that over-expressed the cadmium-tolerant yeast metallothionein gene [22, 23]. Many researchers like French et al. [24], Doty et al. [24], McCutcheon and Schnoor [25], Mackova et al. [26], Meagher [27], Eapen et al. [28], Doty [18] and Macek et al. [29] put forth that tobacco plants were the first to have been genetically modified to remediate organic contaminants like volatile and halogenated ones. In the literature, the worth of transgenic plants for the degradation of chlorinated solvents, explosives, phenolics, etc. has been generally accepted. These plants have been engineered to produce either transgenes responsible for xenobiotic metabolization or transgenes resulting in improved pollutant tolerance. In order to treat wastewater, various plants showed pollutant removal efficiency; *Phragmites australis* showed the capability to reduce the BOD5 value from 1006 mg/L to 57 mg/L in dairy waste [30]. For many decades in the Czech Republic, wetlands which were sub-surfacely constructed with *Phragmites australis* of the size 18 ~ 4,500 m^2 achieved a success rate of 88% reduction in BOD5 and 84% for suspended solids. *Salix atrocinerea* and *Typha latifolia* showed the same results in Spain to treat municipal wastewater [31, 32]. Recent studies by Bodini et al. found that *Cupressus sempervirens* and *Quercus ilex* were tolerant of six months of wastewater from the olive factory, followed by six months of irrigation with water [33]. Low pH, high BOD5 and COD loads, and phytotoxic levels of phenolic compounds characterize the olive mill wastewater and have thus been a waste management issue in the countries producing olive oil. It is important to note that *Glandularia pulchella* phytoremediation of garment effluent has been studied to decolorize structurally distinct dyes. Differential time-related enzyme induction patterns were obtained for lignin peroxidase, tyrosinase and dichlorophenolindophenol reductase during dye mixture decolorization [34].

3.3 Degradation Pathways in Plants

Three major biochemical pathways typically include xenobiotic metabolism in humans, animals and higher plants; as phase I includes transition, conjugation comes under phase II and compartmentalization in phase III [35, 36]. Via N-, O-, and S-dealkylation, aromatic and aliphatic hydroxylation, epoxidation, peroxidation, oxidative desulfurization, sulfoxidation or reduction by cytochrome P450s, hydrophobic contaminants are converted to less hydrophobic metabolites throughout step I [37, 38].

3.4 Cytochrome P450s for Environmental Perspectives

Miscellaneous assemblages of developments, comprising endobiotics such as steroids, xenobiotics, bile acids, fatty acids, prostaglandins and leukotrienes which are crucial for the oxidative, peroxidative and reductive metabolism belonging to the superfamily of heme proteins which is known as Cytochrome P450s enzymes comes under the category of environmental contaminants and therapeutic drugs [39–41]. In 1958, Klingenberg et al. [39] reported for the first time that microsomal carbon monoxide-binding pigment is the CYP enzyme. According to reports of Omura and Sato [42, 43], and Omura [44], hemoprotein nature has been recognized for the enzyme cytochrome P450 due to its absorption peak at 450-nm opticals. These enzymes are found in more than one type in almost all living organisms, thereby representing one class among the principal group of enzymes. The enzyme mechanism involves many cytochrome P450 isoforms which a non-specific NADPH-cytochrome P450 oxidoreductase, and is found in microsomes. A systematic grouping of human types into families and subfamilies has given rise to the notable diversity of CYP enzymes. For a given gene family, the protein sequences are at least 40% the same (e.g. CYP2A6 and CYP2B6), and the sequences within a given subfamily are 55% the same (e.g. CYP2A6 and CYP2A7) [40]. Within various species, the number of families and enzymes varies. While cytochrome P450 (P450 or CYP) monoxygenases play an important role in the oxidative metabolism of endogenous and exogenous liphophilic compounds in higher plants [18, 29, 44], there is very limited molecular knowledge on P450 species metabolizing xenobiotics in plants. There are, on the other hand, a variety of species of P450 that metabolize xenobiotics in human liver microsomes [45]. At least 53 separate CYP genes and 24 pseudogenes were estimated for humans [46]. To date, 11 P450 species in the human liver have been reported to be involved in more than 90% of P450-dependent drug metabolism [45, 47–49]. These P450s oxidize carbon and nitrogen, usually contributing to the formation of a hydroxyl group and often to the eventual elimination of alkyl groups that are close to plant metabolic processes [49]. An analysis of 11 human P450s from CYP1, 2 and 3 families using a recombinant yeast-expressing method has shown that 27 herbicides and 4 insecticides can be metabolized [50]. In addition, another study conducted by the same research group found that 16 herbicides, including triazines, ureas and carbamates, were metabolized by human CYP1A1 and that more than 10 herbicides, including chloroacetanilides, oxyacetamides and 2,6-dinitroanilines, were metabolized by CYP2B6, three insecticides and two industrial chemicals. Seventeen herbicides, including triazine, urea and carbamate, were metabolized by CYP2C19 [50].

3.5 Transgenic Plants for the Rhizoremediation of Organic Xenobiotics

The incorporation of xenobiotics that degrade genes through the root system of sufficient plant species for improved rhizospheric secretion and eventual degradation of contaminants is one of the most promising approaches to developing phytoremediation technology [51–53]. The value of this approach is that the contaminants should not be picked up by plants to detoxify them; instead, the secreted enzymes will degrade the pollutants in the rhizospheric region [53]. In the immediate vicinity of a root that is influenced by root processes, the rhizosphere is the soil. When a root tip joins an amount of soil and disappears sometime after the root is dead and decomposed, it comes into being [54]. Several studies indicate intensely improved dissipation and mineralization at the root–soil interface of chronic organic contaminants [55–59]. Due to the release of plant root exudates, mucigel and root lysates, this rhizosphere effect is typically due to a rise in microbial abundance, diversity and/or metabolic activity (enzymes, amino acids, carbohydrates, low-molecular-mass carboxylic acids, flavonones, and phenolics) [60, 61]. Plants can also strengthen the physical and chemical properties of the polluted soil and increase interaction between the root microbes and the pollutants of the soil.

For the beneficial application of rhizoremediation, transgenic plants secrete detoxifying enzymes which are hydrophobic in nature. It has been reported that Arabidopsis plants being transgenic secrete a root-specific laccase (LAC1) in the rhizospheric region of cotton plants which have shown improved resistance to 2, 4, 6-trichlorophenol and phenolic allelochemicals [62]. Likewise, for the rhizoremediation of bisphenol A and PCP, Sonoki et al. [63] converted tobacco plants with an extracellular fungal laccase from Coriouls versicolar. In addition, the over-expression of *Burkholderia xenovorans*

LB400 bacterial biphenyl dioxygenase (BPDO) resulted in improved oxygenation of 4-chlorobiphenyls [64, 65]. For improved PCB remediation, gene bphC from Pandoraea pnomenusa B-356 was successfully cloned in tobacco plants [66, 67]. Aromatic-cleaving extradiol dioxygenase (DBfB) transgenic Arabidopsis plants resulted in enhanced degradation of 2, 3-dihydroxybiphenyl (2, 3-DHB) [68]. According to Uchida et al. [68], the rhizospheric region of transgenic tobacco plants is where there is enhanced production of 1-chlorobuatne via haloalkane dehydrogenase (DhaA) enzyme activation. Therefore, it is possible that the choice of plants will affect the progress of rhizoremediation technology [65]. For phytoextraction, a plant species with large above-ground biomass is important, whereas for rhizoremediation it is necessary for a plant species with extensive root system or below-ground biomass. The toxins usually impede the development of plants at infected areas. Accelerated production of ethylene is believed to impede root development in response to stress caused by toxins and is considered a significant hindrance to phytoremediation [53]. As per reports of Bernard [69] and Arshad et al. [70], enzymes namely 1-aminocyclopropane-1-carboxylate (ACC) deaminase in bacteria controls ethylene levels in plants by metabolizing its precursor ACC into α-ketobutyric acid and ammonia; and ACC deaminase has been also noticed to regulate the biosynthesis of ethylene in inoculated plant roots (PGPR) [71, 72]. Ethylene levels can be reduced by transgenic plants expressing ACC deaminase genes, resulting in a more extensive root system [70]. It is anticipated that the resulting increase in root growth provided by ACC deaminase may improve the potential for rhizoremediation. In addition, the incorporation of multiple genes associated with various stages of degradation of xenobiotics together with ACC deaminase may improve the remediation potential of transgenic plants [53].

3.6 Transgenic Plants to Be Developed for the Phytoremediation of Some Other Priority Pollutants

The progress from single to multigene engineering, i.e. transgenic technology, has successfully demonstrated the improved degradation and remediation of many persistent hazardous contaminants, although advancement of this technology is yet to be explored for removal of pollutants like hexachlorocyclohexane isomers (HCH) via plants, as this is among the most persistent organochlorine pesticides with high toxicity posing a threat to the environment. Over the last several decades, the systematic and indiscriminate usage of lindane (γ-HCH) and technical HCH has contributed to the pervasive contamination of its four key isomers (α-, β-, γ- and δ-HCH) through separate environmental compartments [73]. Around 1.6 and 4.8 million tonnes of HCH residues worldwide have been discarded or landfilled at the respective processing sites [74, 75]. As per Abhilash [76], various nations have stopped use of HCH for the time being, due to the deleterious effects of residues on habitat. The POPs Evaluation Committee has recently nominated alpha-, β-, and γ-HCH for participation in the Stockholm Convention to resolve HCH pollution at a global level. There is an immediate need, therefore, to end the production of lindane and restore the polluted soil areas. Some plant species developing in an industrial environment capable of accumulating HCH isomers were described by Abhilash and Singh [77]: (*Achyranthes aspera, Calotropis procera, Erinathus munja Dalbergia sisso, Solanum torvum, Withania somnifera, Lantana camera*). A glass house analysis found that *W. somnifera* increased lindane dissipation (73%) in treated soil (20 mg kg−1) [78]. The remediation capacity of this species, in particular rhizoremediation, would therefore be utilized for the decontamination of soil sites polluted with lindane and other HCH. In *Sphingomonas paucimobilis* B90, microbial catabolic profiling shows that dehydrochlorinase LinA1 and LinA2 catalyze the first and second steps in hexachlorocyclohexanes metabolism [79, 80]. Two bacterial strains (*Sphingomonas paucimobilis* UT26 and *S. paucimobilis* B90A) have been documented in several previous studies to degrade both alpha-HCH and gamma isomers [81–83]. *S. paucimobilis* UT26 was used to get purified LinA genes which have the capability of catalyzing the dehydrochlorination via enzymatic activation of dehydrochlorinase of α-HCH, α-pentachlorocyclohexane (α-PCCH) δ-HCH, γ-pentachlorocyclohexane (γPCCH), whereas dehydrochlorination for β-HCH is not attained. γ-HCH is enantioselectively converted to (3 R, 4S, 5S, 6 R)-1,3,4,5,6-PCCH as an initial stage, which is then converted to 1,2,4-trichlorobenzene (1,2,4-TCB) via the presumed but unstable intermediate (3 R, 6 R)-1,3,4,6-tetrachlorocyclohexa-1, 4-diene cyclohexa-1, 4-diene intermediate (3 R, 6 R) [79, 84]. Lin A1 and Lin A2 were 88% identical to each other at the level of amino acids, and Lin A2 was 100% identical to S LinA. UT266 paucimobilis [79]. These genes' further molecular fingerprinting will explain the specifics of their catabolic ability against lindane and other isomers of HCH. The possible incorporation into sufficient HCH resistant/accumulating plant species is one of the possible techniques to harness the catabolic ability of these genes. We have proposed two plant species for the injection of HCH-degrading microbial genes based on our detailed monitoring studies on HCH accumulation capabilities of plant species developing in a lindane generating industrial premise and on glass house experiments. There is a need to evaluate approaches for gene containment in order to analyze the standardization of protocols for genetically transforming *J. curcas* and *W. somnifera* species

3.7 Potential Genes for Phytoremediation

The process of phytoremediation involves accumulation of pollutants along with other essential elements or minerals via roots of plants and undergoes physiological processes. Various report have been published regarding the fate of metal uptake by hyperaccumulating plants in order to understand the mechanisms behind this type of remediation process, and for the production of transgenic plants with enhanced phytoremediation capacity greater insight into the mechanism of hyperaccumulation is necessary. It is possible to envisage at least three separate engineering methods to improved metal uptake [85], which include raising the number of uptake sites, modifying the specificity of the uptake mechanism to decrease unnecessary cation competition and increasing intracellular binding sites. For absorption, transport and sequestration, each metal has a particular molecular mechanism. There has been extensive development in the discovery of genes and proteins implicated in Fe uptake in plants [86, 87]. Membrane transporters with high affinity Zn [88] were also researched. In the production of plants for phytoremediation, the manipulation of metal transporters and vacuolar targeting of metal would have beneficial applications. A good source of genes appropriate for phytoremediation are hyperaccumulators. Specific promotors and regulatory mechanisms offer binding to a few selected pollutants such as heavy metals and radionuclides as Cu and Zn form complexes with organic acids when there is high concentration of malate. Acids and acid anions that have a role in metal storage or in plant internal metal transport are filled with hyperaccumulators. To secrete metal selective ligands into the rhizosphere, transgenic plants could be produced that could precisely solubilize phytoremediation components. It is worthwhile to explore a field to identify simple molecules with selective chelation capacity that plants can produce and secrete into the rhizosphere, and simultaneously to engineer plants with the ability to transport protein for metal chelation. Free histidine in xylem exudates has been found as a metal chelator in the Ni hyperaccumulator. In order to increase the Ni accumulation potential in plants, the concentrations of histidine in xylem exudates can be changed. Other latent metal sequestration and accumulation mediators include family facilitation of cation diffusion (CDF). Table 3.1 summarizes the different genes that have been used for the development of transgenic plants. Heavy metals can be kept in a safe cubicle without troubling the cellular functions in the vacuoles

Table 3.1 List of transgenic plants on behalf of improved plant bioremediation.

Gene manipulation	Source	Host	Effect	References
Pentaerythritol tetranitrate reductase	Bacteria	Tobacco	Degredation of explosives	[24]
Metallothionein II	Human	*Brassica napus and Nicotiana tabacum*	Cd tolerance	[22]
Mercuric reductase	Bacteria	*Liriodendron tullpifera*	Remediation of Hg pollution	[23]
Phytochelatin synthetase	Wheat	*Nicotiana glauca*	Pb accumulation	[89]
Cytochrome P-450	Human	*Populus*	Remediation of trichloromethlene	[19]
Nitroreductase	*E.coli*	*Arabidopsis thalania*	Degradation of 2,4,6-trinitrotoluene (TNT)	[90]
Mercuric reductase Organomercurial lyase	Bacteria	*Arabidopsis thalania*	Detoxification of CH_3Hg	[91]
Cysteine synthase	*Arabidopsis thalania*	Arabidopsis	9x increase in Cd tolerance on medium with 200 mm $CdCl_2$	[92]
Cysteine synthase	*Arabidopsis thalania*	Arabidopsis	Tolerance upto 400 mm $CdCl_2$ with exogenous cysteine supply and 72% with more cadmium accumulation (mature plants grew on 250 mm $CdCl_2$ for 14 days)	[93]

(*Continued*)

Table 3.1 (Continued)

Gene manipulation	Source	Host	Effect	References
ATP sulfurylase	*Arabidopsis thalania*	Indian mustard	2–3 times higher Selenium accumulation in shoots and 1.5 fold higher in roots than WT	[94]
ATP sulfurylase	*Arabidopsis thalania*	Indian mustad	2.5 times higher shoot concentration of different metals both at seedling and mature stages	[95]
Heavy metal transporter	*E.coli*	*Arabidopsis*	Transgenic plants grew better than WT in medium with 0.7 mm Pb(II) or 70 uM Cd(II) and showed higher fresh weight	[96]
Ferric reductase	Yeast	Tobacco	1.5 times higher iron content in transgenic shoots compared to WT plants	[97]
Vacuolar transporter	*Arabidopsis thalania*	Tobacco	15–20% more metal ions in the shoots and higher root tonoplast transport in transgenic plants than in control	[98]

for cellular targets. Therefore, the phytoremediation method of the second generation is the engineering of vacuolar transporters, preferably in particular cell types. By optimizing metal binding points, another solution is to establish artificial metal sinks in the shoot. The creation of transgenic plants for phytoremediation has made great gains, but most genes have been passed from other species to plants. In the transmission of genes from hyperaccumulators to candidate plants, understanding hyperaccumulators can help. Constructing a library of well-characterized genes that work in the acquisition and storing of minerals will help to grow novel plants with enhanced hyperaccumulating characteristics.

3.8 Hitting Transgenics to the Assessment: Plant Bioremediation

In order to create transgenics several diverse approaches have been established for phytoremediation based upon results of hydroponic systems or agar media, followed by testing of polluted substrates which were collected from terrestrial and aquatic environs, and to date none of the results from such studies is available. A greenhouse research has recently been undertaken to assess the phytoextraction potential of transgenic phytochelatin-overproducing mustard plants which were collected from Leadville, Colorado that also metal-polluted soil sites [99]. In their shoots, all kinds of transgenics accumulated considerably higher levels of Cd and Zn than untransformed species. Such findings are promising since they indicate that the results of metal accumulation obtained from hydroponic experiments are a valuable indicator of the metal accumulation capacity of transgenics from environmental soils encompassing metal combinations.

3.9 Potential Risks

While transgenic plants have added advantages over native plant species for successful remediation, many forms of public and science problems have arisen with regard to possible environmental issues. The environmental threats that transgenic plants pose can mainly be classified as direct and indirect risks. Transgenic plants have a high potential to become invasive plants or weeds or have an impact on habitats in terms of direct threats [100]. It is quite obvious that transgenic plants have a wide range of tolerance which makes them stronger rivals for native plant varieties [91]. It has been shown that the transgenes that are particularly applied to plants used for phytoremediation give them a large capacity to survive in extreme conditions that are often found in contaminated areas. Therefore, they have a high capacity to expand through some regions that were previously introduced, causing a possible risk of disruption of natural plant populations. According to Ellstrand and Schierenbeck [101] and Culley and Hardiman [102], in certain ecosystems with the introduction of invasive genotypes, there is an evidence of the adaptation for such new varieties via gene flow temporal variations. The effect of gene flow among transgenic crop cultivars and weeds has been reported to create increased weed problems [103, 104]. In addition, due to frequent application of such herbicides that are tolerable to the transgenic population, the development of herbicide-tolerant weed varieties may occur. The biodiversity in surrounding environments may be influenced by plant communities composed of transgenic plants in many ways. Deleterious transgenes, such as genes implicated in male sterility or fruit maturation, passed by gene flow to relative species, have the ability to extinguish native populations [105]. Highly successful transgenic varieties will replace their wild relatives and make them extinct [106]. Depending on the plant type and the transgenes they carry, transgenic plants may have other immediate potential threats, including impacts on soil microorganisms and nitrogen cycles, local insect ecosystems, food sources for animal species and agricultural practices [107]. Indirect threats can include risks to other species of plants and animals, natural population processes and the functioning of the ecosystem [108].

3.9.1 Risk Assessment Theories and Practices

Risk is always associated with the occurrence of an unexpected event; as per Johnson et al. [109], risk assessment comprises two more aspects of the risk analysis which includes risk control and risk communication [109]. Danger-evaluation programs typically rely on the reasonable speed at which a problem is detected, potential hazards identified, possible exposure mechanisms identified, harm degree assessment and unclear expressions articulated [100]. Government agencies and customers use risk assessment practices for transgenic plants in order to detect likely environmental or human risk from these plant varieties. The key scientific approaches to risk management are qualitative techniques and quantitative approaches, like probabilistic evaluations. Qualitative evaluations are often based on expert judgments, analytical evidence or hybrid approaches. However, because of weaknesses in ecological evidence to construct statistical models, quantitative evaluation methods consume inadequate application for hazard investigation of transgenic plants. By way of the object of dissimilar undesirable results on the ecosystem and the direction or process that leads to such undesirable outcomes, an ecological danger may be explained. According to Hayes [109] and Auer [100], one approach that incorporates current understandings for estimating potential risks from direct or indirect contact to genetically engineered plants, genes for those plants and gene products, is the predictive ecological risk assessment. The most common method of comparing risk evaluations for transgenic varieties, however, is to compare the impacts of historically grown varieties [110]; hence the results of transgenic plants are in most cases similar to those of historically grown varieties.

3.9.2 Contests Aimed at Multifaceted Risk Valuation

Menace-management issues comprise problems through reference to the consequence of transgenic plants on natural plant species, habitats, and non-target organisms; precise and temporal gene flow characterization; and plant fitness under fluctuating environmental conditions. Lack in basic awareness of the biological dimensions of town areas, parks and other natural ecosystems hinder the detection of exposure paths, the prediction of possible dangers and the planning of hypothesis-based research. The assessment of primary and secondary threats from transgenic plants to non-transgenic species, population assemblage, and the functioning of the ecosystem is, however, considered to be challenging. Ecologically relevant information is provided by EIA (Environmental Impact Assessments) experiments for transgenic plants, nonetheless most of these are partial to comparatively short time spans and usually performed on a small scale [111–114]. It is a difficult task to monitor and evaluate long-term ecological influences arising from transgenic plants in a wide range of ecosystems. Masses of novel transgenic plant cultivars are launched worldwide every year and a detailed evaluation of the environmental effects of these plants in a general manner is a monotonous scheme [115].

3.10 Future Research Guidelines

A rigorous long-term field test must be related to the environmental risk evaluation for certain transgenic plants adopted for successful phytoremediation as a crucial phase in research and development. The R&D method needs to explore preventive technologies for the further dissemination of changed genes from transgenic plants to natural plant populations. Therefore it is important to promote the creation of additional mechanisms, such as genetically modified technology to regulate transgenic plant dispersal in natural ecosystems. The phytoremediation process in the case of transgenic plants has an immense potential to rejuvenate contaminated lands.

References

1 Arthur, E.L., et al. (2005). Phytoremediation: an overview. *CRC. Crit. Rev. Plant. Sci.* 24 (2): 109–122.

2 Cunningham, S.D., et al. (1995). Phytoremediation of contaminated soils. *Trends Biotechnol.* 13 (9): 393–397.

3 Singh, O., et al. (2003). Phytoremediation: An overview of metallic ion decontamination from soil. *Appl. Microbiol. Biotechnol.* 61 (56): 405–412.

4 Meagher, R.B. (2000). Phytoremediation of toxic elemental and organic pollutants. *Curr. Opin. Plant. Biol.* 3 (2): 153–162.

5 He, Z.-L. and Yang, X.-E. (2007). Role of soil rhizobacteria in phytoremediation of heavy metal contaminated soils. *J. Zhejiang. Univ. Sci. B.* 8 (3): 192–207.

6 Rafati, M., et al. (2011). Phytoremediation potential of Populus albaand Morus albafor cadmium, chromium and nickel absorption from polluted soil. *Int. J.Environ. Res.* 5 (4): 961–970.

7 Yadav, B.K., et al. (2011). Rhizofiltration of a heavy metal (lead) containing wastewater using the wetland plant Carex pendula. *CLEANSoil, Air, Water* 39 (5): 467–474.

8 Glick, B.R. (2003). Phytoremediation: Synergistic use of plants and bacteria to clean up the environment. *Biotechnol. Adv.* 21 (5): 383–393.

9 Chaney, R.L., et al. (1997). Phytoremediation of soil metals. *Curr. Opin. Biotechnol.* 8 (3): 279–284.

10 McGrath, S.P. and Zhao, F.-J. (2003). Phytoextraction of metals and metalloids from contaminated soils. *Curr. Opin. Biotechnol.* 14 (3): 277–282.

11 Watanabe, M.E. (1997). Phytoremediation on the brink of commericialization. *Environ. Sci. Technol.* 31 (4): 182A–186A.

12 Ellis, D.R., et al. (2004). Production of Se-methylselenocysteine in transgenic plants expressing selenocysteine methyltransferase. *BMC Plant Biol.* 4 (1): 1.

13 Key, S., et al. (2008). Genetically modified plants and human health. *J. R. Soc. Med.* 101 (6): 290–298.

14 Gullner, G., et al. (2001). Enhanced tolerance of transgenic poplar plants overexpressingγ-glutamylcysteine synthetase towards chloroacetanilide herbicides. *J. Exp. Bot.* 52 (358): 971–979.

15 Cherian, S. and Oliveira, M.M. (2005). Transgenic plants in phytoremediation: Recent advances and new possibilities. *Environ. Sci. Technol.* 39 (24): 9377–9390.

16 Sanderman, J.H. (1994). Higher plant metabolism of xenobiotics: The'green liver' concept. *Pharmacogenetics* 4: 225–241.

17 Raskin, I. (1996). Plant genetic engineering may help with environmental clean-up. *Proc. Natl. Acad. Sci. USA* 93: 3164–3166.

18 Eapen, S., Singh, S., and D'Souza, S.F. (2007). Advances in development of transgenic plants for remediation of xenobiotic pollutants. *Biotechnol. Adv.* 25: 442–451.

19 Aken, B.V. (2008). Transgenic plants for phytoremediation: Helping nature to clean up environmental pollution. *Trends Biotechnol.* 26: 225–227.

20 Pilon-Smits, E. (2005). Phytoremediation. *Annu. Rev. Plant Biol.* 56: 15–39.

21 Salt, D.E., Smith, R.D., and Raskin, I. (1998). Phytoremediation. *Annu. Rev. Plant Physiol. Plant Mol. Biol.* 49: 643–668.

22 Misra, S. and Gedamu, L. (1989). Heavy metal tolerant transgenicBrassica napusL. andNicotiana tabacumL. plants. *Theor. Appl. Genet.* 78: 161–168.

23 Rugh, C.L., Wilde, D., Stack, N.M., Thompson, D.M., Summer, A.O., and Meagher, R.B. (1996). Mercuric ion reduction and resistance in transgenic Arabidopsis thalianaplants expressing modified bacterialmerAgene. *Proc. Natl. Acad. Sci. USA* 93: 3182–3187.

24 French, C.E., Nicklin, S., and Bruce, N.C. (1998). Aerobic degradation of 2, 4, 6-trinitrotoluene by Enterobacter cloacae PB2 and by pentaerythritol tetranitrate reductase. *Appl. Environ. Microbiol.* 64: 2864–2868.

25 Doty, S.L., Shang, Q.T., Wilson, A.M., Moore, A.L., Newman, L.A., Strand, S.E., and Gordon, M.P. (2000). Enhanced metabolism of halogenated hydrocarbons in transgenic plants contain mammalian P450 2E1. *Proc. Natl. Acad. Sci. USA* 97: 6287–6291.

26 McCutcheon, S.C. and Schnoor, J.L. (2003). *Phytoremediation: Transformation and Control of Contaminant*. New Jersey, USA: John Wiley & Sons Inc.

27 Mackova, M., Barriault, D., Francova, K., Sylvestre, M., Möder, M., Vrchotova, B., Lovecka, P., Najmanová, J., Demnerova, K., Novakova, M., Rezek, J., and Macek, T. (2006). Phytoremediation of polychlorinated biphenyls. In: *Phytoremediation and Rhizoremediation. Theoretical Background. Series: Focus on Biotechnology* (ed. M. Mackova, D. Dowling, and T. Macek). Springer.

28 Meagher, R.B. (2000). Phytoremediation of toxic elemental and organic pollutants. *Curr. Opin. Plant Biol.* 3: 153–162.

29 Doty, S.L. (2008). Enhancing phytoremediation through the use of transgenic plants and entophytes. *New Phytol.* 179: 318–333.

30 Macek, T., Kotrba, P., Svatos, A., Novakova, M., Demnerova, K., and Mackova, M. (2008). Novel roles for genetically modified plants in environmental protection. *Trends Biotechnol.* 26: 146–152.

31 Ansola, G., González, J.M., Cortijo, R., and De Luis, E. (2003). Experimental and full-scale pilot constructed wetlands for municipal wastewaters treatment. *Ecol. Eng.* 21: 43–52.

32 Kyambadde, J., Kansiimme, F., Gumaelius, L., and Dalhammar, G. (2004). A comparative study of Cyperus papyrusand Miscanthidium violaceum-based constructed wetlands for wastewater treatment in a tropical climate. *Water Res.* 38: 475–485.

33 Bodini, S.F., Cicalini, A.R., and Santori, F. (2011). Rhizosphere dynamics during phytoremediation of olive mill wastewater. *Bioresour. Technol.* 102: 4383–4389.

34 Kabra, A.N., Khandare, R.V., Waghmode, T.R., and Govindwar, S.P. (2012). Phytoremediation of textile effluent and mixture of structurally different dyes by Glandularia pulchella(Sweet) tronc. *Chemosphere* 87: 265–272.

35 Schmidt, B., Faymonville, T., Gembé, E., Joußen, N., and Schuphan, I. (2006a). Comparison of the biotransformation of the 14C-labelled insecticide carbaryl by non-transformed and human CYP1A1, CYP1A2-, and CYP3A4-transgenic cell cultures ofNicotiana tabacum. *Chem. Biodiv.* 3: 878–896.

36 Schmidt, B., Joußen, N., Bode, M., and Schuphan, I. (2006b). Oxidative metabolic profilingof xenobiotics by human P450s expressed in tobacco cell suspension cultures. *Biochem. Soc. Trans.* 34: 1241–1245.

37 Theodoulou, F.L. (2000). Plant ABC transporters. *Biochim. Biophys. Acta* 1465: 79–103.

38 Schroder, P. (2007). Exploiting plant metabolism for the phytoremediation of organic xenobiotics. In: *Phytoremediation: Methods and Reviews* (ed. N. Willey). Totowa, NJ: Humana Press Inc.

39 Klingenberg, M. (1958). Pigments of rat liver microsomes. *Arch. Biochem. Biophys.* 75: 376–386.

40 Nelson, D.R., Koymans, L., Kamataki, T., Stegeman, J.J., Feyereisen, R., Waxman, D.J., et al. (1996). P450 superfamily: Update on new sequences, gene mapping, accession numbers and nomenclature. *Pharmacogenetics* 6: 1–42.

41 Kreuz, K., Tommasini, R., and Martinoia, E. (1996). Old enzyme for a job. *Plant Physiol.* 111: 349–353.

42 Omura, T. and Sato, R. (1964). The carbon monoxide-binding pigment of liver microsomes. I. Evidence for its hemeprotein nature. *J. Biol. Chem.* 239: 2370–2378.

43 Omura, T. and Sato, R. (1962). A new cytochrome in liver microsomes. *J. Biol. Chem.* 237: 1375–1376.

44 Inui, H. and Ohkawa, H. (2005). Herbicide resistance plants with mammalian P450 monooxygenase genes. *Pest Manag. Sci.* 61: 286–291.

45 Inui, H., Kodama, T., Ohkawa, Y., and Ohkawa, H. (2000). Herbicide metabolism and cross tolerance in transgenic potato plants co-expressing human CYP1A1, CyP2B6, and CYP2C19. *Pestic. Biochem. Physiol.* 66: 116–129.

46 Nelson, D.R. (1999). Cytochrome P450 and the individuality of species. *Arch. Biochem. Biophys.* 369: 1–10.

47 Funae, Y., Obata, N., and Kirigami, S. (1998). Multiple function of cytochrome P450. *Kan Tan Sui* 37: 91.

48 Kawahigashi, H., Hirose, S., Ohkawa, H., and Ohkawa, Y. (2007). Herbicide resistance of transgenic rice plants expressing human CYP1A1. *Biotechnol. Adv.* 25: 75–85.

49 Kawahigashi, H., Hirose, S., Ohkawa, H., and Ohkawa, Y. (2008). Transgenic rice plants expressing human P450 genes involved in xenobiotic metabolism for phytoremediation. *J. Mol. Microbiol. Biotechnol.* 15: 212–219.

50 Inui, H., Shiota, N., Motoi, Y., Ido, Y., Inoue, T., Kodama, T., et al. (2001). Metabolism of herbicides and other chemicals in human cytochrome P450 species and in transgenic potato plant co-expressing human CYP1A1, CYP2B6 and CYP2C19. *JPesticSci* 26: 28–40.

51 Glick, B.R. (2003). Phytoremediation: Synergistic use of plants and bacteria to clean up the environment. *Biotechnol. Adv.* 21: 239–244.

52 Gerhardt, K.E., Huang, X.D., Glick, B.R., and Greenberg, B.M. (2009). Phytoremediation and rhizoremediation of organic soil contaminants: Potential and challenges. *Plant Sci.* 176: 20–30.

53 Kawahigashi, H. (2009). Transgenic plants for phytoremediation of herbicides. *Curr. Opin. Biotechnol.* doi:10.1016/j.copbio.2009.01.010.

54 Darrah, P.R., Jones, D.L., Kirk, G.J.D., and Roose, T. (2006). Modeling the rhizosphere: A review of methods for up scaling to the whole plant scale. *Eur J Soil Sci* 57: 13–25.

55 Anderson, T.A., Guthrie, E.A., and Walton, B.T. (1993). Bioremediation in the rhizopshere: Plant roots and associated microbes clean contaminated soil. *Environ. Sci. Technol.* 24: 885–889.

56 Abhilash, P.C. and Singh, N. (2009b). Withania somniferaDunal mediated dissipation of lindane from simulated soil: Implications for rhizoremediation of contaminated soils. *J. Soils Sediments* doi:10.1007/s11368-009-0085-x.

57 Kuiper, I., Lagendijk, E.L., Bloemberg, G.V., and Lugtenberg, B.J. (2004). Rhizoremediation: A beneficial plant-microbe interaction. *Mol. Plant Microbe Interact.* 17: 6–15.

58 Chaudhry, Q., Blom-Zandstra, M., Gupta, S., and Joner, E.J. (2005). Utilizing the synergy between plants and rhizosphere microorganisms to enhance breakdown of organic pollutants in the environment. *Environ. Sci. Pollut. Res.* 12: 34–48.

59 Anderson, T.A. and Coats, J.R. (1995). Screening rhizosphere soil samples for the ability to mineralize elevated concentrations of atrazine and metachlor. *J. Environ. Sci. Health B* 30: 473–484.

60 Curl, E.A. and Truelove, B. (1986). *The Rhizosphere*. Heidelberg: Springer-Verlag.

61 Kidd, P.S., Prieto-Fernandez, A., and Monterroso, C. (2008). Rhizosphere microbial community and hexachlorocyclohexane degradative potential in contrasting plant species. *Plant Soil* 302: 233–247.

62 Wang, G.D., Li, Q.J., Luo, B., and Chen, X.Y. (2004). Ex planta phytoremediation of trichlorophenol and phenolicallelochemicals via engineered secretory laccase. *Nat. Biotechnol.* 22: 893–897.

63 Sonoki, T., Kajita, S., Ikeda, S., Uesugi, M., Tatsumi, K., Katayama, Y., et al. (2005). Transgenic tobacco expressing fungal laccase promotes the detoxification of environmental pollutants. *Appl. Microbiol. Biotechnol.* 67: 138–142.

64 Mohammadi, M., Chalavi, V., Novakova-Sura, M., Laiberte, J.F., and Sylvestre, M. (2007). Expression of bacterial biphenyl-chlorobiphenyl dioxygenase genes in tobacco plants. *Biotechnol. Bioeng.* 97: 496–505.

65 Sylvestre, M., Macek, T., and Mackova, M. (2009). Transgenic plants to improve rhizoremediation of polychlorinated biphenyls (PCBs). *Curr. Opin. Biotechnol.* doi:10.1016/j.copbio.2009.01.006.

66 Francova, K., Sura, M., Macek, T., Szekeres, M., Bancos, S., Demnerova, K., et al. (2003). Preparation of plants containing bacterial enzyme for degradation of poly chlorinated biphenyls. *Fresenius Environ. Bull.* 12: 309–313.

67 Novakova, M., Mackova, M., Chrastilova, Z., Viktorova, J., Szekeres, M., Demnerova, K., et al. (2009). Cloning of the bacterialbphCgene intoNicotiana tabacumto improve the efficiency of PCB phytoremediation. *Biotechnol. Bioeng.* 102: 29–37.

68 Uchida, E., Ouchi, T., Suzuki, Y., Yoshida, T., Habe, H., Vamaguchi, I., et al. (2005). Secretion of bacterial xenobiotic-degrading enzymes from transgenic plants by an apoplastic expressional system: An applicability for phytoremediation. *Environ. Sci. Technol.* 39: 7671–7677.

69 Bernard, R.G. (2005). Modulation of plant ethylene levels by the bacterial enzyme ACC deaminase. *FEMS Microbiol. Let.* 251: 1–7.

70 Arshad, M., Saleem, M., and Hussain, S. (2007). Perspecives of bacterial ACC deaminase in phytoremediation. *Trends Biotechnol.* 25: 356–362.

71 Glick, B.R. (2005). Modulation of plant ethylene levels by the bacterial enzyme ACC deaminase. *FEMS Microbiol. Lett.* 251: 1–7.

72 Glick, B.R., Penrose, D.M., and Li, J. (1998). A model for the lowering of plant ethylene concentrations by plant growth promoting bacteria. *J. Theor. Biol.* 190: 63–68.

73 Abhilash, P.C. and Singh, N. (2009a). Seasonal variation of HCH isomers in open soil and plant– Rhizospheric soil system of a contaminated environment. *Environ. Sci. Pollut. Res.* doi:10.1007/s11356-009-0133-5.

74 Vijgen, J., Li, Y.F., Martin, F., Lal, R., and Weber, R. (2006). The legacy of lindane and technical HCH production. *Organohalog Compd.* 68: 899–904.

75 Weber, R., Gaus, C., Tysklind, M., Johnston, P., Forter, M., Hollert, H., Heinisch, H., Holoubek, I., Lloyd-Smith, M., Masunaga, S., Moccarelli, P., Santillo, D., Seike, N., Symons, R., Torres, J.P.M., Verta, M., Varbelow, G., Vijgen, J., Watson, A., Costner, P., Woelz, J., Wycisk, P., and Zennegg, M. (2008). Dioxin- and POP-contaminated sites: contemporary and future relevance and challenges. *Environ. Sci. Pollut. Res.* 15: 363–393.

76 Abhilash, P.C. (2009). *Monitoring of Organochlorine Pesticide (Lindane) in Soil–plant System of a Contaminated Environment and Its Phytoremediation/bioremediation*. PhD thesis. India: University of Lucknow.

77 Abhilash, P.C. and Singh, N. (2008). Distribution of hexachlorocyclohexane isomers in soil samples from a small scale industrial area of Lucknow, North India, associated with lindane production. *Chemosphere* 73: 1011–1015.

78 Abhilash, P.C. and Singh, N. (2009a). Seasonal variation of HCH isomers in open soil and plant– Rhizospheric soil system of a contaminated environment. *Environ. Sci. Pollut. Res.* doi:10.1007/s11356-009-0133-5.

79 Suar, M., Hauser, A., Poiger, T., Buser, H.R., Muller, M.D., Dogra, C., et al. (2005). Enantioselective transformation ofα-hexachlorocyclohexane by the dehydrochlorinases LinA1 and LinA2 from the soil bacterium Sphingomonas paucimobilis B90A. *Appl. Environ. Microbiol.* 71: 8514–8518.

80 Lal, R., Dadhwal, M., Kumari, K., Sharma, P., Singh, A., Kumari, H., Jit, S., Gupta, S.K., Lal, D., Verma, M., Kaur, J., Bala, K., and Jindal, S. (2008). Pseudomonassp toSphingobium indicum: A journey of microbial degradation and bioremediation of hexachlorocyclohexane. *Indian J. Microbiol.* 48: 3–18.

81 Sahu, S.K., Patnaik, K.K., Sharmila, M., and Sthunathan, N. (1999). Degradation ofα-hexachlorocyclohexane, β-hexachlorocyclohexane, and γ-hexachlorocyclohexane by a soil bacterium under aerobic conditions. *Appl. Environ. Microbiol.* 56: 3620–3622.

82 Imai, R., Nagata, Y., Fukuda, M., Takagi, M., and Yano, K. (1991). Molecular cloning of a Pseudomonas paucimobilis gene encoding a 17-kilodalton polypeptide that eliminates HCH molecules fromγ-hexachlorocyclohexane. *J. Bacteriol.* 173: 6811–6819.

83 Johri, A.K., Dua, M., Tuteja, D., Saxena, R., Saxena, D.M., and Lal, R. (1996). Genetic manipulations of microorganisms for the degradation of hexachlorocyclohexane. *FEMS Microbial. Rev.* 16: 69–84. Joußen N, Heckel D.

84 Trantirek, L., Hynkova, K., Nagata, Y., Murzin, A., Ansorgova, A., Sklenar, V., et al. (2001). Reaction mechanism and stereochemistry ofγ-hexachlorocyclohexane dehydrochlorinase LinA. *J. Biol. Chem.* 276: 7734–7740.

85 Baker, A.J.M., McGrath, S.P., Reeves, R.D., and Smith, J.A.C. (2000). Metal hyperaccumulator plants: A review of the ecology and physiology of a biological resource for phytoremediation of metal-polluted soils. In: *Phytoremediation of Contaminated Soil and Water* (ed. N. Terry and G. Bañuelos), 85–108. Boca Raton, Florida: Lewis.

86 Yi, Y. and Guerinot, M.L. (1996). Genetic evidence that induction of root Fe(III) chelate reductase activity is necessary for iron uptake under iron deficiency. *Plant J.* 10: 835–844.

87 Eide, D., Broderius, M., Fett, J., and Guerinot, M.L. (1996). A novel ironregulated metal transporter from plants identified by functional expression in yeast. *Proc. Natl. Acad. Sci. USA* 93: 5624–5628.

88 Grotz, N., Fox, T., Connolly, E., Park, W., Guerinot, M.L., and Eide, D. (1998). Identification of a family of zinc transporter genes from *Arabidopsis* that respond to zinc deficiency. *Proc. Natl. Acad. Sci. USA* 95: 7220–7224.

89 Gisbert, C., Ros, R., Haro, A.D., Walker, D.J., Bernal, M.P., Serrano, R., and Navarro-Aviñó, J. (2003). A plant genetically modified that accumulates Pb is especially promising for phytoremediation. *Biochem. Biophys. Res. Commun.* 303: 440–445.

90 Aken, B.V. (2009). Transgenic plants for enhanced phytoremediation of toxic explosives. *Curr. Opin. Biotechnol.* 20: 231–236.

91 Bizily, S.P., Rugh, C.L., and Meagher, R.B. (2000). Phytodetoxification of hazardous organomercurials by genetically engineered plants. *Nat. Biotechnol.* 18: 213–217.

92 Dominguez-Solis, J.R., Gutierrez-Alcala, G., Vega, J.M., Romero, L.C., and Gotor, C. (2001). The cytosolic O-acetylserine(thiol)lyase gene is regulated by heavy metals and can function in cadmium tolerance. *J. Biol. Chem.* 276 (33): 9297–9302.

93 Dominguez-Solis, J.R., Lopez-Martin, M.C., Ager, F.J., Ynsa, M.D., Romero, L.C., and Gotor, C. (2004). Increased cysteine availability is essential for cadmium tolerance and accumulation in Arabidopsis thaliana. *Plant Biotechnol. J.* 2: 469–476.

94 Pilon-Smits, E. (2005). Phytoremediation. *Annu Rev Plant Biol* 56: 15–39.

95 Wangeline, A.L., Burkhead, J.L., Hale, K.L., Lindblom, S.D., Terry, N., Pilon, M., and Pilon-Smits, E.A. (2004). Overexpression of ATP sulfurylase in indian mustard: Effects on tolerance and accumulation of twelve metals. *J. Environ. Qual.* 33 (1): 54–60.

96 Lee, J., Bae, H., Jeong, J., Lee, J.Y., Yang, Y.Y., Hwang, I., Martinoia, E., and Lee, Y. (2003). Functional expression of a bacterial heavy metal transporter in Arabidopsis enhances resistance to and decreases uptake of heavy metals. *Plant Physiol.* 133 (2): 589–596.

97 Samuelsen, A.I., Martin, R.C., Mok, D.W.S., and Machteld, C.M. (1998). Expression of the yeast FRE genes in transgenic tobacco. *Plant Physiol.* 118: 51–58.

98 Hirschi, K.D., Korenkov, V.D., Wilganowski, N.L., and Wagner, G.J. (2000). Expression of Arabidopsis CAX2 in tobacco. Altered metal accumulation and increased manganese tolerance. *Plant Physiol.* 124: 125–133.

99 Bennett, L.E., Burkhead, J.L., Hale, K.L., Terry, N., Pilon, M. and Pilon-Smits, E.A.H. (2003) Analysis of transgenic Indian mustard plants for phytoremediation of metal-contaminated mine tailings. *J. Env. Qual.* 32:432–440.

100 Auer, C. (2008). Ecological risk assessment and regulation for genetically-modified ornamental plants. *CRC. Crit. Rev. Plant. Sci.* 27 (4): 255–271.

101 Ellstrand, N.C. and Schierenbeck, K.A. (2006). Hybridization as a stimulus for the evolution of invasiveness in plants? *Euphytica* 148 (12): 35–46.

102 Culley, T.M. and Hardiman, N.A. (2007). The beginning of a new invasive plant: A history of the ornamental Callery pear in the United States. *AIBS Bull.* 57 (11): 956–964.

103 Simard, M.-J., et al. (2006). TransgenicBrassica napusfields and Brassica rapa weeds in Quebec: Sympatry and weed-crop in situ hybridization. *Botany* 84 (12): 1842–1851.

104 Zelaya, I.A., et al. (2007). Transfer of glyphosate resistance: Evidence of hybridization inConyza(Asteraceae). *Am. J. Bot.* 94 (4): 660–673.

105 Hancock, J.F. (2003). A framework for assessing the risk of transgenic crops. *Bioscience* 53 (5): 512–519.

106 Hegde, S.G., et al. (2006). The evolution of California's wild radish has resulted in the extinction of its progenitors. *Evolution* 60 (6): 1187–1197.

107 Wilkinson, M.J., et al. (2003). Risk assessment of GM plants: Avoiding gridlock? Trends. *Plant. Sci.* 8 (5): 208–212.

108 Johnson, K.L., et al. (2007). How does scientific risk assessment of GM crops fit within the wider risk analysis? Trends. *Plant. Sci.* 12 (1): 15.

109 Hayes, K.R. (2004). *Best Practice and Current Practice in Ecological Risk Assessment for Genetically Modified Organisms.* Hobart: CSIRO Division of Marine Research.

110 Conner, A.J., et al. (2003). The release of genetically modified crops into the environment. *Plant J.* 33 (1): 19–46.

111 Snow, A.A. and Palma, P.M. (1997). Commercialization of transgenic plants: Potential ecological risks. *Bioscience* 47 (2): 86–96.

112 Omura, T. (1999). Forty years of cytochrome P450. *Biochem. Biophys. Res. Commun.* 266: 690–698.

113 Clemens, S., Kim, E.J., Neumann, D., and Schroeder, J.I. (1999). Tolerance to toxic metals by a gene family of phytochelatin synthases from plants and yeast. *EMBOJ.* 18: 3325–3333.

114 Anderson, T.A. and Coats, J.R. (1995). Screening rhizosphere soil samples for the ability to mineralize elevated concentrations of atrazine and metachlor. *J. Environ. Sci. Health* 30: 473–484.

115 Pollard, A.J., Dandridge, K.L., and Jhee, E.M. (2000). Ecological genetics and the evolution of trace element hyperaccumulation in plants. In: *Phytoremediation of Contaminated Soil and Water* (ed. N. Terry and G. Bañuelos), 251–264. Boca Raton, Florida: Lewis.

4

Role of RS and GIS in Water Quality Monitoring and Remediation

CHAPTER MENU

4.1 Introduction

Water is one of the prominent precious natural assets that are indispensable for human survival and the ecosystem's well-being. Water purity is acknowledged to authenticate the health of the biosphere, including for humans. Any disturbance in its characteristics poses a threat to human well-being, particularly for production of victuals and biological diversity [1]. Water resources embody marine and inland aquatic systems (freshwater resources). Since the Industrial Revolution, the escalation of anthropogenic conduct, particularly in the industrial domain, has had its detrimental effects on water bodies thus being a worldwide concern.

Aquatic reservoirs and water distribution schemes that gratify the water necessities of the people are frequently polluted by domestic effluents, oil spillage, industrial effluents of myriad recalcitrants, pesticide and fertilizer residues from the agriculture industry, and leachates stemming out from landfill and paving its route into the groundwater aquifers. The general pollutants include heavy metals, nitrates/nitrites, and pathogenic microorganisms (bacteria and viruses). In addition, there are the noxious wastes from pharmaceutical companies, sanitary pads and hormone-disrupting compounds, the consequences of which are yet to be completely comprehended [2].

Aquatic Environmental Bioengineering: Monitoring and Remediation of Contamination, First Edition.
Rouf Ahmad Bhat, Mohammad Yaseen Mir, Gowhar Hamid Dar and Moonisa Aslam Dervash.

Providentially, novel hi-tech progressions have facilitated solutions, and remote sensing (RS) and geographical information systems (GIS) are no exceptions to deal with various natural resource concerns through technological management strategies. The catastrophic transformations in the quality of aquatic systems, conduits of contamination and transportation of noxious streams can be evaluated with synoptic scale coverage via the virtue of RS [3]. Delineation with the advent of satellite data in an oil spill calamity by providing monitoring technology for RS settings [4] is a supportive strategy as far as management and mitigation approaches are concerned. Besides, water remediation through remote sensing and GIS has manifold intentions, ranging from decontamination of aquatic systems to the harmless discarding of residues in an ecofriendly approach. The RS-based modus operandi is known to deal with monitoring and management of water quality; nonetheless, GIS is what makes it achievable (by compilation of data, and communiqué of water quality hazards in real time, through display maps) to impeccably associate water quality monitoring spatiotemporally with remediation approaches at an economical cost. This chapter aims to consider the relevance of RS and GIS in the monitoring of water characteristics and its remediation thereof.

4.2 Scope of RS and GIS in Water Monitoring

RS relies on the averment that various water elements reflect or emit dissimilar energy quanta (of incident radiation) in varied portions of the electromagnetic spectrum. Water pollutants may modify the reflectance based on their chemical, physical and structural attributes [5, 6]. Sensors present in various RS satellites are proficient to record the reflected/emitted wavelengths, which are synchronized to generate an image that makes available important information on the degree of contagion of aquatic resources and quantum of noxious wastes. The images procured from satellites differ in quality on account of dissimilar sensor distinctiveness, such as their "spatial resolution, spectral resolution, radiometric resolution, and temporal resolution" [6]. Maneuvering in the "visible, near-infrared (NIR), middle infrared (MIR), and shortwave infrared (SWIR)" segment of the electromagnetic spectrum (300–3000 nm), it is labeled "Optical RS"; those commissioning in the thermal range (3000–5000 and 8000–14,000 nm) represents "Thermal RS", whereas "Microwave RS" relies on microwaves in the wavelength range of 1 mm to 1 m of the electromagnetic spectrum [6].

Satellite imageries are liberally and commercially obtainable at diverse spatial resolutions, like "Moderate Resolution Imaging Spectroradiometer (MODIS), Medium Resolution Imaging Spectrometer (MERIS), WorldView-2, WorldView-3, Landsat 5 TM, Landsat 7 ETM, Landsat 8, GeoEye, NOAA AVHRR, IKONOS, Nimbus 7, Formosat, Cartosat, Advanced Space borne Thermal Emission and Reflection Radiation (ASTER), Indian Remote Sensing (IRS) LISS-3 and P6, QuickBird, HICO, SeaWIFS, and SPOT HRV." For instance, "SPOT HRV uses radiations in three wavelength bands: the green (500–590 nm), red (610–680 nm), and NIR (790–890 nm) bands, and IKONOS uses four bands in the spectrum: blue, green, red and NIR, while a Landsat TM has seven bands: blue, green, red, NIR bands, two SWIR bands, and a thermal IR band." Supplementary to RS satellites, hyperspectral airborne missions are accomplished to evaluate water characteristics. "AVIRIS, DAIS, CASI 1500, HYDICE, Hymap and MVIS" are some airborne sensors used in water analysis. "Microwave radiometers (MWR)" evaluate microwave emissions from topographic aspects to appraise sea surface temperature (SST) and water characteristics such as salinity. "AQUA, TRMM, Nimbus 5, Nimbus 7, and SEASAT" satellites function in the microwave range of the electromagnetic spectrum [6].

Geographical information systems (GISs) are an authoritative contrivance in natural resource management researches. As per the documented report of the ESRI (1990) [7], "it includes utilization of computer hardware, software, geographic data, and human resources premeditated to work with all forms of geographically referenced information". The incorporation of topographical characteristics with physical factors has facilitated decision-making and has been used broadly to examine alterations in land use-land cover, population density transformations, shifting housing patterns and flood-zone mapping. GIS databases operate with different thematic maps or strata that bear various forms of information, such as "elevation, soil type, land use, and population density, etc.". Using portable global positioning system (GPS) units, coordinates of the spot where water samples are meant to be taken can be documented, and then fed into the GIS database [6]. Information from satellite images, aerial photographs, soil maps, topographic maps and local survey maps can be integrated in GIS. In water quality management and remediation discourses, GIS may be used to explore, for instance, the spatial relationship between the quantum of contaminants and topography [8] or land use-land cover or soil type [6, 9]. Correspondingly, GIS facilitates hazard-zone mapping of agricultural fields in the immediate vicinity of water bodies [6, 10]. GIS enables monitoring alterations in a temporal-lapse context within a watershed/area and relating transformations in water characteristics [6, 10].

4.3 Assessment of Certain Impurities in Water with the Aid of RS and GIS

The spectral signatures [6] composed by sensors endow facts on types of contaminants and their concentration [6]. The spectral reflectance of water is low in all regions of the electromagnetic spectrum [6]. However, reflectance mounts in the visible portion due to various substances suspended in water [6] and aids in differentiating between clear and murky waters [6]. With augmented algae levels, the reflectance will amplify across most of the wavelengths.

4.3.1 Suspended Load

Organic as well as inorganic matter constitutes the suspended load. It undertakes an imperative task in water quality management because of its relevance with total primary production and fluxes of nuisance and recalcitrant pollutants [11]. Suspended loads are the most frequent contaminants in surface waters and affect the passage of light, thus reducing the photosynthetic aptitude of submerged plants. Besides, it is well documented that suspended loads possess a well-established relationship with radiance/reflectance which relies upon varied factors like color, water depth, sediments type, texture, sensor view and sun angles [6]. In order to estimate suspended matter in water, remote sensing techniques are available. Various airborne platforms such as, "Landsat, SPOT, IRS, Coastal Zone Color Scanner (CZCS) and Sea- viewing Wide Field of View Sensor (SeaWiFS)" reveal the considerable affiliation between suspended load and radiance/reflectance of satellite or aerial vehicles.

The absorption of radiation by suspended particulate matter is usually less compared to chlorophyll, however the scattering is greatly prominent. Taking into cognizance optical wavelength, the RS of terrestrial components is extensive compared to the RS of aquatic systems which is constrained to a comparatively contracted range. It has been also documented in scientific reports that "TM4 (Landsat image)" has a fair relationship with "TSM", and would be appropriate for an estimation of suspended particulate matter in aquatic systems [6].

4.3.2 Phytoplankton

Nutrient enrichment of aquatic systems could possibly enumerate the quantity of chlorophyll enclosed in the fundamental units of algal planktons. Various scientific methodologies rely upon RS for mapping chlorophyll a, an algal indicator and an explicit consideration for water quality evaluation predominantly in inland waters [12]. "Landsat, SPOT, SeaWiFS and CZCS" have been using the manifold algorithms and wavelengths to map the chlorophyll of marine and freshwaters. As a key rule, chlorophyll absorption occurs in short wavelengths (hyperspectral sensors; "reflectance height at 690 nm above the base line from 670 nm to 850 nm"). Information regarding the quantity of phytoplankton has imperative insinuations for "*primary production and carbon cycle models*" as well as for assessing the status of aquatic environs [6]. A huge improbability in recognition of the concentration of chlorophyll is predominant due to cyanobacterial proliferation. In order to deal with such a scenario, broadband sensors like "AVHRR", multispectral sensors such as, "CZCS and SeaWiFS", and synthetic aperture radars have been utilized to map the coverage of the cyanobacterial populace [6, 13].

Correspondingly, plant canopy light reflectance attributes could be used to record the existence of weeds in aquatic environs. Compared to other plants, *Eichhornia crispes* has elevated NIR reflectance [6], whereas *Hydrilla* possess feeble NIR reflectance but elevated NIR reflectance compared to water. This makes them contrastingly noticeable in color infrared imagery with bright orange-red and reddish-brown image responses, respectively [6, 12].

4.3.3 Turbidity

Water turbidity is the outcome of fluvial suspended sediment concentrations. It is related to the optical attributes of water, which results in the scattering and absorbance of light. It somehow hinders the transmittance of light, thus is an adversary of clarity. Satellite data procured from "IKONOS" [14], "Landsat 7 ETM-F bands" [14] using "Landsat 7 ETM + Band 3 (red portion of the electromagnetic spectrum) and Band 4 (near-infrared portion of the electromagnetic spectrum), Landsat 5 TM Band 2, Band 3/Band 2, Band 4/Band 3, and a multivariate regression analysis using Bands 3 and 2" [15] are best to forecast turbidity of inland water bodies.

4.4 Benefits of RS in Assessment of Water Quality

One foremost benefit of RS observations over conventional water quality assessment techniques offers spatiotemporal information of surface water characteristics. With the advent of novel airborne sensors, huge information on chlorophyll a, color index, turbidity, sediment load, surface water and temperature wave height can be evaluated regularly [16]. Remotely procured data possess the prospect of endowing an understanding of extensive scale alterations, the relation between offshore and nearshore waters and the capacity to acquire a durable, near-daily observation of the study area signifying seasonal and interannual changeability. Contrarily, superior RS technology is an economical and cyclical quantitative technique for quantifying water quality to permit comprehensive management. Success in RS to evaluate water quality is well documented in the literature [15].

Compared to limnological studies, *in situ* quantification is frequently constrained to chosen sampling stations. But RS data offers the water body's synoptic panorama by obtaining the attributes of an area rather than a point, and can integrate several characteristics with one composite measurement and capability to improve models with an incessant response from satellite quantification.

4.4.1 Soil Moisture Mapping for Floods and Droughts

Soil moisture is a function of instructive state variables of the water balance. Soil moisture is basically the availability of water in the soil's unsaturated zone, whereas the surface soil moisture is availability of water in the top few centimeters of soil. Distribution of downpour into runoff and permeation is determined by surface soil moisture. Soil moisture has an imperative function in forecasting drought, floods and crop yields.

It is basically impractical to accomplish soil moisture information at large spatial scales with *in-situ* sensors. For this reason, it is not usual to contain soil moisture in decision-making. Passive microwave sensors on board satellites like "AMSR-E, SMOS and Feng Yung" provide global-scale estimations of daily surface soil moisture free of cost, which endows continuous soil moisture estimation without being affected by weather circumstances. That is also the primary reason for applying novel evapotranspiration (*ET*) algorithms based on soil moisture measurements [17].

4.4.2 Spatially Distributed Crop Water Use Estimation

Water paucity is an international concern. Consequently, monitoring and conservational strategies are requisite for its competent management. Consumptive impacts of agriculture, urban greens, forests, soil and natural vegetation are major factors which consume water through evaporation (E) and transpiration (T) putting a toll on fiscal, societal and environmental development [18]. Evapotranspiration (ET) is taken into cognizance by several practices ranging from customary techniques to novel "*modeling and spatially distributed RS evaluation*". Bowen ratio, eddy correlation, heat pulse velocity, lysimeters, and surface renewal are frequently used to evaluate ET at individual plant and field levels with < 90% accuracy [19]. Moreover, extensive labor, apparatus expenditure, and applicability problems are viewed as giant obstacles to utilize this modus operandi at a large scale.

The definite rate of evapotranspiration cannot be calculated precisely from everyday weather data. Rainfall can be estimated effortlessly with inexpensive gauges but not *ET* from land surfaces (except at sites outfitted for systematic energy balance apparatus). As a result, a novel technique ("ETLook") has been tested to generate a spatial estimation of evapotranspiration from satellite imagery and a surface energy balance.

"ETLook" is a binary source model which can generate a spatial estimation of evapotranspiration from satellite imagery and a surface energy balance. Likewise, "SWAT" is a popular distributed hydrological model that illustrates many appropriate hydrological attributes, and soil physical and biophysical processes. It is liberally obtainable, internationally adoptable and can compute atmospheric land surface, soil, groundwater and stream-flow processes. The "SWAT" model parameterization and calibration can be done using remote sensed data on land use, rainfall, topography and *ET*.

4.4.3 Surface Water Quality Monitoring and Remediation

The use of multispectral satellite sensors such as "Landsat TM5", "Landsat ETM 7", "ASTER" [20] and "ALOS AVNIR" are essential to assess water characteristics and sediment problems in water bodies [21]. Similarly, "WorldView-2 satellite imagery" can be exercised to obtain water quality parameters in a lotic water system [22]. Images from "Hyperspectral Imager for the Coastal Ocean (HICO)" on the "International Space Station (ISS)" can be exercised to derive chl-a, CDOM absorption and turbidity in marine waters [23]. Likewise, as per the documentation of Fichot et al. 2015, an airborne sensor, "portable remote imaging spectrometer (PRISM)" can be exploited to examine water characteristics, viz.: chl-a, DOC, methylmercury concentrations, and turbidity at high spatial resolution (2.6 m grid) across coastal wetlands.

Pertinent to an assessment of water quality, RS has consummated landmark achievements in surface water remediation. GIS is exercised to process spatial data to investigate surface water contamination risk [24]. The powerful consortium of software and modeling tools such as "ArcGIS" [7], "Chemicals, Runoff, and Erosion from Agricultural Management Systems (CREAMS)" [25], "Soil and Water Assessment Tool (SWAT)" [26] and "SPAtially Referenced Regressions On Watershed attributes (SPARROW)" [27, 28] are significant in surface water quality assessment.

Several hydrological models are available to approximate runoff and sediment yield from a watershed. "RUSLE, EPIC, ANSWERS, CORINE, ICONA, MIKE SHE, CASC 2D, AGNPS, CREAMS, SWAT and WEPP" are a few among the models [29–36]. These hydrological models are usually site-specific, and hence they need to be appropriately calibrated before applying them to a watershed in a particular region. The calibration of any model is a cumbersome job as it requires a lot of time. Several factors are to be taken into account while calibrating a hydrological model and a lot of input data is also required. Hence, research was conducted to ascertain whether the WEPP model is needed for calibration every time it is applied to a new watershed in the region

RS data can appreciably pave the way for meeting the remediation targets of inland waters in case of cyanobacterial proliferation and the presence of toxins (cyanotoxins); the algorithms for the retrieval of chl-a and phycocyanin (C-PC) from data can be acquired by the "CASI-2 and the Airborne Imaging Spectrometer for Applications (AISA) Eagle Sensor". By combining hyperspectral RS, GPS and GIS technologies, mapping of nuisance and exotic species invasion over big and remote aquatic environs can be executed to restore freshwater, estuarine and floodplain habitats [37]. Likewise, the "HyMap images" could be converted to perceptible surface reflectance for recognition of exotic species and submerged aquatic plants. Using GIS software, mapping their allocation in the aquatic setup can be achieved [38]. Active airborne/spaceborne "Synthetic-Aperture Radar (SAR)" systems can offer a swift rejoinder in case of oil spills [39, 40]. The consortium of varied and sophisticated RS and GIS modus operandi will also facilitate appraisal of the efficacy of the endeavors that are espoused for alleviation.

4.4.4 Groundwater Quality Monitoring and Remediation

Even though remote sensing is principally appropriate for assessing surface water characteristics, with novel progressions in the sensor installation and physical hydrologic models, it has been feasible to accomplish larger measures in groundwater investigations. Remote sensing and GIS technology by exercising image data from "IRS ID (LISS III) and Landsat 5 TM" in conjunction with "Integrated Land and Water Information System (ILWIS)" software possess the potential to prepare groundwater quality maps of TDS, hardness, incrustation problems and magnesium concentration, in addition to the mapping of groundwater prospectives [8]. While investigating the studied rock–water interface, residence time of water in aquifers, and human and natural activities, in the context of groundwater quality, "IRS 1D, LISS III, PAN, and PAN + LISS III images" and GIS software can aid in mapping the spatial variation of groundwater quality [41]. "Geostatistics and ArcGIS" have the potential to be exercised in order to generate spatial distribution maps of groundwater characteristics such as pH, electrical conductivity, chloride, hardness and nitrate concentrations [42]. "Landsat and LISS III imagery" in a GIS framework can be helpful in the assessment of weathering and leaching actions of feldspars and magnesium calcite in addition to estimations of nitrate, manganese and chromium in groundwater reservoirs [43]. On account of this assessment strategy, the groundwater can be rated into "very good, good, moderate, poor, very poor, and unfit classes based on groundwater quality index (GWQI) values" [43].

Groundwater monitoring is intimately pursued by the progression of remediation strategies to restore the resource. For example: "PESTRAS for derivation of spatial patterns of the pesticide leaching potential" [44], "Nitrate Leaching and Economic Analysis package (NLEAP) model" [45], "Agricultural Pressures and Impacts on European waters (Ag–PIE) model" [10] with the widespread agenda to tackle the apprehensions coupled with scattered water pollution based on "DPSIR framework (Driving Force—Pressure—State— Impact—Response)".

"DRASTIC" is a competent overlay and index model which can tackle groundwater susceptibility to impending contamination based on geological and hydrological factors [46]. The "DRASTIC index" takes into cognizance aquifer type, the aquifer's hydraulic conductivity, depth of water, recharge by rainfall, soil properties, topography and the impact of the vadose zone. The coupling of "DRASTIC and GIS" is useful in risk appraisal of groundwater and aquifers [47]. To evaluate the groundwater vulnerability to nitrate concentration, the "RSIVL model (net recharge R, soil type S, impact of vadose zone I, groundwater velocity V, and land use type L)" is the modification of the "DRASTIC model" [48]. In all these scientific assessments, GIS has assisted in the mapping of vulnerable locations and detection of polluted zonations.

4.5 Future Prospectus of RS and GIS Applications in Water Quality Studies

- A RS and GIS approach in water quality studies is proficient in several aspects. Water quality monitoring and remediation through progressions in the field would probably help manage water resources on a extensive scale. Concerning the budding interest in GIS in water quality assessment, GPS and GIS linkages are expected to reinforce research in the future, making real-time monitoring more specific.
- Water quality monitoring and remediation through the "Sentinel-2 and Landsat 8 imagery" [49] seems quite beneficial.
- The "Multispectral Imager (MSI)" onboard the "Sentinel 2A satellite" has better quality capabilities for mapping chl-a, TSS and CDOM [50].
- The "Sentinel-2 mission" offers quality information for monitoring agricultural practices, land use/land cover changes and pollution in lakes and coastal waters.
- "Sentinel 2B images" retrieved recently made it feasible to study Australia's Great Barrier Reef's bleaching because of its high resolution and frequent revisits.
- The "Sentinel-3 satellites" carry contemporary instruments to measure oceans, land, ice and atmosphere and can supervise aquatic biological productivity and oceanic contamination.
- Among the upcoming missions, "Sentinel-6" is equipped with a radar altimeter to measure sea surface height to assist in oceanography and climate studies.
- The "Landsat Data Continuity Mission (LDCM)" is the next-generation technology [24].
- Moreover, missions such as the "Surface Water and Ocean Topography (SWOT)" are expected to better understand the world's oceans and their terrestrial surface waters.
- Further, "Environmental Mapping and Analysis Program (EnMAP)", the "German hyperspectral satellite", has one of its missions to monitor the quality of inland and coastal waters, vis-à-vis mapping of algae [51].
- "EnMAP" and other upcoming imaging spectroscopy missions, such as "NASA's HyspIRI [52]", "Japan's Hyperspectral Imager Suite (HISUI)", "PRISMA", "HYPXIM" and "SHALOM" may thus unveil novel viewpoints in RS-based water quality researches [51].

References

1 Tebbs, E.J., Remedios, J.J., and Harper, D.M. (2013). Remote sensing of chlorophyll-a as a measure of cyanobacterial biomass in Lake Bogoria, a hypertrophic, saline–alkaline, flamingo lake, using Landsat ETM+. *Remote Sens Environ* 135: 92–106.
2 Petrovic, M., Eljarrat, E., Lopez de Alda, M.J., and Barceló, D. (2004). Endocrine disrupting compounds and other emerging contaminants in the environment: A survey on new monitoring strategies and occurrence data. *Anal Bioanal Chem*. 378, 549. https://doi.org/10.1007/s00216-003-2184-7.
3 Kallio, K., Koponen, S., Ylöstalo, P., Kervinen, M., Pyhälahti, T., and Attila, J. (2015). Validation of MERIS spectral inversion processors using reflectance, IOP and water quality measurements in boreal lakes. *Remote Sens Environ* 157: 147–157.
4 Jha, M.N., Levy, J., and Gao, Y. (2008). Advances in remote sensing for oil spill disaster management: State-of-the-art sensors technology for oil spill surveillance. *Sensors* 8 (1): 236–255.
5 Kirk, J.T.O. (1983). *Light and Photosynthesis in Aquatic Ecosystems*. Cambridge: Cambridge University Press. 509.
6 Ramadas, M. and Samantaray, A.K. (2018). Applications of Remote Sensing and GIS in Water Quality Monitoring and Remediation: A state-of-the-Art review. In: *Water Remediation. Energy, Environment, and Sustainability*, (S. Bhattacharya, A. Gupta, A. Gupta, and A. Pandeyeds.), Singapore: Springer. https://doi.org/10.1007/978-981-10-7551-3_13.
7 ESRI. (1990). *Understanding GIS the Arc/info Method*. Redlands, CA, USA: Environmental Systems Research Institute.
8 Srinivasa Rao, Y. and Jugran, D.K. (2003). Delineation of groundwater potential zones and zones of groundwater quality suitable for domestic purposes using remote sensing and GIS. *Hydrol Sci J* 48 (5): 821–833.
9 Everitt, J.H., Yang, C., Escobar, D.E., Webster, C.F., Lonard, R.I., and Davis, M.R. (1999). Using remote sensing and spatial information technologies to detect and map two aquatic macrophytes. *J Aquat Plant Manag* 37: 71–80.
10 Oikonomidis, D., Dimogianni, S., Kazakis, N., and Voudouris, K. (2015). A GIS/remote sensing-based methodology for groundwater potentiality assessment in Tirnavos area, Greece. *J Hydrol* 525: 197–208.
11 Ritchie, J.C. and Charles, M.C. (1996). Comparison of measured suspended sediment concentration estimated from landsat MSS data. *Int J. Remote Sensing* 9 (3): 379–387.
12 Svejkovsky, J. and Shandley, J. (2001). Detection of offshore plankton blooms with AVHRR and SAR imagery. *Int. J. Remote Sensing* 22: 471–485.
13 Ferdi, L., Miller, W.O., and Kehinde, S. (2007). Mapping turbidity in Charles River, Boston using a high resolution satellite. *Environmental Monitoring and Assessment* 132 (1–3): 311–320.
14 Liza, K.L. (2007) *Turbidity mapping and prediction in ice marginal lakes at the bering glacier system, Alaska*. M.Sc Thesis, University of Michigan.50p.
15 Wang, F., Han, L., Kung, H.T., and Van Arsdale, R.B. (2006). Applications of Landsat5 TM imagery in assessing and mapping water quality in Reelfoot Lalce, Tennessee. *Int. J. Of Remote Sensing* 27 (23): 5269–5283.

16 Zhang, Y., Pulliainen, J., Koponen, S., and Hallikainen, M. (2002). Application of an empirical neural network to surface water quality estimation in the Gulf of Finland using combined optical data and microwave data. *Remote Sensing of Environment* 81 (2–3): 327–336.

17 Ines, A.V.M. and Mohanty, B.P. (2008). Near-surface soil moisture assimilation for quantifying effective soil hydraulic properties under different hydroclimatic conditions. *Vadose Zone J* 7: 39–52.

18 Bastiaanssen, W.G.M., Van Der Wal, T., and Visser, T.N.M. (1996). Diagnosis of regional evaporation by remote sensing to support irrigation performance assessment. *Irrig. Drain. Systems* 10: 1–23.

19 Wang, X. and Xie, H. (2018). A Review on Applications of Remote Sensing and Geographic Information Systems (GIS) in Water Resources and Flood Risk Management. *Water*, 10(5):608. https://doi.org/10.3390/w10050608

20 Abdelmalik, K.W. (2016). Role of statistical remote sensing for Inland water quality parameters prediction. *The Egyptian Journal of Remote Sensing and Space Science* 21 (2): 193–200.

21 Volpe, V., Silvestri, S., and Marani, M. (2011). Remote sensing retrieval of suspended sediment concentration in shallow waters. *Remote Sens Environ* 115 (1): 44–54.

22 El Saadi, A.M., Yousry, M.M., and Jahin, H.S. (2014). Statistical estimation of rosetta branch water quality using multi-spectral data. *Water Sci* 28 (1): 18–30.

23 Keith, D.J., Schaeffer, B.A., Lunetta, R.S., Gould, R.W., Jr, Rocha, K., and Cobb, D.J. (2014). Remote sensing of selected water-quality indicators with the hyperspectral imager for the coastal ocean (HICO) sensor. *Int J Remote Sens* 35 (9): 2927–2962.

24 Lim, J. and Choi, M. (2015). Assessment of water quality based on Landsat 8 operational land imager associated with human activities in Korea. *Environ Monit Assess* 187 (6): 384.

25 Frere, M.H., Ross, J.D., and Lane, L.J. (1980). The nutrient submodel. In: *CREAMS: A Field Model for Chemicals, Runoff and Erosion from Agricultural Management Systems*, (W. Kniseled.), Vol. 26. USDA Conserv Res Rep, 340.

26 Arnold, J.G., Williams, J.R., Srinivasan, R., and King, K.W. (1996). *The Soil and Water Assessment Tool (SWAT) User's Manual*. Temple, TX.

27 Alexander, R.B., Smith, R.A., Schwarz, G.E., Boyer, E.W., Nolan, J.V., and Brakebill, J.W. (2007). Differences in phosphorus and nitrogen delivery to the Gulf of Mexico from the Mississippi River Basin. *Environ Sci Technol* 42 (3): 822–830.

28 Robertson, D.M., Schwarz, G.E., Saad, D.A., and Alexander, R.B. (2009). Incorporating uncertainty into the ranking of SPARROW model nutrient yields from Mississippi/Atchafalaya River basin watersheds. *J Am Water Resour Assoc* 45 (2): 534–549.

29 Yousuf, A., Bhardwaj, A., Tiwari, A.K., and Bhatt, V.K. (2015). Simulation of runoff and sediment yield from a forest micro-watershed in *Shivalik* foot-hills using WEPP model. *Int J Agri Sci Res* 5: 67–78.

30 Han, F., Ren, L., Zhang, L., and Li, Z. (2016). The WEPP model application in a small watershed in the Loess Plateau. *PLoS One*. 11: e0148445. DOI:10.1371/journal.pone.0148445.

31 Yousuf, A. and Singh, M.J. (2016). Runoff and soil loss estimation using hydrological models, remote sensing and GIS in *Shivalik* foothills: A review. *Journal of Soil and Water Conservation* 15: 205–210.

32 Yousuf, A., Bhardwaj, A., Tiwari, A.K., and Bhatt, V.K. (2017). Simulation of runoff and sediment yield from a forest micro-watershed in *Shivalik* foot-hills using WEPP model. *Indian J Soil Cons* 45: 21–27.

33 Sharma, S.P. and Bhardwaj, A. (2017). Runoff quantification from small non-arable rangeland watershed in Shivalik foot-hills using WEPP model. *J Indian Water Resour Soc* 37: 25–36.

34 Yousuf, A. and Bhardwaj, A. (2018). WEPP model: Present simulation capabilities, field performance and future enhancement needs. *Res J Agri Sci* 9: 221–230.

35 Melaku, N.D., Renschler, C.S., Holzmann, H., Strohmeier, S., Bayu, W., Zucca, C., Ziadat, F., and Klik, A. (2018). Prediction of soil and water conservation structure impacts on runoff and erosion processes using SWAT model in the northern Ethiopian highlands. *J Soil Sediment* 18: 1743–1755.

36 Yousuf, A., Bhardwaj, A., and Prasad, V. (2021). Simulating impact of conservation interventions on runoff and sediment yield in a degraded watershed using WEPP model. *ECOPERSIA* 9: 3.

37 Hunter, P.D., Tyler, A.N., Carvalho, L., Codd, G.A., and Maberly, S.C. (2010). Hyperspectral remote sensing of cyanobacterial pigments as indicators for cell populations and toxins in Eutrophic Lakes. *Remote Sens Environ* 114 (11): 2705–2718.

38 Hestir, E.L., Khanna, S., Andrew, M.E., Santos, M.J., Viers, J.H., Greenberg, J.A., and Ustin, S.L. (2008). Identification of invasive vegetation using hyperspectral remote sensing in the California Delta ecosystem. *Remote Sens Environ* 112 (11): 4034–4047.

39 Shen, L., Xu, H., and Guo, X. (2012). Satellite remote sensing of harmful algal blooms (HABs) and a potential synthesized framework. *Sensors* 12 (6): 7778–7803.

40 Leifer, I., Lehr, W.J., Simecek-Beatty, D., Bradley, E., Clark, R., Dennison, P., Hu, Y., Matheson, S., Jones, C.E., Holt, B., and Reif, M. (2012). State of the art satellite and airborne marine oil spill remote sensing: Application to the BP Deepwater Horizon oil spill. *Remote Sens Environ* 124: 185–209.

41 Basavarajappa, H.T. and Manjunatha, M.C. (2015). Groundwater quality analysis in Precambrian rocks of Chitradurga district, Karnataka, India using Geo-informatics technique. *Aquat Procedia* 4: 1354–1365.

42 Nas, B. and Berktay, A. (2010). Groundwater quality mapping in urban groundwater using GIS. *Environ Monit Assess* 160 (1): 215–227.

43 Singh, C.K., Shashtri, S., Mukherjee, S., Kumari, R., Avatar, R., Singh, A., and Singh, R.P. (2011). Application of GWQI to assess effect of land use change on groundwater quality in lower Shiwaliks of Punjab: Remote sensing and GIS based approach. *Water Resour Manag* 25 (7): 1881–1898.

44 Tiktak, A., van der Linden, A., and Leine, I. (1996). Application of GIS to the modeling of pesticide leaching on a regional scale in the Netherlands. In: *Applications of GIS to the Modeling of Non-point Source Pollutants in the Vadose Zone* (ed. L.C. Dennis and K. Loague). 259–281.

45 Shaffer, M.J., Hall, M.D., Wylie, B.K., and Wagner, D.G. (1996). NLEAP/GIS approach for identifying and mitigating regional nitrate-nitrogen leaching. In: *Applications of GIS to the Modeling of Non-Point Source Pollutants in the Vadose Zone* (ed. L.C. Dennis and K. Loague). 283–294.

46 Knox, R.C., Sabatini, D.A., and Canter, L.W. (1993). *Subsurface Transport and Fate Processes*. Lewis Publishers.

47 Kazakis, N. and Voudouris, K.S. (2015). Groundwater vulnerability and pollution risk assessment of porous aquifers to nitrate: Modifying the DRASTIC method using quantitative parameters. *J Hydrol* 525: 13–25.

48 Huan, H., Wang, J., and Teng, Y. (2012). Assessment and validation of groundwater vulnerability to nitrate based on a modified DRASTIC model: A case study in Jilin City of northeast China. *Sci Total Environ* 440: 14–23.

49 Toming, K., Kutser, T., Laas, A., Sepp, M., Paavel, B., and Nõges, T. (2016). First experiences in mapping lake water quality parameters with Sentinel-2 MSI imagery. *Remote Sens* 8 (8): 640.

50 Dörnhöfer, K., Göritz, A., Gege, P., Pflug, B., and Oppelt, N. (2016). Water constituents and water depth retrieval from Sentinel-2A—A first evaluation in an Oligotrophic Lake. *Remote Sens* 8 (11): 941.

51 Foerster, S., Carrère, V., Rast, M., and Staenz, K. (2016) Preface: The Environmental Mapping and Analysis Program (EnMAP) Mission: Preparing for its Scientific Exploitation

52 Devred, E., Turpie, K.R., Moses, W., Klemas, V.V., Moisan, T., Babin, M., Toro-Farmer, G., Forget, M.H., and Jo, Y.H. (2013). Future retrievals of water column bio-optical properties using the Hyperspectral Infrared Imager (HyspIRI). *Remote Sensing* 5 (12): 6812–6837.

5

Advancement on Bioaugmentation

Strategies for Processing Industry Wastewater

CHAPTER MENU

5.1 Introduction

Major environmental issues such as global warming, and an imbalance in the ecological processes of the soil which leads to lower agricultural yields have occurred as a result of the undesirable and unwanted activities by humankind. Among all these, it has been seen that xenobiotic compounds are increasingly being disposed of into the environment, and these are responsible for causing tremendous hazard because of their very

Aquatic Environmental Bioengineering: Monitoring and Remediation of Contamination, First Edition.
Rouf Ahmad Bhat, Mohammad Yaseen Mir, Gowhar Hamid Dar and Moonisa Aslam Dervash.

high stability as well as complexity. Several methods have been proposed for getting rid of these materials but the most important method which has proven to be both efficient as well as having great disposal strategy is the process of bioremediation. This technique of remediation is executed through the introduction of very efficient strains of microbes (bioaugmentation) and also by the process of addition of rate limiting nutrients to the soil (biostimulation) in order to enhance the process of remediation significantly. Previous literature has provided a very broad landscape with regard to the efficiency of some of the important microbial strains, but lacks basic information with regard to their degree of efficacy, as well as eco-friendliness, when compared with other disposal strategies.

The environment is currently at a tipping point, or we can say at the point of ecological damage, which is mainly due to a very rapid increase in the human populationm as well as urbanization and industrialization [1]. We can say in other words that the environment has become endangered as a result of increased human activities which are constantly ruining it [2]. The industrial production of chemicals and the use of petroleum and its derivatives are of serious concern when it comes to the contamination of the aquatic and terrestrial ecosystems [3–6]. The most common source of pollutants that are found in the soil and water resources have been found to be oil refineries, and also from the seeping of water from agricultural lands that are treated with pesticides. Most of these pollutants remain present in the environment for long periods of time [7]. Among all the pollutants that have been discussed, the most toxic are polythene bags, because their toxic nature is mainly due to their high molecular weight as well as their complex structure [8]. Every year 140 million tonnes of synthetic polymers are produced and each year an estimated 500 billion to nearly 1 trillion plastic bags are used [9–11]. The polythene bags are then dumped into landfills after use, which results in severe environmental pollution because they are non-biodegradable under natural conditions [12, 13].

When it comes to the ecosystem services, soil plays an important role and also harbours complex biological communities that are present on the earth [14]. Therefore, the preservation of the quality of the soil is a matter of great priority while also maintaining the ecosystem attributes that are of great relevance [15]. One of the most important causes of soil degradation is the process of contamination, and in Europe there are about 2.5 million sites that are potentially contaminated which have an estimated annual management cost of about 6 billion euros [16]. When it comes to the problem of soil contamination, different anthropogenic activities such as the combustion of fossil fuels along with agricultural practices and waste disposal methods have contributed significantly [17]. Also the presence of contaminants in soil cause a very negative impact on the soil biota which has resulted in altered diversity of biological communities that are present in the soil [18]. Bioaugmentation is an *in situ* bioremediation technique in which there is exogenous addition of microorganisms to the contaminated soil. It is a technique in which microbial strains are introduced into the contaminated zone in order to increase the contaminant biodegradation. Also, biological activity may result in degrading the contaminants besides reducing their mobility as well as toxicity.

5.2 Present Disposal Techniques and Their Limitations

Several techniques have been developed for the efficient removal of hazardous toxic pollutants that are present in the soil, and some of the methods that are commonly employed include incineration and recycling. However, the process of incineration results in the emission of large amounts of toxic emissions which mainly include carbon monoxide and methane. This has resulted in changes in the climate of the region, and also the disposal of the materials that mainly include polythene bags into the landfills results in posing serious threats not only to the soil microflora but also proves to be very dangerous for mankind. Besides this, there are other techniques that are under development, which mainly include organic solvent pollutant extraction and also the process of chemical reduction of contaminants and plasma torch techniques. Of the above techniques, use of biological means for the process of removal of soil contaminants was considered to be the most environmentally friendly approach. This approach is talked about in terms of bioremediation and it mainly involves the use of microorganisms in order to destroy waste materials. The main reason behind this is their biodiversity and most importantly the catabolic potential, and among all the bioremediation approaches, *in situ* bioremediation is considered to be the most suitable. It involves the detoxification of harmful chemicals through the process of mineralization. It has been seen that for improving the rate of waste degradation through the process of bioremediation, *in situ* bioremediation is mainly coupled with bioaugmentation and biostimulation.

5.3 Bioaugmentation as an Emerging Strategy

Bioaugmentation is a method where there is application of both autochthonous and allochthonous-wild as well as GMO to waste sites so as to increase the rate of removal of undesired compounds. Clean water is required by industries, and at the same time they are also responsible for the generation of large amounts of wastewater which is contaminated with toxic compounds. Situations like this were already faced in the developed world, but are now also faced by developing countries where growing industrialization is the main problem. China, which is considered as one of the fastest growing developing countries with regards to growing industrial areas, has been found to generate more than 20 billion m^3/year of wastewater [19]. The increasing need for large amounts of clean water that is mainly used for industrial activities has been the area of main concern because of the ever-increasing human population and also because the supply of freshwater is very limited. It therefore becomes necessary to reuse both industrial as well as domestic water in order to overcome this long-standing problem [20].

Before discharging contaminated wastewater into the environment, it needs treatment in order to maintain the concentration of pollutants at acceptable levels. With growing awareness in the population about the consequences of pollutants on human health, legislation has been tightened with regard to the discharge of the pollutants into the environment. As a result of all this, strategies are being developed to increase the efficiency of the treatment plants for cleaning of industrial wastewater. Figure 5.1

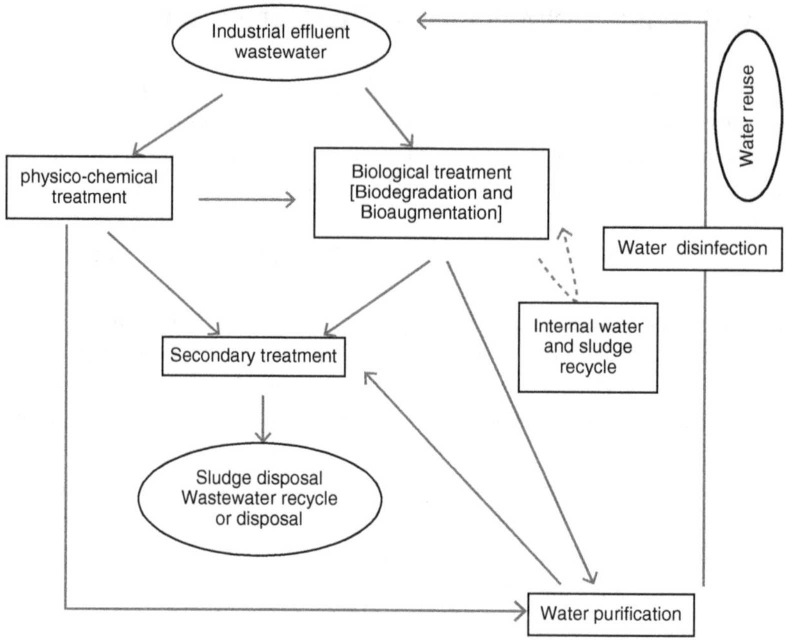

Figure 5.1 Industrial wastewater treatment plan (adapted and modified from Fung et al., 2012) [21].

describes a typical industrial wastewater treatment plant. As a result of employing this technique the products of various treatments, particularly physicochemical as well as biological treatments, will be subjected to purification and disinfection before they are reused, and will be based on the type of the contaminations.

Various approaches are used in the in the physicochemical treatment which mainly include advanced oxidation and reverse osmosis, but these techniques are costly especially when it comes to full-scale treatment [22–24]. Also, these approaches lead to the production of toxins. Biodegradation of organic pollutants through the process of biological treatment is carried out with the help of microorganisms that are present in the wastewater. However, some pollutants are not efficiently biodegraded by the microorganisms, especially the highly complex compounds and that ultimately results in compromising the quality of the water. As a result of all these limitations, strategies based on bioaugmentation are employed. This approach is cheaper and more environmentally friendly when compared to other approaches. A large number of studies cite this approach as a potential tool for the removal of contaminants from the soil [3, 25–31]. The current chapter tries to focus on the applications of bioaugmentation in relation to wastewater and the emphasis is particularly on the microbial aspects of bioaugmentation. It will also try to identify knowledge gaps with respect to future research efforts. In addition to this, we will focus on the limitations of bioaugmentation and will try to have a look at the parameters that mainly affect the efficiency of biodegradation. And we will focus on new areas which mainly include nanotechnology and quorum sensing, which have the potential to be used in order to increase the success of wastewater bioaugmentation.

5.3.1 Bioaugmentation Principle

The aim is to increase the rate of degradation of pollutants, which is mainly achieved by adding microorganisms that assist in the degradation of pollutants. Increasing the population of microorganisms in a contaminated site will not only increase the process of removal of pollutants but will also help in increasing the genetic capacity of that very particular site. It can therefore be concluded that the process of bioaugmentation results in an increase in the gene pool and thereby helps in increasing the genetic diversity of that site. In a nutshell, we can say that genetic diversity can be increased by helping the augmentation of the diversity of microbes.

5.3.2 Cell Bioaugmentation

The process of cell bioaugmentation mainly depends on the catabolic activity of microbial strains that are inoculated [32]. The process of degradation of the target contaminants can be increased by the removal of the contaminants and it will ultimately result in decreasing the time that is required for intended bioremediation [33]. The microbial strains that are inoculated then need to compete for the energy and resources with the population of microbes that are already present in the soil. However, the main drawback for the success of cell bioaugmentation is the high mortality rate of the microbial strains which is caused both due to biotic as well as abiotic stress. Also, their limited dispersal in the soil matrix also happens to be the major drawback [34, 35]. It has also been observed that many factors which mainly include adhesion of the cell to the soil organic matter results in decreasing the abundance of the bacteria. In order to overcome this problem, different authors [36, 37] have suggested that surfactants and foams should be used. In spite of all these limitations, it has been concluded that cell bioaugmentation has great potential when it comes to the bioremediation of soil that is contaminated with organic compounds.

5.3.3 Biological Augmentation as a Tool for Improving the Wastewater Treatment Efficiency

Bioaugmentation has received a great amount of attention due to its capability of working on both metabolic as well as catabolic pathways [38]. In this technique, compounds are targeted for which there has been prior identification and isolation of the microbial consortia. These help in degrading the target compounds through metabolic and catabolic pathways and ultimately result in the biotransformation of the target pollutants. Two types of biologically mediated reactions are found; namely metabolic and cometabolic. Whereas metabolic reactions are associated with processes that are linked to growth in which micro pollutants are mineralized, in the case of catabolic processes there is no sustainable growth of the microorganisms and it leads to the formation of transformed products [39]. A diagram depicting these two processes is shown in Figure 5.2.

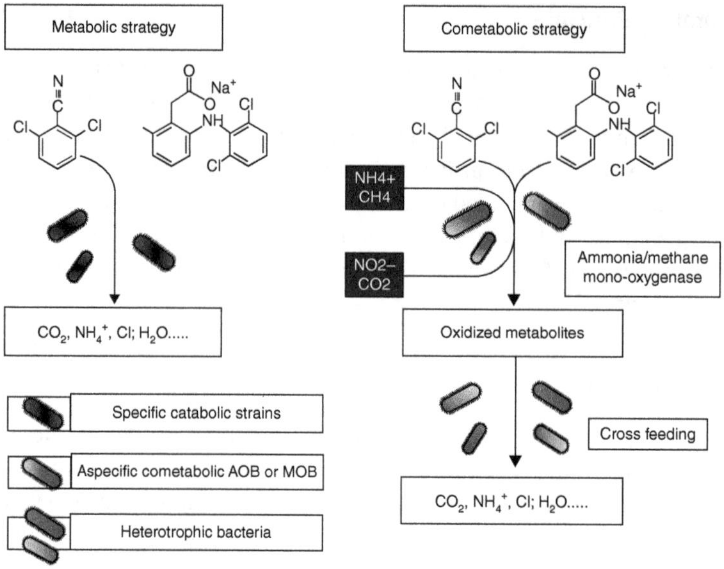

Figure 5.2 Metabolic and cometabolic strategies.

5.3.4 Role of Bioaugmentation in Removing Recalcitrant Pollutants from Industrial Wastewater

As has been seen, some pollutants are resistant to the process of biodegradation and wastewater sludge is known to contain microorganisms that help in degrading these pollutants. A number of factors are responsible for this resistance which include low bioavailability and low biodegradability. Some pollutants having a definite chemical structure are so complex that some microorganisms are absolutely necessary for their biodegradation. In a number of cases some compounds, being new, are not used by the microorganisms as their substrate [40]. In order to overcome these challenges, bioaugmentation has proven to be very useful because it can be targeted against a specific pollutant that is otherwise very dominant. It has therefore been concluded that this approach is very useful for both emerging pollutants and also in cases of those pollutants that are present in large concentrations. During the last decades, a large number of investigations have been used to test bioaugmentation strategies in order to clean wastewater and majority of these have focussed on recalcitrant molecules.

5.4 Bioaugmentation Applications

5.4.1 Removal of Compounds

In plastic components, degreasing agents and wood preservatives, halogenated compounds find their applications [41]. In 2012 the amount of chlorinated solvents used was 764 000 metric tonnes [42]. It is this very extensive use that leads to contamination of wastewater and bioaugmentation has proven to be very useful for the successful elimination of these pollutants. The biodegradation of 2,4-dichlorophenol with the help of bacteria through the process of bioaugmentation has recently been reported in a laboratory setup by using wastewater that is enriched with AS [43]. Recently, it has been reported that using a fluidized biofilm reactor leads to an increase in the biodegradation of 2,4,6-trichlorophenol which is followed by bioaugmentation with Desulfitobacterium sp [44]. These studies were carried out at laboratory-scale, and are very suitable for the removal of chlorinated molecules. However, it still remains to be evaluated when considered at the level of full-scale wastewater treatment plants.

5.4.2 Removal of Lignin

In the paper industry, the process of bioaugmentation has been successfully employed for the treatment of wastewater. Large amounts of wastewater are generated in these industries and contain a major volume of lignin which is commonly called black liquor. For every ton of pulp that is generated seven tonnes of black liquor are produced [45]. A number of compounds like lignin and polysaccharides are present in the black liquor. Biological treatment that is completely natural with AS is not able to remove the compounds because lignin biodegrading microorganism are generally absent in wastewater [46]. It has therefore been concluded that the addition of lignin biodegrading microorganisms proves to be an effective strategy for the removal of pollutants that are originated from black liquor. Zhang et al. [47] in an experiment tested the lignocellulose biodegrading microorganisms and found that bioaugmneted AS helped in the elimination of lignin by more than 50%. All these results showed that bioaugmentation is a very good alternative when it comes to enhancing the biological treatment of wastewater that has lignin present in very high concentrations [47]. However, this process still needs much more evaluation.

5.4.3 Pyridine and Quinoline

These are N-heterocyclic aromatic compounds which are used as solvents and are commonly present in industrial raw materials, and it is because of their use that they find their way into industrial wastewater. Quinolines are also present in petroleum products and have the ability to persist in the environment because of their low biodegradability. A recent study was carried out which showed quinoline in wastewater is biodegraded when bioaugmented with *Burkholderia pickettii* [48], and another study has evaluated that there is good success when quinoline is biodegraded using wastewater bioaugmented with *Paracoccus* sp. and *Psedomonas* sp [49]. In the study discussed later, it was seen that although there was lower concentration of quinoline, the nitrogen concentration was found to be high. In order to overcome this limitation, the same bacteria were tested in a SBR reactor that contained a modified zeolite. Zeolites have been known to help in the removal of nitrogen by the process of adsorption. The results were very promising and showed a reduction in the concentration of quinoline and pyridine [50]. A number of studies have further been carried out which have showed an increase in the pyridine removal when industrial wastewater was bioaugmented with *Rhizobium* sp [51]. Till date, nothing has been confirmed on the application of this approach when it comes to being applied in field conditions.

5.4.4 Cyanides

One of the most toxic compounds released from coal during the coking process is cyanide [52]. It therefore becomes necessary that this wastewater released from the industries must be suitably treated before being released into the environment. Bioaugmentation was applied to wastewater treatment processes that were producing coke in order to increase the efficiency of the removal of cyanides by biological means, and this was done by using a cyanide-degrading yeast, namely *Cryptococcus humicolus*. However, it was less efficienct because it was found to have a lower setting performance as well as a slow biodegradation rate [52]. This technique was one of the first of its kind where bioaugmentation was evaluated for a full-scale treatment plant, and it was concluded that further investigations are required in order to make this approach more efficient when it comes to the process of cyanide removal.

5.4.5 Nicotine

A large amount of toxic substances are associated with tobacco industry wastewater [53], and out of these nicotine is found to be a possible carcinogen [54]. Approximately 60 tonnes of contaminated wastewater are produced for every ton of cigarettes and it was estimated that more than 5 trillion cigarettes were produced throughout the world in 2009 [55, 56]. It has been reported that bioaugmentation will prove a successful strategy in removing these polluatnts. Various studies have been carried out which have reported that there are many bacteria which are capable of degrading nicotine, which mainly include *Acinetobacter* sp. and *Sphingomonas* sp [57]. Wang et al. [58] carried out an experiment in which they tested the bioaugmentation effect on the degradation of nicotine with the help of *Acinetobacter* sp. and it was seen that there was a tremendous increase in the removal of the nicotine; this removal was found to be associated with a significant increase in the bacterial content [58]. Nicotine is very toxic to bacteria and therefore its removal has been found to be associated with increased bacterial growth which in turn helps in augmenting the biodegradation process. Similar results were also reported while using another strain named Pseudomonas sp. HF-1 [59]. All these studies show the potential of bioaugmentation in the elimination of nicotine. However, it should be kept in mind that these experiments were carried out on a small scale and till date there has been no evidence on the applications of this approach in terms of tobacco wastewater treatment plants.

5.5 Bioaugmentation Technologies and Their Limitations

So far we have seen the role of bioaugmentation in wastewater on a small laboratory scale and we have found that the results are very encouraging. However, to date this success has not been achieved on a large scale. On one hand we have seen the benefits of bioaugmentation but a number of failures of bioaugmentation have also been reported [60, 61]. One such study which has been carried out on a large scale is the *in situ* removal of chlorinated solvents that are present in groundwater [62]. Studies have reported that there is a decrease in the number of exogenous microorganisms shortly after they are added to a site. There have been a number of explanations which have documented the death of these introduced microorganisms which mainly include abiotic and biotic stress. There are a number of reasons for this stress which mainly include temperature changes, pH and the competition between introduced and indigenous microorganisms [62, 63].

5.5.1 Grazing of Protozoans

Studies have reported the potential causes which are responsible for the failure of bioaugmentation [63]. At a small scale, *Microvirgula aerodenitricans* was added to a medium that contained acetate in order to remove N_2 and it was found that there was no difference in the removal of N_2 between the nonbioaugmented and bioaugmented reactors. Further studies revealed that the added bacteria disappeared within two days from the reactor for which the main reason was the growth of the protozoa which was found to destroy the bacteria, a phenomenon which was called protozoan grazing. As a result of the phenomenon of protozoan grazing, failure in the technique of bioaugmentation has been reported particularly in the removal of recalcitrant pollutants. It has also been reported in AS studies through the use of *Psedomonas cepacia* [64] that was engineered by transfecting it with a green floutescent protein gene [65]. It was concluded that before the process of bioaugmentation is carried out, studies should be carried out on whether the microorganisms that are bioaugmented are really in a position to grow efficiently in the tested environment or not.

5.5.2 Inoculum Size

The size of the inoculums is yet another important factor that determines the success of the process of bioaugmentation. A study was carried out in which the ability of *Pseudomonas cepacia* was evaluated for biodegrading p-nitrophenol in lake waters, and it was seen that the bioaugmented bacteria was not in a position to biodegrade the pollutants at a concentration of < 400 cells/mL. However, there were encouraging results when the bacterial concentration was between 104–105 cells/mL [66]. This was associated with the phenomenon of protozoan grazing and it was concluded that inoculum size is essential for the success of the process of bioaugmentation. Studies have indicated that an inoculums size of 106–107 cells/mL is very important for the bioaugmentation of groundwater, and this can be adapted in the process of wastewater treatment also [60].

5.5.3 Bacteriophage Infection

Bacteriophages are considered to be the most vast and diverse biological entities on the planet and are known for their ability to infect and destroy bacteria. It has been reported that there are ten phages for every bacterial cell in most of the studied ecosystems including wastewater [67, 68]. Evidence suggests that the reason for the failure of bioaugmentation has been the infection of bacteria by these bacteriophages. For example, a study has reported the effect of phage infection on *Lutimonas* sp. which is a nitrifying bacteria, and it has greatly affected the removal of ammonia from wastewater [69]. The bioaugmentation failed because of the disappearance of the bacterial strains that were found to be associated with an increase in the population of phages in the environment. Similar results have been reported when it was done for the removal of the phosphates by *Candidatus accumulibacter* that was bioaugmented in a wastewater plant [70, 71]. It was therefore concluded that phage infection has a tremendous impact on bioaugmented bacterial growth. Therefore, prior to the initiation of the process of bioaugmentation, the ability of the bacteria to become adapted to the new environment should be established.

5.6 Strategies for Improving the Effectiveness of Bioaugmentation

5.6.1 Immobilizing the Cells in Bioaugmentation

In order to overcome the limitations that we have mentioned, immobilization of the microorganisms can be one of the methods. It is a process in which the living microorganisms are entrapped with the help of a semi-permeable gel which has many advantages over the process of free cell bioaugmentation. For instance, it can help in protection against protozoan grazing as well as against bacteriophjage infections, besides helping in enhancing the biological and physical stabilities. Overall, it is associated with a very high biomass concentration and has been found to show successful results in wastewater when compared to the free cell systems [72]. A study has been carried out which has reported that there is an increase in the removal of nitrogen as well as phosphate from the wastewater by employing encapsulated *Chlorella* sp [73]. Yet another study has worked on the removal of azo dyes by using *Bacillus firmus* [74]. Some other promising results have also been reported elsewhere [75–78]. Although this technique is associated with an improvement in the bioaugmentation results, it is a costly technique especially when large quantities of wastewater need to be treated.

5.6.2 Quorum Sensing

Bioaugmented bacterial colonization is an important factor that determines the success of the process of bioaugmentation. In a majority of the cases, it is ensured by the fact that there is formation of biofilms which happen to be the structures through which the bacterial communities get attached by a self-produced polymeric matrix. A process known as quorum sensing (QS) helps in the formation of these biofilms. In this process, certain chemicals are released by the bacteria which are known as auto inducers that are essential for bacterial communication and ultimately lead to the biofilm formation as well as bacterial cooperation which results in an increase in the bacterial population. The auto inducers which are most commonly reported are acylated-L-homoserine lactones [79]. In case of the pathogenic bacteria, the cooperation between the bacteria leads to the phenomenon of virulence which ultimately results in the development of the disease. A new area of research has opened up which is mainly based on the inhibition of the QS. It uses it as a strategy in order to treat the bacterial infections and in this process a number of QS inhibitors are being investigated [79, 80]. Another way of improving the process of bioaugmentation for the treatment of the wastewater is to modulate the process of QS [81]. A number of studies have been carried out which have showed the presence of QS in wastewater bacteria although the production of the AHL auto inducers were found to help in regulating the dynamics of microbes in the bioreactors [82, 83]. These studies have thrown light on the fact that microbial communities exhibit the process of QS during the process of wastewater treatment, and has led to the increase in biofilms and ultimately improved the process of biodegradation of pollutants. In spite of all these facts, there is need for further research in order to establish the proper dynamics between bioaugmentation and quorum sensing (QS).

5.6.3 Gene Transfer and Genetically Modified Microorganisms

In order to increase the applications of bioaugmentation and to improve its success rate, genetically modified microorganisms have also been used. These microorganisms are modified by inserting some genes into them that are responsible for encoding catabolic enzymes that are then used in the biodegradation of pollutants, thereby helping to increase the efficiency of biodegradation. The genetic manipulation of *Pseudomonas* bacterium which helps in the biodegradation of oil was reported by [84] which proved to be a breakthrough. *Pseudomonas* sp. has been modified genetically with the help of plasmids that contained genes which were coding for catabolic enzymes and ultimately helped in the biodegradation of monoaromatic compounds. It was seen that there was a marked increase in the biodegradation of the benzoate derivatives [85, 86]. Also the use of genetically modified *E. coli* in the process of biodegradation of atrazine has been tested successfully [87, 88]. Similar studies have been carried out in wastewater for the removal of heavy metals [89–92].

5.7 Bioaugmentation and Nanotechnology

Nanotechnology is a new field that is increasingly becoming a technology that has a number of applications in every field of science and technology [93]. A number of nanomaterials have been developed and then tested in a number of varying conditions. The nanomaterials which are most commonly used include titanium dioxide and silver, as well as gold nanoparticles, and have been found to have a size ranging between 1–100 nm thereby helping in providing a large surface area which can be further increased by attaching of the functional groups. The overall need is to adsorb as many target molecules as possible. This phenomenon has been found to be very useful especially in removing both organic and inorganic pollutants from the environment [94–97]. Naonoparticles have been found to inhibit the growth of bacteria and as a result they have been exploited for their use as antibiotics [98, 99]. It has been concluded that the role of nanoparticles as far as the process of bioaugmetation is concerned is limited to only inhibiting the populations of microbes in the contaminated environment.

5.8 Future Prospects

Industries can consume large amounts of water in their production processes and during this production process they generate huge volumes of wastewater which are highly complex in nature. Aquatic life, both flora and fauna, are at risk because of these contaminants. Organic compounds are so refractory in nature that they remain largely unnoticed, and result in various types of toxicities such as carcinogenicity, as well as having mutagenic effects. In order to make it possible that there is production of high quality wastewater generated from process industries, bioaugmentation has proven to be the most useful technique. Bioaugmentation is very successful in removing refractory toxic compounds especially while using specific microorganism that are adapted to a particular toxic compound. These mainly include the utilization of the enzymes as well as using microbial consortium that have the capability to degrade a particular type of target compound.

5.9 Conclusion

The process of bioaugmentation has tremendous potential when it comes to removing pollutants from wastewater. As we have discussed, there is great scope for bioaugmentation when it comes to its applications at the laboratory scale, but some challenges are still present for the large-scale implementation of bioaugmentation. Up to this date, bioaugmentation has been successfully tested for the removal of chlorinated compounds. One of the most common drawbacks that we have witnessed is the problem in maintaining adequate numbers of biodegrading microorganisms during the process of bioaugmentation. Factors such as inoculum size, bacteriophage infections and phenomena like protozoan grazing help in determining the success of the process of bioaugmentation. It is therefore necessary that much attention should be paid to these parameters while applying the process of bioaugmentation at a large scale. QS has also been shown to affect the bioaugmentation process; it mainly affects microorganism growth and researches have been carried out which have shown that QS can be controlled through the addition of activators as well as inhibitors. However, this concept has not received much attention when compared to biodegradation of the organic pollutants. Nanomaterials have also been used in the process of remediation but the applications of this approach have not been explored at a large scale. We can conclude by saying that there is still much more scope for improving the biodegradation of pollutants that are present in contaminated wastewater.

References

1 Cherniwchan, J. (2012). Economic growth, industrialization and the environment. *Resource and Energy Economics* 34: 442–467.

2 Goudie, A.S. (2013). *The Human Impact on the Natural Environment: Past, Present and Future*. John Wiley and Sons. 424.

3 Mrozik, A. and Piotrowska-Seget, Z. (2010). Bioaugmentation as a strategy for cleaning up of soils contaminated with aromatic compounds. *Microbiol. Res.* 165: 363–375.

4 Tyagi, M., da Fronseca, M.M., and de Carvalho, C.C. (2010). Bioaugmentation and biostimulation strategies to improve the effectiveness of bioremediation processes. *Biodegradation* 22: 231–241.

5 Federici, E., Giubilei, M., Santi, G., zanaroli, G., Negroni, A., Fava, F., Petruccioli, M., and D'Annibale, A. (2012). Bioaugmentation of a historically contaminated soil by polychlorinated biphenyls with *Lentinus tigrinus*. *Microb. Cell. Fact.* 11: 35.

6 Schultz-Jansen, N., Aamand, J., and Sorenson, S.R. (2016). Bioaugmentation potential of free and formulated 2,6-dichlorobenzamide (BAM) degrading *Aminobacter* MSH1 in soil, sand and water. *AMB Express* 6: 33.

7 Srirangan, K., Akawi, L., Moo-Young, M., and Chou, C.P. (2012). Towards sustainable production of clean energy carriers from biomass resources. *Applied Energy* 100: 172–186.

8 Harshavardhan, K. and Jha, B. (2013). Biodegradation of low-density polyethylene by marine bacteria from pelagic waters, Arabian Sea, India. *Mar. Pollut. Bull.* 77: 100–106.

9 Shimao, M. (2001). Biodegradation of plastics. *Current Opinion in Biotechnology* 12: 242–247.

10 Leja, K. and Lewandowicz, G. (2010). Polymer biodegradation and biodegradable polymers-a review. *Polish. J. Environ. Stud.* 19: 255–266.

11 Roy, P.K., Titus, S., Surekha, P., Tulsi, E., Deshmukh, C., and Rajagopal, C. (2008). Degradation of abiotically aged LDPE films containing pro-oxidant by bacterial consortium. *Polym. Degrad. Stab.* 93: 1917–1922.

12 Burd, D. (2008). Plastic not fantastic. *Reports of the Canada Wide Science Fair* 1: 1–5.

13 Jalil, M.A., Mian, M.N., and Rahman, M.K. (2013). Using plastic bags and its damaging impact on environment and agriculture: An alternative proposal. *International Journal of Learning and Development* 3: 1–4.

14 Barrios, E. (2007). Soil biota, ecosystem services and land productivity. *Ecol. Econ.* 64: 269–285.

15 Garbisu, C., Alkorta, I., and Epelde, L. (2011). Assessment of soil quality using microbial properties and attributes of ecological relevance. *Appl. Soil Ecol.* 49: 1–4.

16 Panagos, P., Van Liedekerke, M., Yigini, Y., and Montanarella, L. (2013). Contaminated sites in Europe: Review of the current situation based on data collected through a European network. *J. Environ. Public Health* 2013: 1–11.

17 Besser, J.M., Brumbaugh, W.G., Allert, A.L., Poulton, B.C., and Schmitt, C.J. (2009). Ecological impacts of lead mining on Ozark streams-toxicity of sediment and pore water. *Ecotoxicol. Environ. Saf.* 72: 516–526.

18 Burges, A., Epelde, L., and Garbisu, C. (2015). Impact of repeated single-metal and multi-metal pollution events on soil quality. *Chemosphere* 120: 8–15.

19 Zheng, X., Zhang, Z., Yu, D., Chen, X., Cheng, R., Min, S., Wang, J., Xiao, Q., and Wang, J. (2015). Overview of membrane technology applications for industrial wastewater treatment in China to increase water supply. *Resour. Conserv. Recycl.* 105: 1–10.

20 Norton-Brandao, D., Scherrenberg, S.M., and van Lier, J.B. (2013). Reclamation of used urban waters for irrigation purposes: A review of treatment technologies. *J. Environ. Manag.* 122: 85–98.

21 Fung, K.Y., Lee, C.M., Ng, K.M., Wibowo, C., Deng, Z., and Wei, C. (2012). Process development of treatment plants for dyeing wastewater. *AIChe J.* 58: 2726–2742.

22 Subba Rao, A.N. and Venkatarangaiah, V.T. (2014). Metal oxide-coated anodes in wastewater treatment. *Environ. Sci. Pollut. Res. Int.* 21: 3197–3217.

23 Bagal, M.V. and Gogate, P.R. (2014). Wastewater treatment using hybrid treatment schemes based on cavitation and Fenton chemistry: A review. *Ultrason. Sonochem.* 21: 1–14.

24 Siegrist, H. and Joss, A. (2012). Review on the fate of organic micropollutants in wastewater treatment and water reuse with membranes. *Water Sci. Technol.* 66: 1369–1376.

25 Semrany, S., Favier, L., Djelal, H., Taha, S., and Amrane, A. (2012). Bioaugmentation: Possible solution in the treatment of Bio-Refractory Organic Compounds (Bio-ROCs). *Biochem. Eng. J.* 69: 75–86.

26 Tyagi, M., da Fonseca, M.M., and de Carvalho, C.C. (2011). Bioaugmentation and biostimulation strategies to improve the effectiveness of bioremediation processes. *Biodegradation* 22: 231–241.

27 Lebeau, T., Braud, A., and Jezequel, K. (2008). Performance of bioaugmentation-assisted phytoextraction applied to metal contaminated soils: A review. *Environ. Pollut.* 153: 497–522.

28 Thompson, I.P., van der Gast, C.J., Ciric, L., and Singer, A.C. (2005). Bioaugmentation for bioremediation: The challenge of strain selection. *Environ. Microbiol.* 7: 909–915.

29 Singer, A.C., van der Gast, C.J., and Thompson, I.P. (2005). Perspectives and vision for strain selection in bioaugmentation. *Trends Biotechnol.* 23: 74–77.

30 El Fantroussi, S. and Agathos, S.N. (2005). Is bioaugmentation a feasible strategy for pollutant removal and site remediation? *Curr. Opin. Microbiol.* 8: 268–275.

31 Herrero, M. and Stuckey, D.C. (2015). Bioaugmentation and its application in wastewater treatment: A review. *Chemosphere* 140: 119–128.

32 Singh, A. and Ward, O.P. (2004). *Biotechnology and Bioremediation an Overview*. Berlin: Springer-Verlag. 1–17.

33 Nowak, A. and Mrozik, A. (2016). Facilitation of co-metabolic transformation and degradation of monochlorophenols by Pseudomonas sp. CF600 and changes in its fatty acid composition. *Water Air Soil Pollut.* 227: 83.

34 Pepper, I.L., Gentry, T.J., Newby, D.T., Roane, T.M., and Josephson, K.L. (2002). The role of cell bioaugmentation and gene bioaugmentation in the remediation of co-contaminated soils. *Environ. Health Perspect.* 110: 943–946.

35 Quan, X., Tang, H., Xiong, W., and Yang, Z. (2010). Bioaugmentation of aerobic sludge granules with a plasmid donor strain for enhanced degradation of 2,4-dichlorophenoxyacetic acid. *J. Hazard Mater.* 179: 1136–1142.

36 Wang, S. and Mulligan, C.N. (2004). An evaluation of surfactant foam technology in remediation of contaminated soil. *Chemosphere* 57: 1079–1089.

37 Franzetti, A., Caredda, P., Ruggeri, C., La Colla, L., Tamburini, E., Papacchini, M., and Bestetti, G. (2009). Potential application of surface active compounds by *Gordonia* sp. strain BS29 in soil remediation technologies. *Chemosphere* 75: 801–807.

38 Hai, F.I., Modin, O., Yamamoto, K., Fukushi, K., Nakajiam, F., and Nghiem, L.D. (2012). Pesticide removal by a mixed culture of bacteria and white-rot fungi. *J. Taiwan Inst. Chem. Eng. 43:* 459–462.

39 Tran, N.H., Urase, T., Ngo, H.H., Hu, J., and Ong, S.L. (2013). Insight into metabolic and cometabolic activities of autotrophic and heterotrophic microorganisms in the biodegradation of emerging trace organic contaminants. *Bioresour Technol.* 146: 721–731.

40 Providenti, M.A., Lee, H., and Trevors, J.T. (1993). Selected factors limiting the microbial degradation of recalcitrant compounds. *J. Ind. Microbiol.* 12: 379–395.

41 Mattes, T.E., Alexander, A.K., and Coleman, N.V. (2010). Aerobic biodegradation of the chloroethenes: Pathways, enzymes, ecology, and evolution. *FEMS Microbiol. Rev.* 34: 445–475.

42 Loffler, F.E., Ritalahti, K.M., and Zinder, S.H. (2013). Dehalococcoides and reductive dechlorination of chlorinated solvents. In: *Bioaugmentation for Groundwater Remediation*, (ed. C.H. Ward), 39–88. New York, NY, USA: Springer Science and Business Media New York.

43 Quan, X., Shi, H., Liu, H., Lv, P., and Qian, Y. (2004). Enhancement of 2,4-dichlorophenol degradation in conventional activated sludge systems bioaugmented with mixed special culture. *Water Res.* 38: 245–253.

44 Puyol, D., Monsalvo, V.M., Sanchis, S., Sanz, J.L., Mohedano, A.F., and Rodriguez, J.J. (2015). Comparison of bioaugmented EGSB and GAC-FBB reactors and their combination with aerobic SBR for the abatement of chlorophenols. *Chem. Eng. J.* 259: 277–285.

45 Biermann, C.J. (1993). *Essentials of Pulping and Papermaking*. San Diego, CA, USA: Academic Press, Inc.

46 Wu, J., Xiao, Y.Z., and Yu, H.Q. (2005). Degradation of lignin in pulp mill wastewaters by white-rot fungi on biofilm. *Bioresour. Technol.* 96: 1357–1363.

47 Zheng, Y., Chai, L.Y., Yang, Z.H., Tang, C.J., Chen, Y.H., and Shi, Y. (2013). Enhanced remediation of black liquor by activated sludge bioaugmented with a novel exogenous microorganism culture. *Appl. Microbiol. Biotechnol.* 97: 6525–6535.

48 Tuo, B.H., Yan, J.B., Fan, B.A., Yang, Z.H., and Liu, J.Z. (2012). Biodegradation characteristics and bioaugmentation potential of a novel quinoline-degrading strain of Bacillus sp. isolated from petroleum-contaminated soil. *Bioresour. Technol.* 107: 55–60.

49 Bai, Y., Sun, Q., Zhao, C., Wen, D., and Tang, X. (2009). Simultaneous biodegradation of pyridine and quinoline by two mixed bacterial strains. *Appl. Microbiol. Biotechnol.* 82: 963–973.

50 Bai, Y., Sun, Q., Xing, R., Wen, D., and Tang, X. (2010). Removal of pyridine and quinoline by bio-zeolite composed of mixed degrading bacteria and modified zeolite. *J. Hazard. Mater.* 181: 916–922.

51 Liu, X., Chen, Y., Zhang, X., Jiang, X., Wu, S., Shen, J., Sun, X., Li, J., Lu, L., and Wang, L. (2015). Aerobic granulation strategy for bioaugmentation of a sequencing batch reactor (SBR) treating high strength pyridine wastewater. *J. Hazard. Mater.* 295: 153–160.

52 Park, D., Lee, D.S., Kim, Y.M., and Park, J.M. (2008). Bioaugmentation of cyanide-degrading microorganisms in a full-scale cokes wastewater treatment facility. *Bioresour. Technol.* 99: 2092–2096.

53 Talhout, R., Schulz, T., Florek, E., van Benthem, J., Wester, P., and Opperhuizen, A. (2011). Hazardous compounds in tobacco smoke. *Int. J. Environ. Res. Public Health* 8: 613–628.

54 Russo, P., Cardinale, A., Margaritora, S., and Cesario, A. (2012). Nicotinic receptor and tobacco-related cancer. *Life Sci.* 91: 1087–1092.

55 Zhong, W., Zhu, C., Shu, M., Sun, K., Zhao, L., Wang, C., Ye, Z., and Chen, J. (2010). Degradation of nicotine in tobacco waste extract by newly isolated *Pseudomonas* sp. ZUTSKD. *Bioresour. Technol.* 101: 6935–6941.

56 Eriksen, M., Mackay, J., and Ross, H. (2012). *The Tobacco Atlas*. Atlanta, GA, USA: American Cancer Society.

57 Wang, M.Z., Yang, G.Q., Wang, X., Yao, Y.L., Min, H., and Lv, Z.M. (2011). Nicotine degradation by two novel bacterial isolates of *Acinetobacter* sp. TW and *Sphingomonas* sp. TY and their responses in the presence of neonicotinoid insecticides. *World J. Microbiol. Biotechnol.* 27: 1633–1640.

58 Wang, J.H., He, H.Z., Wang, M.Z., Wang, S., Zhang, J., Wei, W., Xu, H.X., Lv, Z.M., and Shen, D.S. (2013). Bioaugmentation of activated sludge with *Acinetobacter* sp. TW enhances nicotine degradation in a synthetic tobacco wastewater treatment system. *Bioresour. Technol.* 142: 445–453.

59 Wang, M., Yang, G., Min, H., Lv, Z., and Jia, X. (2009). Bioaugmentation with the nicotine-degrading bacterium *Pseudomonas* sp. HF-1 in a sequencing batch reactor treating tobacco wastewater: Degradation study and analysis of its mechanisms. *Water Res.* 43: 4187–4196.

60 Lyon, D.Y. and Vogel, T. (2013). Bioaugmentation for Groundwater Remediation: An Overview. In: *Bioaugmentation for Groundwater Remediation*, (Ed. C.H. Ward), 1–37. New York, NY, USA: Springer Science and Business Media.

61 Kalogerakis, N. (2016) Bioaugmentation: is it really needed for the bioremediation of contaminated sites? (accessed on 23 June 2016).

62 Stroo, H.F., Leeson, A., and Ward, C.H. (2013). *Bioaugmentation for Groundwater Remediation*. New York, NY, USA: Springer Science and Business Media. 389.

63 Bouchez, T., Patureau, D., Dabert, P., Juretschko, S., Dore, J., Delgenes, P., Moletta, R., and Wagner, M. (2000). Ecological study of a bioaugmentation failure. *Environ. Microbiol.* 2: 179–190.

64 Goldstein, R.M., Mallory, L.M., and Alexander, M. (1985). Reasons for possible failure of inoculation to enhance biodegradation. *Appl. Environ. Microbiol.* 50: 977–983.

65 Eberl, L., Schulze, R., Ammendola, A., Geisenberger, O., Erhart, R., Sternberg, C., Molin, S., and Amann, R. (1997). Use of green fluorescent protein as a marker for ecological studies of activated sludge communities. *FEMS Microbiol. Lett.* 149: 77–83.

66 Ramadan, M.A., el-Tayeb, O.M., and Alexander, M. (1990). Inoculum size as a factor limiting success of inoculation for biodegradation. *Appl. Environ. Microbiol.* 56: 1392–1396.

67 Suttle, C.A. (2007). Marine viruses: Major players in the global ecosystem. *Nat. Rev. Microbiol.* 5: 801–812.

68 Shapiro, O.H. and Kushmaro, A. (2011). Bacteriophage ecology in environmental biotechnology processes. *Curr. Opin. Biotechnol.* 22: 449–455.

69 Fu, S., Fan, H., Liu, S., Liu, Y., and Liu, Z. (2009). A bioaugmentation failure caused by phage infection and weak biofilm formation ability. *J. Environ. Sci.* 21: 1153–1161.

70 Barr, J.J., Slater, F.R., Fukushima, T., and Bond, P.L. (2010). Evidence for bacteriophage activity causing community and performance changes in a phosphorus-removal activated sludge. *FEMS Microbiol. Ecol.* 74: 631–642.

71 Lee, S.H., Otawa, K., Onuki, M., Satoh, H., and Mino, T. (2007). Population dynamics of phage-host system of *Microlunatus phosphovorus* indigenous in activated sludge. *J. Microbiol. Biotechnol.* 17: 1704–1707.

72 Shi, S., Qu, Y., Ma, F., and Zhou, J. (2014). Bioremediation of coking wastewater containing carbazole, dibenzofuran and dibenzothiphene by immobilized naphthalene-cultivated *Arthrobacter* sp. W1 in magnetic gellan gum. *Bioresour. Technol.* 166: 79–86.

73 Zeng, X., Danquah, M.K., Zheng, C., Potumarthi, R., Chen, X.D., and Lu, Y. (2012). NaCS–PDMDAAC immobilized autotrophic cultivation of *Chlorella* sp. for wastewater nitrogen and phosphate removal. *Chem. Eng. J.* 187: 185–192.

74 Ogugbue, C.J., Sawidis, T., and Oranusi, N.A. (2011). Evaluation of colour removal in synthetic saline wastewater containing azo dyes using an immobilized halotolerant cell system. *Ecol. Eng.* 37: 2056–2060.

75 Surkatti, R. and El-Naas, M.H. (2014). Biological treatment of wastewater contaminated with p-cresol using Pseudomonas putida immobilized in polyvinyl alcohol (PVA) gel. *J. Water Process Eng.* 1: 84–90.

76 Pakshirajan, K. and Kheria, S. (2012). Continuous treatment of coloured industry wastewater using immobilized *Phanerochaete chrysosporium* in a rotating biological contactor reactor. *J. Environ. Manag.* 101: 118–123.

77 Khondee, N., Tathong, S., Pinyakong, O., Powtongsook, S., Chatchupong, T., Ruangchainikom, C., and Luepromchai, E. (2012). Airlift bioreactor containing chitosan-immobilized *Sphingobium* sp. P2 for treatment of lubricants in wastewater. *J. Hazard. Mater.* 213–214: 466–473.

78 Del Castillo, I., Hernández, P., Lafuente, A., Rodríguez-Llorente, I.D., Caviedes, M.A., and Pajuelo, E. (2012). Self-bioremediation of cork-processing wastewaters by (chloro) phenol-degrading bacteria immobilised onto residual cork particles. *Water Res.* 46: 1723–1734.

79 Kalia, V.C. (2013). Quorum sensing inhibitors: An overview. *Biotechnol. Adv.* 31: 224–245.

80 Jiang, T. and Li, M. (2013). Quorum sensing inhibitors: A patent review. *Expert Opin. Ther. Pat.* 23: 867–894.

81 Valle, A., Bailey, M.J., Whiteley, A.S., and Manefield, M. (2004). N-acyl-l-homoserine lactones (AHLs) affect microbial community composition and function in activated sludge. *Environ. Microbiol.* 6: 424–433.

82 Chong, G., Kimyon, O., Rice, S.A., Kjelleberg, S., and Manefield, M. (2012). The presence and role of bacterial quorum sensing in activated sludge. *Microb. Biotechnol.* 5: 621–633.

83 Yeon, K.M., Cheong, W.S., Oh, H.S., Lee, W.N., Hwang, B.K., Lee, C.H., Beyenal, H., and Lewandowski, Z. (2009). Quorum sensing: A new biofouling control paradigm in a membrane bioreactor for advanced wastewater treatment. *Environ. Sci. Technol.* 43: 380–385.

84 Chakrabarty, A.M. (1985). Genetically-manipulated microorganisms and their products in the oil service industries. *Trends Biotechnol.* 3: 32–39.

85 McClure, N.C., Fry, J.C., and Weightman, A.J. (1991). Survival and catabolic activity of natural and genetically engineered bacteria in a laboratory-scale activated-sludge unit. *Appl. Environ. Microbiol.* 57: 366–373.

86 Nusslein, K., Maris, D., Timmis, K., and Dwyer, D.F. (1992). Expression and transfer of engineered catabolic pathways harbored by *Pseudomonas* spp. introduced into activated sludge microcosms. *Appl. Environ. Microbiol.* 58: 3380–3386.

87 Wei, M.J., Wang, H., Liu, C., and Ning, D.L. (2008). Bioaugmentation with immobilized genetically engineered microorganism (GEM)/CAS process for treatment of atrazine wastewater. *Huan Jing Ke Xue* 29: 1555–1560.

88 Jin, R., Yang, H., Zhang, A., Wang, J., and Liu, G. (2009). Bioaugmentation on decolorization of C.I. Direct blue 71 by using genetically engineered strain *Escherichia coli* JM109 (pGEX-AZR). *J. Hazard. Mater.* 163: 1123–1128.

89 Okino, S., Iwasaki, K., Yagi, O., and Tanaka, H. (2002).Removal of mercuric chloride by a genetically engineered mercury-volatilizing bacterium *Pseudomonas putida* PpY101/pSR134. *Bull. Environ. Contam. Toxicol.* 68: 712–719.

90 Deng, X. and Wilson, D.B. (2001). Bioaccumulation of mercury from wastewater by genetically engineered. *Escherichia Coli. Appl. Microbiol. Biotechnol.* 56: 276–279.

91 Almaguer-Cantu, V., Morales-Ramos, L.H., and Balderas-Renteria, I. (2011). Biosorption of lead (II) and cadmium (II) using *Escherichia coli* genetically engineered with mice metallothionein I. *Water Sci. Technol.* 63: 1607–1613.

92 Deng, X., Li, Q.B., Lu, Y.H., Sun, D.H., and Huang, Y.L. (2003). Uptake of nickel from industrial wastewater by genetically engineered Escherichia coli JM109. *Sheng Wu Gong Cheng Xue Bao.* 19: 343–348.

93 de Morais, M.G., Martins, V.G., Steffens, D., Pranke, P., and da Costa, J.A. (2014). Biological applications of nanobiotechnology. *J. Nanosci. Nanotechnol.* 14: 1007–1017.

94 Ghasemzadeh, G., Momenpour, M., Omidi, F., Hosseini, M.R., Ahani, M., and Barzegari, A. (2014). Applications of nanomaterials in water treatment and environmental remediation. *Front. Environ. Sci. Eng.* 8: 471–482.

95 Garner, K.L. and Keller, A.A. (2014). Emerging patterns for engineered nanomaterials in the environment: A review of fate and toxicity studies. *J. Nanopart. Res.* 16: 1–28.

96 Crane, R.A. and Scott, T.B. (2012). Nanoscale zero-valent iron: Future prospects for an emerging water treatment technology. *J. Hazard. Mater.* 211-212: 112–125.

97 Fu, F., Dionysiou, D.D., and Liu, H. (2014). The use of zero-valent iron for groundwater remediation and wastewater treatment: A review. *J. Hazard. Mater.* 267: 194–205.

98 Beyth, N., Houri-Haddad, Y., Domb, A., Khan, W., and Hazan, R. (2015). Alternative antimicrobial approach: Nano-antimicrobial materials. Evid.-Based Complement. *Altern. Med.* 2015: 246012.

99 Zhang, C., Li, M., Xu, X., and Liu, N. (2015). Effects of carbon nanotubes on atrazine biodegradation by *Arthrobacter* sp. *J. Hazard. Mater.* 287: 1–6.

6

Photocatalysis in Relation to Water Remediation

CHAPTER MENU

6.1 Introduction

The word "photocatalysis" means a chemical reaction which is caused by the photoabsorption of matter, also known as a "photocatalyst"—a semiconductor (a branch of chemistry that deals with chemical reactions in presence of light and photocatalysts)—which enhances the rate of reaction and remains unchanged during the reaction. During the process, water molecules get adsorbed over the surface of photocatalysts (TiO2) in normal conditions and are used to participate in the reaction to decompose pollutants in water (Figure 6.1). Photocatalysis has become the most explored method of obtaining potable water in remote areas [1]. Photocatalysis is used in a broad range of research areas especially environmental- and energy-related fields, as shown in Figure 6.2 [2].

Millions of people suffer and die every year due to lack of access to safe drinking water, mostly in under-developed nations, because there is no technology to treat the huge amounts of wastewater discharged from industries and sewage treatment plants, leading to the risk for pathogens inhabiting water. Malato and coworkers [4] reported that nearly 4 billion people are at risk and may lose their lives [5, 6]. According to reports many people across the globe lack access to safe drinking water, and annually large numbers of people lose their lives due to water pathogenecity. It has been estimated that natural

Aquatic Environmental Bioengineering: Monitoring and Remediation of Contamination, First Edition.
Rouf Ahmad Bhat, Mohammad Yaseen Mir, Gowhar Hamid Dar and Moonisa Aslam Dervash.

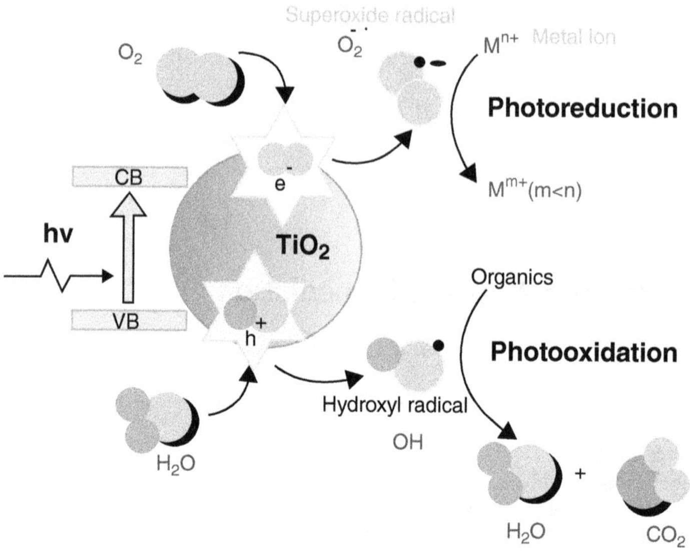

Figure 6.1 Mechanism of photocatalysis [3].

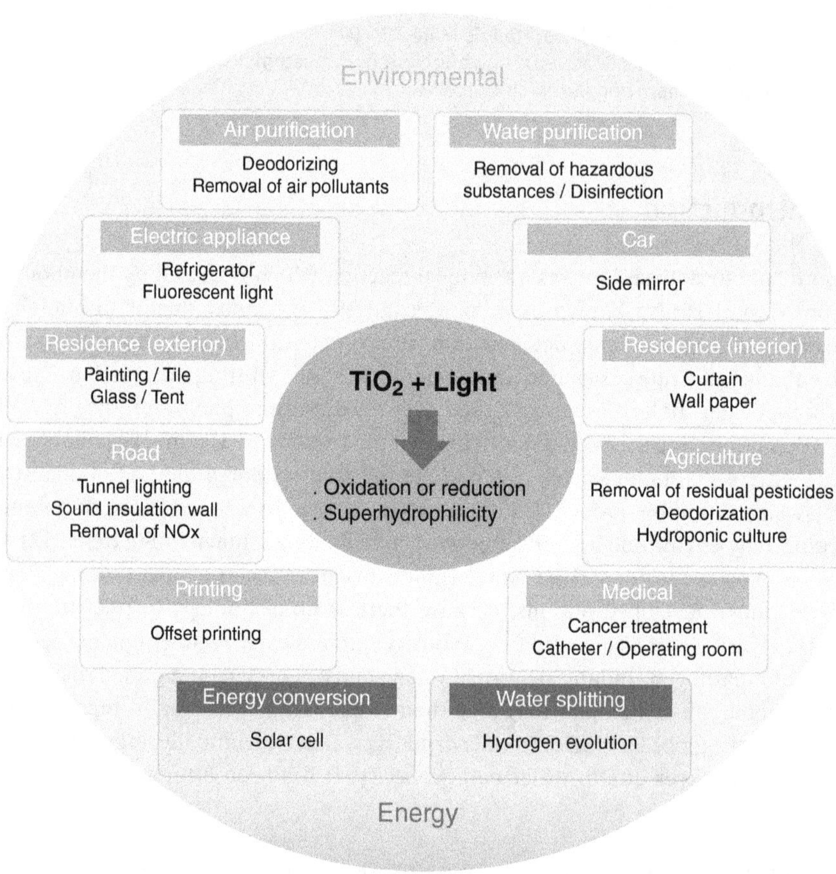

Figure 6.2 Applications of TiO2 photocatalysis [2].

waters are being continuously contaminated due to the addition of pollutants to the natural water systems of the planet. In order to counter the deterioration of safe-water shortages, it is beneficial to develop innovative H_2O management systems with cheap and advanced quality for the treatment of wastewater. The future reclamation or recycling, or providing onset pastoral discarded H_2O or treated urban effluent from management plants for agronomic and manufacturing processes is one of the few appealing alternatives [7, 8]. While we know about providing advanced treatments like removing pathogenic coliforms, inorganic as well as organic molecules are hard to remove because they interact with others to form complex compounds, and are costly to handle and are typically correlated with treating wastewater [9]. Inland surface water treatment methods presently available, such as adsorption or coagulation, simply concentrate the toxins existing by pointing them to supplementary phases, but they continue to exist and are not "eradicated" or "demolished" entirely, as per Padmanabhan and coworkers in [10]. Sedimentation, filtration, chemical and membrane systems come under the category of traditional or conventional treatments methods which requires high operating charges and could lead to the formation of toxic secondary contaminants [11]. The most popular and commonly used method of decontamination has been chlorination. Chlorination-generated disinfectant by-products remain cancer-posing to life forms [12–14]. According to the US Environmental Protection Agency (EPA) concern has been raised over secondary contaminants rather than primary ones which are more potent for human health, thus regulations were given for general awareness regarding their presence across the nations. A frequently used method of disinfection has been chlorination. Chlorination-generated disinfection by-products are mutagenic and carcinogenic to human health [12–14]. These also contributed to accelerated research and development in the area of Advanced Oxidation Processes (AOPs) as an emerging technology for water treatment. These AOPs are based on the *in situ* generation of highly reactive transient species (i.e. H_2O_2, OH, O_2, O_3) for the mineralization of organic refractory materials, water contaminants and by-products of disinfection (Table 6.1) [3, 15].

Heterogeneous photocatalysis using semiconductor catalysts (TiO_2, ZnO, Fe_2O_3, CdS, GaP and ZnS) among these AOPs showed their efficacy in degrading a wide variety of unclear refractory organics into readily biodegradable compounds and finally mineralizing them into harmless carbon dioxide and water. Titanium dioxide (TiO_2) has earned the greatest interest in the R&D of photocatalysis technology among semiconductor catalysts. Titanium dioxide is the most active photocatalyst and is stable following repeated catalytic cycles of 300 nm < l < 390 nm *hv* energy, whereas cadmium-selenium or gallium-polonium are degraded to produce toxic content [4]. A variety of essential features of heterogeneous photocatalysis have enhanced their feasible use in the treatment of water, including (1) environmental operating pressure and temperature, (2) total parent and non-secondary compounds of mineralization, and (3) cheap machinery operations. Due to photoinduction on TiO_2 surfaces can generate highly reactive oxygen species (ROS) which can be used for organic mineralization and microbial inactivation. To date a wide range of technological obstacles are present for using TiO_2 catalysts for water treatment purposes. As an industrial process, the post-separation of the semiconductor TiO_2 catalyst after water treatment remains the major barrier to practicality. Together with its high surface-area-to-volume ratio and surface energy, the fine particle

Table 6.1 Photocatalytic disinfection of microorganisms using different catalysts Based on [3, 15].

S.No.	Microorganism	Catalyst material
Disinfection in presence of UV radiation		
1	*Legionella pneumophila*	TiO_2
2	*Bacillus anthracis*	TiO_2
3	*E.coli, Lactobacillus helveticus*	TiO_2 and ZnO
4	*E.coli, Bacillus megaterium*	TiO_2, V doped TiO_2, Au capped TiO_2 composites
5	*E.coli*	Ag impregnated TiO_2
Disinfection in presence of visible radiation		
1	*E.coli, Staphylococcus aureus*	AgI/TiO_2
2	*E.coli*	Montmorillonite supported Ag/TiO_2
3	*E.coli, Pseudomonas aeruginosa, Staphylococcus aureus*	PdO and N doped TiO_2
4	*E. coli*	Graphene oxide-TiO_2 thin films
5	Microcystin-LR	N-F codoped TiO_2

size of TiO_2 produces a strong propensity for catalyst agglomeration during activity. In terms of particle size protection, surface-area reduction and its reuse lifetime, such particle agglomeration is extremely counterproductive. Such technical snags are the production of catalysts with a wider spectrum of photoactivity and their combination with a viable photocatalytic reactor device. Furthermore, understanding the theory behind the typical operating parameters of the reactors and their interactions is also insufficient and poses a daunting process-optimization challenge. The impurity of fresh aquatic environs by unpurified micro and macro pollutants from various man-made activities, including aquatic degradation of toxic microbes, have an important role in the effects of anthropogenesis. Various environmental factors, mostly abiotic parameters, are involved in the oxidation of toxicants and the creation of new compounds, particularly those that are more toxic. There are a variety of physicochemical processes in the hydrosphere alongside biological processes as well as chemical processes which play important roles in new compound synthesis, etc. [16–18]. For the environmental importance, solar light plays an immense role in the self-purification of the natural waters to initiate the reaction of transition metals (TM) like (Fe, Cu, Cr, etc.). When rocks and minerals as well as anthropogenic sources like metal, steel wastes, oil refineries, mines, etc., undergo dissolution, they leads to the formation of TM compounds. In the surface waters, the concentration of TM compounds varies from a few to several hundred micrograms per liter, with most common occurrence of C and Fe which play a role in the metabolic pathways like enzymatic activations, and these are present in their thermostable states of Fe(III) and Cu(II). The necessary step in the conversion of organic compounds to transitional elements is important in the inland waters by involving photochemical reactions. A comparatively recent zone that will be applied in the future is the use of photocatalytic processes in aquatic management [17, 19, 20]. Among the large amount of physicochemical transformations occurring in the hydrosphere, photocatalytic oxidative oxidation of anthropogenic and natural organic pollutants (pesticides, surfactants, organic fertilizers, humic matter, etc.) in the photocatalytic surface water layer is of considerable significance.

6.2 Characteristics of Material

A reasonable band gap is necessary in any photocatalytic substantial so that from electromagnetic radiation a particular wavelength (UV-visible light) must strike the semiconducting photocatalyst surface to excite the electron and thus create electron–hole pairs by excitation to a valence band to produce conduction. The e^*-hole pair combination must be blocked to conduct the photocatalytic effect or the combination time should be long enough to perform the photocatalytic reaction continuum. It is possible to use the photocatalysis approach to break down a broad range of organic materials, including clothing dyes, microorganisms, inorganic substances, etc. Filtration and precipitation are some methods for coupled photocatalysis which have the ability to leach out harmful metals like Hg. The most common methods are discussed below:

6.2.1 Homogeneous Photocatalysis

In the natural self-purification of water, homogeneous photocatalytic processes play a vital role as UV light makes the transformation of hydrocomplexes of transition metal complexes (Fe, Cu, Cr, etc.) in advanced oxidation states [21–23]. Hence, on reaction of OH* radicals with substrates, various organic radicals will be formed. The oxidation of the substrate hydroxyl radicals due to charge transfer via one way occurs by taking a MeOH example into consideration:

$$n + h\mathrm{V} \rightarrow \mathrm{Me}(n-1)^{+} + \mathrm{OH} \tag{6.1}$$

$$\mathrm{MeL}n+ \ + h\mathrm{V} \rightarrow \mathrm{Me}(n-1)^{+} + \mathrm{L} \tag{6.2}$$

Molecular oxygen is important for the creation of photoreactive hydroxyl radicals (Figure 6.1) in the organic pollutants via H_2O_2. For the creation of carbon-based molecules and the removal of transition metal ions, the photochemical procedure can be approved out by photo-induced exchange of e* from ligand (Figure 6.2). The reduced shape of the metal, carbon-based radicals, and radical ions produced at a prior stage of photo-induced catalytic process are involved in the resulting heat-based reactions. In the presence of molecular oxygen, the production of H_2O_2 in the photo-oxidation of alcohols is independent of the procedure of producing hydroxyl radicals, inducing the procedure of regeneration underneath deliberation. Fe^{3+}ions are being photo-induced via inner-aqua in the aq. salts of iron (III) where OH radicals are produced during photo-oxidation of C_2H_5OH to observe hydrogen peroxide accumulation [24–26]. UV is responsible for the initiation of the Fenton reaction by involving organic matter to conversion of other oxidized forms thus, holding wide scope to treat wastewaters [27].

A greater percentage of catalytic processes involving various agents can completely eradicate the pollutants or complex compounds from the effluents, including ultra violet light, ozone, titanium dioxide and hydrogen peroxide, and have many benefits with the advent of Fenton reactions over conventional methods [22, 28].

Different oxidative reactions are involved in the removal of total organic compounds including 4-chlorophenicol which is the component of dyeing industries found in waste effluents needing treatment. The association of pyridine carboxylic and dichloropicolinic acids with H_2O_2 at acidic to alkaline pH levels does not induce oxidation of this compound in the absence of iron ions. Dyeing and finishing plant wastewater is a harmful source of environmental pollution of toxins, marked by a substantial demand for oxygen for use in degradation by both the microbe and chemical reactions involved. To enhance the conversion of colored to uncolored solutions of azo-colorants by using ferric nitrate and hydrogen peroxide reagents due to exposure to photons was studied. Consequently, the intensification of the process occurs as formerly seen with the H_2O_2-Fe_2+ sunlight system, where the drainage wastewater components are the substratum [28]. So far, this has been observed in the substratum-H_2O_2-Fe^{2+} sunlight method where the substratum is the sewage portion of textile manufacturing. Therefore, in this situation, the consequence of photons over the substrate of hydrogen peroxide involves conversion via reduction methods of iron from higher to lower oxidizing states. These studies indicate that by choosing the required initial concentration of the oxidizing agent, the precise intake of H_2O_2 in bleach azo dye solutions can be decreased. The application of aeration at some stages of the process decreases the use of hydrogen peroxide and thereby lowers the economic expense of this procedure for the treatment of water. The initial quantity of the catalyst can be reduced drastically due to the regeneration of Fe^{2+} ions in the reaction. The Cu^{2+} ions can also be photoreduced in the substrate-H_2O_2-Cu^{2+-} light method. Shaped Cu(I) ions can also catalyze the formation of reactive hydroxyl radicals by decomposing hydrogen peroxide through the reaction of the Fenton kind. The primary vulnerability of the Photo-Fenton process is due to the inclusion of reagents used to adjust the pH, and production of hydroxide slurries that need to be used or buried are high costs for using H_2O_2, and for the use of UV and energy external acidificing and neutralization systems as well as the high salinity of water under treatment. Via the on-site synthesis of H_2O_2 by electrochemical oxygen reduction and the substitution of UV light with natural photon interactions to oxidize the ferric iron state, for example, the cost of handling by the approaches under deliberation can be reduced.

6.2.2 Heterogeneous Photocatalysis

The use of solar energy seems to be a promising method for the future with its efficacy to play a vital role in cumbersome photocatalytic water management in the initial stages. Conferring to Herrmann and coworkers [29] and Soboleva and coworkers [30] for commercial water treatment purposes, human-induced photosynthesis is used to induce photochemical reactions in order to treat pollutants of both inland water surfaces as well as effluents, with the advent of using semiconductors to achieve the heterogeneous process. TiO_2 is commercially available, nontoxic and has synthetic stability for industrial commodity. As per reports of Gaya et al. [11] and Stroyuk et al. [31], e* structure is used

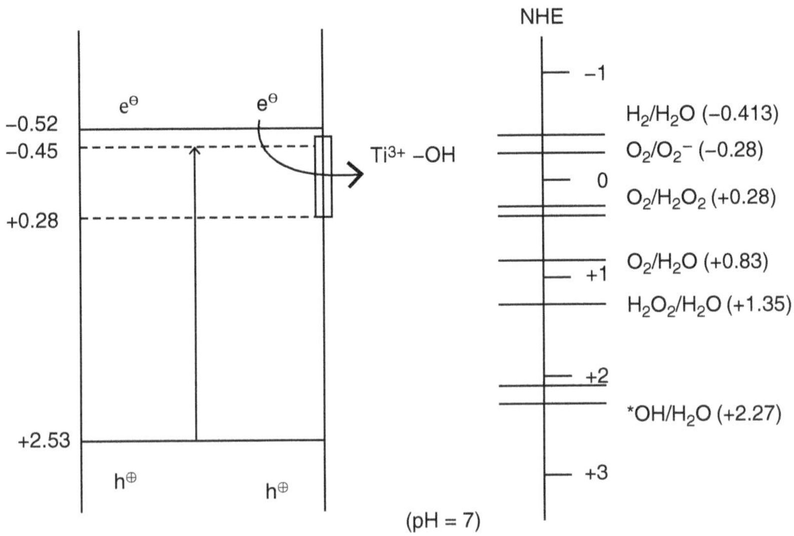

Figure 6.3 Schematic diagram of photogenerating initial oxidizing agents on the TiO_2 surface in the presence of e*-donor and e* acceptor compounds. (*Source*: Prihod'ko and Soboleva [32]).

to include the presence of free or valence electrons and to create holes in the semiconductor materials used. Various factors like quantum energy and crystal lattices are involved in the production of deficient centres (Figure 6.3).

$$TiO_2 + hV \rightarrow TiO_2(e- + h+)$$

6.3 Consequence of Ultra Violet/Titanium Dioxide/Hydrogen Peroxide

The outcome of fraternization of ultra violet radiation, photocatalyst TiO_2 and oxidant H_2O_2 on TC elimination was successively studied. The greater elimination of TC cannot be achieved by mixing only titanium dioxide and hydrogen peroxide. There is no consensus on the efficacy of the mixture of TiO_2 and oxidant H_2O_2 under ultra violet radiation in previously documented work. Under such a combo, the reported efficiencies of UV were attached mutually by oxidant and photocatalysts [33], although as rivalry for ultra violet incident light interactions for the oxidant and the photocatalyst [34] or the hydrogen peroxide adsorptionon TiO_2 that decreases catalyst activity [35].

6.3.1 Chlorophenol

2-Chlorophenol(2-CP) is another waste component added to treated sample water to analyze the removal efficiency of the UV/TiO_2 method with the advent of hydrogen peroxide, though it took ninety minutes to remove TC from the effluent, and so is not worthwhile. After treatment the concentration of chlorophenol was decreased with 98% of removal efficacy.

6.3.2 2,4-Dichlorophenol

Results for 2,4-DCP revealed that the findings obtained for 2,4-dichlorophenol(2,4-DCP) had little effect on TC removal at the end of the process, while H_2O_2 was once again beneficial for carbon mineralization and after 120 minutes, the 2,4-DCP conversion was above 95%.

6.3.3 2,4,6-Trichlorophenol

For 2,4,6-trichlorophenol, the findings obtained remained comparable to those with CH_2OH attained for phenol. There was no discrepancy among TC elimination of 5mgL-1 and 10mgL-1interms, although the inclusion of H_2O_2 increased wastewater mineralization. Rendering to findings of high throughput characterization techniques including high performance liquid chromatography, the addition of hydrogen peroxide caused an improvement in the rate of degradation, as a conversion of 100% 2,4,6-TCP was achieved in 30 minutes (Figure 6.4)

The photocatalytic method was used primarily to treat synthetic wastewater by the natural carbon alloy. Various ultra violet light, TiO_2, H_2O_2 and Fe^{2+} were measured and carbon output was investigated for the alteration of the exclusion and target compound. The key outcomes were:

1) We achieved a success rate of 58% and 56%, using ultra violet light along with hydrogen peroxide and ultra violet light with titanium dioxide respectively.
2) Adding of ferric ions greatly improved the total carbon elimination obtained (84%) for ferric ions using ultra violet/hydrogen peroxide. However, there was no further addition of titanium dioxide as a photocatalyst.

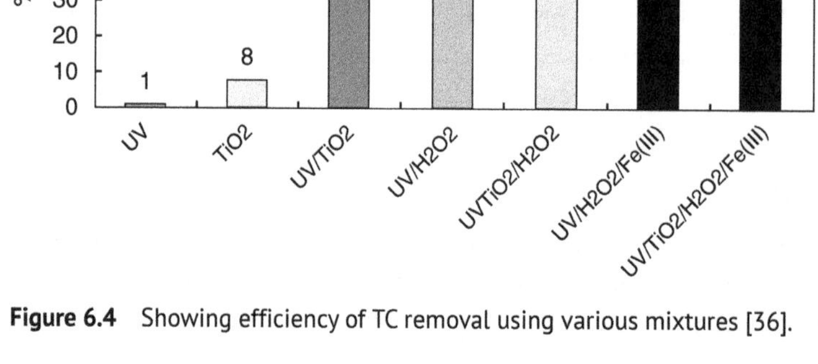

Figure 6.4 Showing efficiency of TC removal using various mixtures [36].

3) If either 5mg/L or10mg/L was partly replaced with the carbon of the wastewater the final TC elimination of phenol or 2,4,6-trichlorophenol in the UV/TiO_2 phase was reduced.
4) The wastewater comprising 2-chlorophenol was usually the simplest to mineralize in the UV/TiO_2 phase, while the unique comprising 2,4,6-TCP was the most resistant. The substitute H_2O_2 reduced the discrepancies found with respect to the elimination of TC.

6.4 Obstacles for Applicability

To achieve success for devised photocatalysts, more challenges with use of various combo or mixtures to attain greater removal efficiency for TC removal are required in order to treat either inland waters or effluents from various sources. These experiments, generally carried out on a bench scale, have a tendency to show up more opportunities and also the disadvantages of the expertise once it is implemented on the ground. There is a gap between the applicability of photocatalysis in industrial application, and the scope in academic literature regarding the practicality to overlay emphasis on substantial project and automatous appraisals. Skepticism has therefore developed, causing us to wonder whether, in the next two decades, photo-induced mechanism drives turn out to be a standard H_2O management system. Low photoconversion competence remains a main test among the leading barriers found.

6.4.1 Advancement of Photocatalytic Materials

With the advent of material science and nanotechnology, significant efforts were made to formulate the modeling of new photocatalysts, but there are various limitations in manufacturing industries to pursue the growth of photocatalytic materials. Of increasing importance is research aimed at developing stronger, development-driven catalyst materials. Early attempts have started with TiO_2 anatase enhancements since the fusion of the popular P25 TiO_2 flame pyrolysis powder is not only industrially economical, but it also demonstrated that the rutile fraction of the minority phase contributes to improved photocatalytic activity. For the scientific community, the photocatalyst response of TiO_2 is based primarily on improving catalyst absorption and the high rate of recombination for the primary animal's ecb − and hvb + produced by photographs. Surface hydrogenation is used to create distortion-induced midgap states, and to create blue or black TiO_2 crystals absorbed near the IR spectrum area [37]. Although it helps to enhance visible-light absorption, unnecessary recombination and an unnecessarily limited band structure can result in too many midgap states, limiting ecb-/hvb + pair redox potential and influencing the form of ROS produced [38]. Heterojunctions with smaller band gaps [39] or sensitization with organic chromophors [40] can also expand the absorption to involve further midgap situations.

6.4.2 Photocatalytic Reactor Design and System Evaluation

Instead of reactor or systems engineering, the focus of academic research is on material production. It is certainly unlikely that photocatalytic water treatment systems could increase performance by developing a reactor configuration. Despite numerous studies in this field, slim realistic architectures have been largely restricted to light reactor systems with a slurry of well-mixed TiO particulate matter to maximize photon absorption and mass transfer of redox species [41]. In the following selected niche implementations such systems can compete with other AOPs, but they remain unoptimized and have drawbacks, such as catalyst fouling or photoaggregation [41, 42]. Instead, it is possible to immobilize photocatalysts on support substrates, removing the need for isolation from ultrafiltration and reducing shear stress on catalyst particles. But when using a darkened catalyst surface area, a decreased light catalyst surface area per volume water processed, and higher photon dispersal speeds, the performance of immobilized systems can also be limited. The immobilization substrates investigated include optical light fibers for enhanced light control [43] and fluidized mediums [44], porous foam substrates [45], membranes [46] and electrospun fibers. The bulk of these immobilized reactor projects remain at the benchmark. The advancement in LED technology is primed to spur a new wave of revolutionary reactor designs because of its advantages over gas discharge lights, which involve less prone cartridges, lack of harmful materials, compact size and fast warm-up time [47, 48].

6.5 Strategies for Improving Research Outcomes

Photocatalysis is the prominent technique for the breakdown of residual compounds of wastewaters of different industries that are difficult to biodegrade [49]. Combined with recent developments in the field of materials science and reactor engineering, prospects may arise for better treatment of photocatalytic water. Based on our combined knowledge, we suggest subsequent techniques for investigators with a view to encouraging the photocatalysis of H_2O management systems lacking capitulating to the traps that have achieved success over the previous three periods. Therefore, a combination of photocatalysis with biological treatment is suggested to minimize treatment costs and improve the overall treatment efficiency as shown in proposed wastewater treatment systems.

Researchers can extend performance requirements to contain considerations as to the cost and availability of goods, the feasibility of large production and the tendency to foul the suggested water content and, under future operational conditions, long-term reliability and efficiency, in order to build innovations that are more likely to draw market interest [50]. They need to assess the efficiency of photo-induced methods in definite circumstances. It is important to use existing methods to measure an LVRPA computational assessment, in which analysis constituents do not influence the absorption of light or otherwise modify the catalyst [51]. As photocatalyst contenders progress to practical use, testing should take place under standardized conditions using characteristic testing compounds and solution conditions (e.g., pH, ionic composition, ROS scavenger). When comparing the efficacy of different AOPs a suitable figure of merit, such as EE/O [52], should be published.

Although developing constituents after similar grounds may be a successful method for developing novel photocatalysts for water treatment, blind search can also contribute to the preferment of resources that are not appropriate for realistic applications, such as materials produced for water splitting or photovoltaics. Hoffmann et al. [53]found the most desirable application of the material has not been described in several studies on photocatalyst efficiency. The photocatalyst's efficiency depends on the treatment properties of a contaminant [54], or the relative concentrations. Investigators may however create photocatalysts for applications that are impractical or of lack of interest to the industry [55–58]. In order to tackle this obstacle, scientists will need to address the main obstacles to conventional uses, as potable water and industrial wastewater treatment, or to find niches that can compete with current photocatalytic water treatment systems [59–62].

References

1 Gupta, A., Mondal, K., and Kumar, S. (2018). Role of photo-catalysis in water remediation. In: *Water Remediation*, 117–134. Singapore: Springer.

2 Nakata, K. and Fujishima, A. (2012). TiO2 photocatalysis: Design and applications. *J. Photochem. Photobiol. C Photochem. Rev.* 13 (3): 169–189.

3 Vinu, R. and Madras, G. (2010). Environmental remediation by photocatalysis. *J. Ind. Inst. Sci.* 90 (2): 189–230.

4 Malato, S., Ferna´ndez-iba´n˜ez, P., Maldonado, M.I., Blanco, J., and Gernjak, W. (2009). Decontamination and disinfection of water by solar photocatalysis: Recent overview and trends. *Catal. Today* 147: 1e59.

5 Wintgens, T., Salehi, F., Hochstrat, R., and Melin, T. (2008). Emerging contaminants and treatment options in water recycling for indirect potable use. *Water Sci. Technol.* 57: 99e107.

6 Sua´rez, S., Carballa, M., Omil, F., and Lema, J.M. (2008). How are pharmaceutical and personal care products (PPCPs) removed from urban wastewaters? *Rev. Environ. Sci. Biotechnol.* 7: 125e138.

7 Bradley, B.R., Daigger, G.T., Rubin, R., and Tchobanoglous, G. (2002). Evaluation of onsite wastewater treatment technologies using sustainable development criteria. *Clean Technol. Environ. Policy* 4: 87e99.

8 Lapen˜a, L., Cerezo, M., and Garcı´a-augustin, P. (1995). Possible reuse of treated municipal wastewater for Citrusspp. *Plant irrigation.Bull. Environ. Contam. Toxicol.* 55: 697e703.

9 Viessman, W., Jr. and Hammer, M.J. (1998). *Water Supply and Pollution Control*, 6th e. California USA: Addison Wesley Longman Inc.

10 Padmanabhan, P.V.A., Sreekumar, K.P., Thiyagarajan, T.K., Satpute, R.U., Bhanumurthy, K., Sengupta, P., Dey, G.K., and Warrier, K.G.K. (2006). Nano-crystalline titanium dioxide formed by reactive plasma synthesis. *Vacuum* 80: 11e12.

11 Gaya, U.I. and Abdullah, A.H. (2008). Heterogeneous photocatalytic degradation of organic contaminants over titanium dioxide: A review of fundamentals, progress and problems. *J. Photochem. Photobiol. C* 9 (1): 1–12.

12 Yang, H. and Cheng, H. (2007). Controlling nitrite level in drinking water by chlorination and chloramination. *Sep. Purif. Technol.* 56: 392e396.

13 Lu, J., Zhang, T., Ma, J., and Chen, Z. (2009). Evaluation of disinfection by-products formation during chlorination and chloramination of dissolved natural organic matter fractions isolated from a filtered river water. *J. Hazard. Mater.* 162: 140e145.

14 Coleman, H.M., Marquis, C.P., Scott, J.A., Chin, S.S., and Amal, R. (2005). Bactericidal effects of titanium dioxide-based photocatalysts. *Chem. Eng. J.* 113: 55e63.

15 Esplugas, S., Gime´nez, J., Conteras, S., Pascual, E., and Rodrı´guez, M. (2002). Comparison of different advanced oxidation processes for phenol degradation. *Water Res.* 36: 1034e1042.

16 Goncharuk, V.V. (1997). Photooxidation degradation of the organic compounds in water. *Khimiya V Interesakh Ustoichivogo Razvitiya* 4 (5): 345–355.

17 Goncharuk, V.V., Soboleva, N.M., and Nosonovich, A.A. (2003). Physico-chemical aspects of soil contamination with heavy metals and hydrosphere. *Khimiya V Interesakh Ustoichivogo Razvitiya* 11 (6): 795–809.

18 Soboleva, N.M., Saprykina, M.N., Kosinova, V.N., Nosonovich, A.A., and Goncharuk, V.V. (2012). Inactivation ofCandida albicansin the photo-fenton system. *J. Water Chem. Technol.* 34: 96–102.

19 Solozhenko, E.G., Soboleva, N.M., and Goncharuk, V.V. (1995). Decolourization of azodye solutions by Fenton's oxidation. *Wat. Er. Res. Earch* 29 (9): 2206–2210.

20 D¨ukkanci, M., G¨und¨uz, G., Yilmaz, S., and Prihod'ko, R.V. (2010). Heterogeneous Fenton-like degradation of Rhodamine 6G in waterusing CuFeZSM-5 zeolite catalyst prepared by hydrothermal synthesis. *J. Hazardous Mater.* 181 (1–3): 343–350.

21 Prihod'ko, R., Stolyarova, I., G¨und¨uz, G., et al. (2011). Fe-exchanged zeolites as materials for catalytic wet peroxide oxidation. Degradation of Rodamine G dye. *Appl. Catal. B* 104 (1-2): 201–210.

22 Hofman-Caris, C.H.M., Harmsen, D.J.H., and Beerendonk, E.F. (2003). Degradation of priority compounds by UV and UVoxidation. *Adv. Oxidation Process. TECHNEAU.*

23 Arslan, I. and Balcioglu, I.A. (1999). Degradation of commercial reactive dyestuffs by heterogenous and homogenous advanced oxidation processes: A comparative study. *Dyes Pigment.* 43 (2): 95–108.

24 Nosonovich, A.A., Soboleva, N.M., and Goncharuk, V.V. (2011). Generation of hydrogen peroxide in oxidative degradation process of the organic compounds. *J. Water Chem. Technol.* 33 (5): 465–476.

25 Goncharuk, V.V., Soboleva, N.M., and Nosonovich, A.A. (2010). Photooxidative destruction of organic compounds by hydrogen peroxide in water. *J. Water Chem. Technol.* 33 (5): 456–476.

26 Nosonovich, A.A., Soboleva, N.M., and Goncharuk, V.V. (1995). Photodestructive oxidation of cetylpyridinium bromide in water. *Khimiya I Tekhnologiya Vody* 17 (5): 451–456.

27 Solozhenko, E.G., Soboleva, N.M., and Goncharuk, V.V. (2004). Using of H2O2-Fe2+(Fe3+) catalytic system in wastewater treatment. *J. Water Chem. Technol.* 26 (3): 1–16.

28 Goncharuk, V.V. (2005). *Environmental Aspects of Modern Technologies Protect the Aquatic Environment.* Ukraine: Naukova Dumka, Kiev.

29 Herrmann, J.-M. (2005). Heterogeneous photocatalysis: State of the art and present applications. *Topics Catal.* 34: 149–465.

30 Soboleva, N.M., Nosovich, A.A., and Goncharuk, V.V. (2007). Theheterogenic photocatalysis in water treatment processes. *J. Water Chem. Technol.* 29 (2): 72–89.

31 Stroyuk, A.L., Kryukov, A.I., Kuchmii, S.Y., and Pokhodenko, V.D. (2005). Quantum size effects in semiconductor photocatalysis. *Theor. Exper. Chem.* 41 (4): 207–228. Pic source Photocatalysis: Oxidative Processes in Water Treatment.

32 Prihod'ko, R.V. and Soboleva, N.M. (2013). Hindawi publishing corporation. *J. Chem.*, Article ID 168701, 8 pages. http://dx.doi.org/10.1155/2013/168701.

33 Dixit, A., Mungray, A.K., and Chakraborty, M. (2012). Photochemical oxidationofphenolicwastewatersanditskinetic study. *Desalin Water Treat* 40 (1–3): 56–62.

34 Poulopoulos, S.G. and Philippopoulos, C.J. (2004). Photo assisted oxidation of chlorophenols in aqueous solutions using hydrogen peroxide and titanium dioxide. *J. Environ. Sci. Heal—PartA Toxic / Hazardous Subst Environ. Eng.* 39 (6): 1385–1397.

35 Tanaka, K., Hisanaga, T., and Harada, K. (1989). Efficient photocatalytic degradation of chloralhydrate inaqueous semiconductor suspension. *J. Photochem. Photo Biol. A Chem.* 48 (1): 155–159.

36 Poulopoulos, S.G., Yerkinova, A., Ulykbanova, G., and Inglezakis, V.J. (2019). Photocatalytic treatment of organic pollutants in a synthetic wastewater using UVlight and combinations of TiO2, H2O2 and Fe(III). *PLoS ONE* 14 (5): e0216745. https://doi.org/10.1371/journal.pone.0216745.

37 Chen, X.B., Liu, L., Yu, P.Y., and Mao, S.S. (2011). Increasing solar absorption for photocatalysis with black hydrogenated titanium dioxide nanocrystals. *Science* 331 (6018): 746–750.

38 Zhao, C., Pelaez, M., Dionysiou, D.D., Pillai, S.C., Byrne, J.A., and O'Shea, K.E. (2014). UV and visible light activated TiO2 photocatalysis of 6-hydroxymethyl uracil, a model compound for the potent cyanotoxin cylindrospermopsin. *Catal. Today* 224: 70–76.

39 Shiraishi, Y., Sakamoto, H., Sugano, Y., Ichikawa, S., and Hirai, T. (2013). Pt–Cu bimetallic alloy nanoparticles supported on anatase TiO2; Highly active catalysts for aerobic oxidation driven by visible light. *ACS Nano* 7 (10): 9287–9297.

40 Chen, C., Ma, W., and Zhao, J. (2010). Semiconductor-mediated photodegradation of pollutants under visible-light irradiation. *Chem. Soc. Rev.* 39 (11): 4206–4219.

41 Benotti, M.J., Stanford, B.D., Wert, E.C., and Snyder, S.A. (2009). Evaluation of a photocatalytic reactor membrane pilot system for the removal of pharmaceuticals and endocrine disrupting compounds from water. *Water Res.* 43 (6): 1513–1522.

42 Li Puma, G. and Brucato, A. (2007). Dimensionless analysis of slurry photocatalytic reactors using two-flux and six-flux radiation absorption–scattering models. *Catal. Today* 122 (1): 78–90.

43 Ling, L., Tugaoen, H., Brame, J., Sinha, S., Li, C., Schoepf, J., Hristovski, K., Kim, J.H., Shang, C., and Westerhoff, P. (2017). Coupling light emitting diodes with photocatalyst-coated optical fibers improves quantum yield of pollutant oxidation. *Environ. Sci. Technol.* 51 (22): 13319–13326.

44 Karches, M., Morstein, M., Rudolf Von Rohr, P., Pozzo, R.L., Giombi, J.L., and Baltanás, M.A. (2002). Plasma-CVD-coated glass beads as photocatalyst for water decontamination. *Catal. Today* 72 (3): 267–279.

45 Plesch, G., Gorbár, M., Vogt, U.F., Jesená, K.K., and Vargová, M. (2009). Reticulated macroporous ceramic foam supported TiO2 for photocatalytic applications. *Mater. Lett.* 63 (3): 461–463.

46 Athanasekou, C.P., Romanos, G.E., Katsaros, F.K., Kordatos, K., Likodimos, V., and Falaras, P. (2012). Very efficient composite titania membranes in hybrid ultrafiltration/photocatalysis water treatment processes. *J. Membr. Sci.* 392–393: 192–203.

47 Lee, C.G., Javed, H., Zhang, D., Kim, J.H., Westerhoff, P., Li, Q., and Alvarez, P.J.J. (2018). Porous electrospun fibers embedding TiO2 for adsorption and photocatalytic degradation of water pollutants. *Environ. Sci. Technol.* 52 (7): 4285–4293.

48 Matafonova, G. and Batoev, V. (2018). Recent advances in application of UV light-emitting diodes for degrading organic pollutants in water through advanced oxidation processes: A review. *Water Res.* 132: 177–189.

49 Oller, I., Malato, S., and Pérez, J.S. (2011). Combination of advanced oxidation processes and biological treatments for wastewater decontamination: A review. *Sci. Total Environ.* 409: 4141–4166.

50 Fujishima, A. and Honda, K. (1972). Electrochemical photolysis of water at a semiconductor electrode. *Nature* 238: 37.

51 Henan Bright Commercial Co., Ltd. Factory sales nano-titanium dioxide TiO2 degussa P25. https://www.alibaba.com/product-detail/Factory-sales-Nano-Titanium-DioxideTiO2_60601130450.html?spm=a2700.7724838.2017115.1.6d0263f8Q9brPJ (accessed 5 February 2018).

52 Bolton, J.R., Bircher, K.G., Tumas, W., and Tolman, C.A. (2001). Figuresof-merit for the technical development and application of advanced oxidation technologies for both electric-and solar-driven systems (IUPAC Technical Report). *Pure Appl. Chem.* 73: 627.

53 Hoffmann, M.R., Martin, S.T., Choi, W.Y., and Bahnemann, D.W. (1995). Environmental applications of semiconductor photocatalysis. *Chem. Rev.* 95 (1): 69–96.

54 Ryu, J. and Choi, W. (2008). Substrate-specific photocatalytic activities of TiO2 and multiactivity test for water treatment application. *Environ. Sci. Technol.* 42 (1): 294–300.

55 Richardson, S.D. (2008). Environmental mass spectrometry: Emerging contaminants and current issues. *Anal. Chem.* 80: 4373e4402.

56 Solozhenko, E.G., Soboleva, N.M., and Goncharuk, V.V. (2004). Using of H2O2-Fe2+(Fe3+) catalytic system in wastewater treatment. *J. Water Chem. Technol.* 26 (3): 1–16.

57 Deng, Y. (2009). Advanced Oxidation Processes (AOPs) for reductiono for ganicpollutantsinl and filll eachate: A review. *Int. J. Environ. Waste Manag.* 4 (3/4): 366.

58 Stolyarova, I.V., Kovban', I.B., Prikhod'ko, R.V., Kushko, A.O., Sychev, M.V., and Goncharuk, V.V. (2007). Relationship between the catalytic behavior of FeZSM-5 zeolites in oxidative degradation of dyes and the nature of their active centers. *Russian J. Appl. Chem.* 80 (5): 746–753.

59 Gaya, U.I. and Abdullah, A.H. (2008). Heterogeneous photocatalytic degradation of organic contaminants over titanium dioxide: Areview of fundamentals, progress and problems. *J. Photochem. Photobiol. C: Photochem. Rev.* 9: 1e12.

60 Safarzadeh-Amiri, A., Bolton, J.R., and Cater, S.R. (1997). Ferrioxalate-mediated photo degradation of organic pollutants in contaminated water. *Water Res.* 31 (4): 787–798.

61 Van Gerven, T., Mul, G., Moulijn, J., and Stankiewicz, A. (2007). A review of intensification of photocatalytic processes. *Chem. Eng. Process.* 46 (9): 781–789.

62 Goncharuk, V.V., Vakulenko, V.F., Sitnichenko, T.N., Yu.Shvadchina, O., and Sovab, A.N. (2013). Photocatalytic destruction of fulvic acids by ozone and oxygen. *J. Water Chem. Technol.* 35: 62–70.

7

Biochemical Systems

Cathode Advanced Wastewater Treatment

CHAPTER MENU

7.1 Introduction

Bioelectrochemical systems (BESs) are complex systems that use the interaction of microbes with solid electron acceptors/donors (for example, an electrode) to remove contaminants from wastewater while still producing electricity [1]. In a typical BES, such as a microbial fuel cell (MFC), microorganisms oxidize organics in the anode and pass electrons to an anode electrode, which then transfers electrons to a cathode to reduce a terminal electron acceptor like oxygen [2]. When a load is applied to the circuit, this electron movement generates an electrical current or electricity. Microbial electrolysis cells (MECs) have been used to generate hydrogen (via the hydrogen evolution reaction) [3], drive desalination in microbial desalination cells (MDCs) [4] and synthesize value-added compounds in microbial electro synthesis cells (MES) [5, 6]. BESs have progressed in recent years, especially in terms of system scaling, and several large-scale systems have been investigated [2, 7]. There is also a lot of enthusiasm about using BESs to recycle valuable wastewater resources like nutrients, electricity and water.

Wastewater treatment is considered to play a crucial role in the development of BES and is critical to the future potential use of this technology. Organic removal is a primary

Aquatic Environmental Bioengineering: Monitoring and Remediation of Contamination, First Edition.
Rouf Ahmad Bhat, Mohammad Yaseen Mir, Gowhar Hamid Dar and Moonisa Aslam Dervash.

goal of wastewater treatment, and the efficiency of organic removal in BESs varies greatly depending on a number of factors, including wastewater composition and strength, hydraulic retention time, reactor configurations, temperature and so on. The BES anode is where the majority of electrons are extracted for generation [8]. However, nutrients such as nitrogen and phosphorus can hardly be removed in a BES anode because their removal (if through biological pathways) would need to involve aerobic conditions. As a consequence, the BES anode effluent is an untreated source that still retains residues like organic substances and nutrients. This has been demonstrated in a number of previous studies of MFCs treating real domestic wastewater, in which cathodic treatment significantly improved organic and ammonium compound removal [9, 10]. In certain cases, such a stream cannot be discharged or reused for nonpotable purposes right away. To boost wastewater treatment in BESs while avoiding the installation of more reactors, we are considering using the cathode of a BES for further treatment. A cathode may have more physical space for longer treatment of host biological processes, and a cathodic reduction reaction can be used to remove certain contaminants that aren't soluble in the anode. A wide range of contaminants, including pathogens, can be rejected by inserting the membrane into the cathode, resulting in high-quality effluent. The aim of this chapter is to provide a fast overview of cathode-promoted wastewater treatment in BESs, to discuss and evaluate several major treatments that are stimulated or aided by cathodes, and to encourage further research into this approach to improve BES effluent performance.

7.2 Cathodic Catalysis in BES and Implications for Catalyst Design

7.2.1 Cathodic Catalysis Characteristic in BES

BES is a biocatalytic electrochemical device that uses biocatalysts (either microbial cells or enzymes) to catalyze electrochemical reactions at one or both electrodes [11, 12]. Electrons flow from the anode to the cathode through an external circuit in such systems, which are followed by substrate oxidation at the anode and reduction reactions at the cathode. Energy can be derived from this process in the form of electricity or chemical fuels. Notably, electron flow can occur spontaneously based on the functional electrode potential of the anode and cathode, i.e. without the need for additional power, making it a microbial fuel cell (MFC), or it can be powered by an external power source, i.e. electrical power is needed, making it a microbial electrolysis cell (MEC) [11]. Although electrical energy can be directly harvested in an MFC, a MEC can generate a large range of high energy density products (e.g. H_2, CH_4, or ethanol) with very little power consumption [13]. As a result, while net energy recovery is possible in either BES operating mode, it is limited to a small number of redox reaction processes and heavily dependent on the real energy recovery mechanism. ORR, HER and other hydrocarbon synthesis processes are currently the most important cathodic BES reactions that could result in net energy production. Several of these cathodic reactions often occur in conventional electrochemical systems. Many cathodic electrocatalysts used in chemical fuel cells or electrolysis cells may also be used directly in BES, which is not surprising. However, they usually have lower catalytic efficiency than conventional electrochemical systems in terms of operation, selectivity and stability, owing to a suboptimal operating condition in the BES. To meet the requirements of the BES operation, these existing catalysts must either be upgraded or new catalysts must be produced.

7.2.2 Operation Environment

When compared to conventional electrochemical systems, BES makes extensive use of a biocatalyst, most commonly microorganisms. As a result, unlike ordinary electrochemical systems, which are normally operated under well-defined or even harsh environmental/solution conditions to allow for optimum catalysis, BES operating conditions are typically restricted to preserve bioactivity. Neutral pH, low electrolyte ionic strength and ambient temperature and atmospheric pressure are all needed to keep microorganisms' metabolic activities going. Unfortunately, because of the limited reactions, certain mild conditions are normally undesirable for most metal catalysts.

7.2.3 Wastewater Electrolyte

Wastewater is the commonest electrolyte in most BESs. Unlike conventional synthesized electrolytes which have well-defined structure and solution properties, real wastewater is characterized by high complexity and uncertainty in composition and typically lacks adequate conductivity and buffering capacity to support ion transportation [14]. Furthermore, the presence of organic substrates is usually needed for microbial growth in a microbial anode, which increases the risk of cathodic catalyst fouling due to substrate crossover [15]. All of these characteristics make it difficult for BESs to run efficiently and reliably. Owing to the presence of numerous organics nutrients, and metal ions, multiple competitive reactions can occur at the cathode, which is one of the major issues associated with wastewater complexity. Any reductive material, such as organic substrate, ammonia or sulphide, interacts directly with oxygen, decreasing the amount of oxygen accessible to the cathode and slowing the kinetics of electron uptake at an MFC cathode [15]. Additionally, for MECs, such side reactions would not only reduce the rate of development of desired products but also reduce the purity of the product due to the formation of side products [13].

Another significant issue is that certain aqueous contaminants (e.g. dissolved organic matter [16–18] and metal ions [19]) or reaction intermediates (e.g. chloride [20] and sulfide [16, 21] can cause catalyst deactivation during long-term activity [22–25] as a result of dissolution, oxidation, poisoning or fouling. OH accumulation near the ORR cathode, in particular, is known to cause a significant drop in kinetic and thermodynamic efficiency [26]. As a result of these difficulties, selecting and developing cathodic catalysts for BES is problematic as they are needed not only for catalytic activity but also for high selectivity and catalyst durability.

7.2.4 Cathode Over-Potential and Catalysis in BES

One of the primary goals of BES cathodic catalysis is to reduce cathode energy losses (also known as over-potential) and thus increase energy performance. A BES can work in MFC or MEC mode, depending on the thermodynamics of electrode reactions and energy losses. In theory, electrical energy can be harvested if the equilibrium potential of the involved redox pair at the anode is lower than that at the cathode; otherwise, additional energy must be invested, and the recoverable or input energy should be determined by the potential difference between the electrode reactions [27]. However, due to a significant over-potential in electrode reactions, less energy is typically harvested or more power is needed. The over-potential cathode in a BES, in particular, can be important, lowering the potential achieved over the BES and thus lowering the energy efficiency [28]. Although the thermodynamic potential for reducing O_2 to water at neutral pH 7 is around + 0.81 V (in comparison to SHE, hereinafter the same), the initial potential for reducing current on non-catalyzed carbon is often below 0.1 V. Kinetic losses (i.e. resistance to charge and mass transfer) and thermodynamic losses can be approximately divided in a cathodic over-potential.

7.2.5 Photo-Aided Cathodic Catalysis

Increased generation of electricity or fuel output can be achieved by introducing an external power source, ideally a renewable energy source such as solar energy, to provide an additional driving force for electron transfer in addition to reducing the cathode's over-potential. As a result, photocathode and photoelectrocatalysts have been developed, which have greatly aided cathodic electrocatalytic reactions in BESs [29, 30]. In the photocathode metabolism of the BES, there are strong synergies between cathodic photoelectrocatalysis and anodic metabolism, as well as photocatalytic and electrocatalytic processes. During the process, electron-hole pairs are produced upon photo irradiation at a semiconductor, then the photogenerated charges (either electrons or holes) are transferred to a catalytically active site for electrocatalytic reactions, while electrons are delivered from the anode to combine with holes or participate directly in cathodic reactions. To allow such synergies, the photoelectrocatalyst must have the three properties described below:

1) The material used in electrocatalytic reactions should be conductive and active.
2) Light absorption (ideally visible light) and photo-excited electron transport should be possible. These two properties do not have to be provided by the same material. This means that a photoelectrocatalyst can be made up of either a catalytically active absorber or a structure that separates light absorption from electrocatalytic action (i.e. a photocatalyst combined with a normal electrochemical co-catalyst) [31].
3) The band structure is extremely important. To extract microbial electrons from the anode, the photoelectrocatalyst should have a more positive valence band edge than the anode potential.

Meanwhile, to transfer photo-excited electrons to a terminal electron acceptor, it is necessary to have a conduction band edge that is more negative than the redox potential of the target cathode reactions. As a consequence, BES photoelectrocatalysts should be rationally designed to enable efficient interaction between photoelectrocatalysts, anodic electricigens and catholytes.

7.3 Wastewater Treatment

One of the most important biological factors that can affect the efficiency of BES treatment, and thus the recovery of electrons, is the substrate's existence. BES can work with a wide range of substrates, including both inorganic and organic molecules, as electron donors and acceptors. The efficiency of electron recovery, on the other hand, is determined by the electron donor's oxidation state and its ratio to the microbe that can oxidize it. The most widely used anodic fuels among the basic substrates are glucose and acetate, but other simple substrates, such as sucrose, starch, butyrate, dextran, peptone, ethanol and others, were also evaluated in BES with a primary slogan of power generation. In addition to these basic substrates, BESs demonstrated versatility in the use of a wide range of organic waste, from simple to complex. In BESs, waste from different sources, such as factories, manufacturing areas, residential areas and so on, was regarded as a potential electron donor. Domestic wastewater, dairy-based wastewater, food wastewater, vegetable waste, and other waste with high biodegradability will generate a lot of electricity, while industrial wastewater with low biodegradability will generate less. Wastewater remains a viable MFC substrate due to its dual benefits of transforming low-value waste into bioenergy.

7.3.1 Highly Biodegradable Wastewater

The characteristics of wastewater vary, depending on the raw materials used as the foundation for its production. Domestic wastewater with a low substrate load is considered simple and highly biodegradable in nature, so treatment with BES is more effective and faster, but power generation is only for a few hours [32]. Dairy wastewater, which is high in lactate, proteins and other milk-based waste components, is also simple in nature and has a higher treatment efficiency [33].

Kitchen waste, food waste, vegetable waste, cheese waste, potato wastewater and other wastewaters are all highly biodegradable [34]. These wastewaters are primarily composed of organic material, especially carbohydrates and proteins, which are easily degraded by almost all bacteria. This form of waste produces a lot of energy and has a high treatment quality. Furthermore, this form of wastewater can operate efficiently at higher organic loading rates, reducing the need for feed dilution.

7.3.2 Complex/Low Biodegradable Wastewater

BES can also handle wastewater that is highly complex and low-biodegradable, such as wastewater from distilleries, pharmaceutical wastewater, lignin-based wastewater and so on. These wastewaters' complex nature and low biodegradability make conversion to reducing equivalents difficult, and the pollutants/components of wastewater (intermediate acceptors) can accept the electrons and protons released for further oxidation, resulting in lower current densities. The removal of color from industrial wastewater, such as distillery and pharmaceutical wastewater, is a vital aspect of wastewater treatment, but BES can easily remove color at the anode. Similarly, toxic halogens and other hydrocarbons are resistant to aerobic remediation, but they can serve as electron acceptors in BES under anaerobic respiration. For efficient treatment and power generation, BES may use solid waste such as kitchen waste, food waste, vegetable waste and so on without dilution. Lignocellulosic biomass [35], dye wastewater [36], landfill leachates [37], cellulose and chitin [38] and reed mannagrass [39] were all investigated as electron donors in manufactured food corporations (MFCs).

7.3.3 Integrated Process for Additional Treatment

BES is also said to be used in the decomposition of fermentation effluent and preliminary treatment methods that involve first-process acid and solvent metabolites, as well as residual organic carbon. Few studies have been published in the literature on the use of organic acids (pure/mixed) and effluents from various processes as primary substrates for power generation in MFC. The comparative MFC findings published in various studies are shown in Table 7.1. All of these studies were carried out in single/dual-chambered membrane-based fuel cell configurations. The conversion efficiencies of the system were comparable to that of conventional fuel cells, implying that the system is more efficient. All studies have found that coulombic efficiency (CE) ranges from 12 to 75%, but studies with actual fermentation effluents have found that the CE ranges from 12 to 45%, which is comparable to normal wastewater.

The enriched biocatalyst was found to have higher treatment efficiencies and power output in the presence of acid metabolites such as acetate and butyrate, suggesting that it could effectively oxidize higher metabolite concentrations in effluents [40]. The treatment obtained using this approach, in particular, is complementary to the first step, which increases the waste's recovery capacity.

7.4 Current Bottlenecks and Challenges for BES

In terms of materials and engineering research, as well as microbial and biotechnological perspectives, progress in the field of BESs has gone hand-in-hand. While advances in design engineering have resulted in increased power output and improved efficiency, the components and mechanisms involved in the transfer of electrons from bacteria to the electrode surfaces are becoming better understood. Although there is not yet a consensus on the final electron transfer mechanism, researchers are optimistic that one will emerge soon. Due to several significant limitations, such as slow transport across cell membranes, bio fuel cells have a significant disadvantage from the start: low power densities, or the power generated per unit electrode surface area. Ohmic voltage losses (caused by resistance to the transport of charges, including both ionic and electronic resistances), over-potential activation (caused by energy barriers to the transfer of charges from bacteria to electrodes), over-potential concentration (caused by resistance to mass transport), and finally coulombic losses are all

Table 7.1 List of wastewaters studied in BES for their treatment.

Wastewater	MFC configuration	Removal efficiency (%)	Reference
Domestic wastewater	Single chamber	66.7	[32]
Domestic wastewater	Double chamber	85	[41]
Dairy wastewater	Single chamber	95.5	[33]
Dairy wastewater	Double chamber	90	[42]
Cattle dung	Single chamber	—	[43]
Pharmaceutical wastewater	Single chamber	85.8	[44]
Swine waste	Single chamber	86	[45]
Meat packing wastewater	Single chamber	87	[46]
Molasses wastewater	Single chamber cuboid MFC	53.2	[47]
Mixed volatile fatty acids	Double chamber	39	[48]
Primary effluent	Single chamber	84	[49]
Dark fermentation effluent	Single chamber	72	[40]
Anaerobic food waste leachate	Single chamber	91	[50]

significant losses associated with these devices. Coulombic losses are caused by the accumulation of biomass, the occurrence of side reactions that do not contribute to the current yield, and the crossover of the substrate from the cathode to the anode and vice versa [51]. pH concerns, high ohmic cell resistance and high over-potentials are all barriers to BESs being used in industry. Environmental variables such as oxidant and proton flux towards the biocathodes are critical in the development of high-performance biocathodes.

Another big concern with these devices is upscaling. Scaling up is generally acknowledged as a major and difficult challenge [52], and there are currently few viable options for achieving successful and cost-effective size. Large-scale reactor efficiency must be at least equal to that of bench-scale reactors today, with production costs that are both economically and environmentally feasible. It has been suggested that the majority of MFC units be used to power real-world applications as a stack. Biocatalyzed reactions, on the other hand, are highly vulnerable to cell reversal in a stacked configuration. Furthermore, since both the anode and the cathode are biologically catalyzed, a stacked operation will be difficult. Unfortunately, both of these techniques are expected to increase the reactor's expense, and upscaling BESs may be a long-term project. The maximum power density provided by the MFC has previously been shown to be proportional to the logarithm of the anode's surface area rather than the anode's surface area [53]. In other words, the power density in MFCs decreases as the surface area of the current-limiting electrode increases, and it cannot be assumed that power density will remain constant as the electrode surface area increases when scaling up these systems. However, it was later discovered that raising the electrode surface area increases the overall reaction rate, resulting in more collected current [54]. Later, it was discovered that increasing the anode bed thickness and the approximate anode area increased the current production in a single chamber MFC with an anode made of a packed bed of irregular graphite granules.

However, due to current distribution problems and mass transport constraints, scaling up from a flat sheet to a higher surface area packed bed did not result in a corresponding rise in current [55]. Some of the problems associated with reactor scale-up for MFCs were recently addressed by Fornero et al. [56] and identified three major obstacles that must be addressed when scaling up MFC reactors. Maintaining low internal resistance while increasing the amount of electrochemically active biomass, optimizing reactor design and finding new ways to separate anode and cathode are just a few of them. In addition to the aforementioned limitations, comparing documented results is a common and frequently discussed issue with BESs, as key experimental parameters are frequently not provided, and significant comparative electrical performance measurements are rarely reported [57]. Furthermore, a comparison is difficult due to a wide variety of designs ranging from two-chambered to single-chambered, mediator or without mediator, membrane or membrane-less. Apart from the design itself, a range of materials are used, such as electrodes ranging from graphite foil, tubing, granules, fiber brush, carbon paper, fabric, felt, foam, activated carbon cloth, reticulated vitreous carbon, conductive polymer modified electrodes and metals such as aluminum, nickel or stainless steel, making it technically difficult to compare the set-up performance [57]. There are a variety of approaches that can be used to address these flaws. Background research to describe and illustrate electrochemical reaction mechanisms, the effects of electrode materials, biofilm, substrate and metabolites, reproducibility and repeatability measurement experiments, and the use of a reference electrode are just a few of them [29]. It has been suggested in recent years that the growth of biocatalyst power density has reached a plateau, and that the next major growth will come from better materials used in these systems [11]. This includes improved anode and cathode electrodes, as well as separators [58] and newer cell designs [59]. Logan [60] recently addressed the importance of new materials in designing the next generation of bioelectrochemical systems [60] and also discussed the recent development of anode/cathode materials and filling materials as three-dimensional electrodes for MFCs. It was discovered that different electrodes behaved differently, and electrode modification proved to be a viable option for increasing MFC performance. Pt is the most commonly used catalyst on the cathode, but its high cost prohibits it from being used in industrial MFCs [61]. The most expensive components of a BES have been identified as electrodes with a precious metal catalyst and a membrane as a separator [62]. The costs associated with these systems are expected to decrease as design changes and new, less costly materials are created. Some figures for MFCs for wastewater electricity production have already been cited [62], and they are expected to be even better for MECs when hydrogen production is factored in [63]. According to Foley et al., MFC's efficiency must surpass 500 W m23 in order to be commercially viable and environmentally compatible with existing anaerobic treatment technology. Although it has been suggested that current bioelectrochemical reactors are costly due to the need for electrode materials, current collectors, membranes and other components, the advantage of microbial electro synthesis lies in the on-site use of electricity for bioproduction and its freedom from the availability of arable land [64, 65].

7.5 Future Directions

Treatment of recalcitrant pollutants in wastewater is one of the many applications of BES, and a few studies using both synthetic and real field substrates have already been published. The integrated role of electrochemical microbial metabolism in a single reactor gives these systems their unique ability to manage complex contaminants that are difficult to treat in conventional processes. Importantly, the use of BES to eliminate harmful pollutants and xenobiotics is currently being researched extensively in order to enhance treatment efficiency. Experiments with real-field wastewater differ greatly from those with synthetic pollutants, especially in terms of energy recovery. The application of BES for the treatment of real-field wastewater should be more focused, with energy recovery as one of the targets, in order to make the system/process economically viable. The disposal of petroleum-based chemicals such as aromatic hydrocarbons and pharmaceutical wastewater are two of the industrial sector's most pressing issues. Applying BES to the treatment of complex structures with simple carbon chains would be very important (breaking aromatic rings). Similarly, using BES for solid waste disposal, such as kitchen, vegetable, slaughterhouse, municipal and so on, would be innovative and reduce pretreatment costs. A few BES researchers are also looking at how to handle chlorinated aliphatic hydrocarbons like trichloroethene (TCE) and perchloroethylene, which are widely used solvents and degreasing agents. It will be important to perform comprehensive research into the full removal of these extremely toxic (and carcinogenic) compounds from soil and groundwater by BES treatment. BES, on the other hand, can be blended into effluents from conventional treatment methods (which are high in acid metabolites) to create value-added chemicals and solvents with minimal applications. The difficulties in upgrading, especially with design issues, are primarily due to a number of BES advantages. To make BES competitive with current conventional processes, it is critical to focus on making it capable of handling large volumes and faster loading speeds.

References

1 Pant, D.et al. (2012). Bioelectrochemical systems (BES) for sustainable energy production and product recovery from organic wastes and industrial wastewaters. *RSC Adv.*2: 1248–1263.

2 Gude, V.G. (2016). Wastewater treatment in microbial fuel cells – An overview. *J. Clean. Prod.*122: 287–307.

3 Zhang, Y. and Angelidaki, I. (2014). Microbial electrolysis cells turning to be versatile technology: Recent advances and future challenges. *Water Res.*56: 11–25.

4 Sevda, S., Yuan, H., He, Z., and Abu-Reesh, I.M. (2015). Microbial desalination cells as a versatile technology: Functions, optimization and prospective. *Desalination*371: 9–17.

5 Rabaey, K. and Rozendal, R.A. (2010). Microbial electrosynthesis: Revisiting the electrical route for microbial production. *Nat. Rev. Microbiol.*8: 706.

6 Sadhukhan, J., et al. (2016). A critical review of integration analysis of microbial electrosynthesis (MES) systems with waste biorefineries for the production of biofuel and chemical from reuse of CO2. *Ren. Sust. Energ. Rev.*56: 116–132.

7 Escapa, A., Mateos, R., Martínez, E.J., and Blanes, J. (2016). Microbial electrolysis cells: An emerging technology for wastewater treatment and energy recovery. From laboratory to pilot plant and beyond. *Ren. Sust. Energ. Rev.*55: 942–956.

8 Pandey, P.et al. (2016). Recent advances in the use of different substrates in microbial fuel cells toward wastewater treatment and simultaneous energy recovery. *Appl. Energy*168: 706–723.

9 Zhang, F., Ge, Z., Grimaud, J., Hurst, J., and He, Z. (2013). Long-term performance of literscale microbial fuel cells treating primary effluent installed in a municipal wastewater treatment facility. *Environ. Sci. Technol.*47: 4941–4948.

10 Ge, Z. and He, Z. (2016). Long-term performance of a 200 liter modularized microbial fuel cell system treating municipal wastewater: Treatment, energy, and cost. *Environ. Sci.*2: 274–281.

11 Rabaey, K. and Rozendal, R.A. (2010). Microbial electrosynthesis — revisiting the electrical route for microbial production. *Nat. Rev. Microbiol.*8: 706–716.

12 Freguia, S., Virdis, B., Harnisch, F., and Keller, J. (2012). *Electrochim. Acta*82: 165–174.

13 Logan, B.E., Call, D., Cheng, S., Hamelers, H.V.M., Sleutels, T.H.J.A., Jeremiasse, A.W., and Rozendal, R.A. (2008). Microbial electrolysis cells for high yield hydrogen gas production from organic matter. *Environ. Sci. Technol.*42: 8630–8640.

14 Pant, D., Van Bogaert, G., Diels, L., and Vanbroekhoven, K. (2010). A review of the substrates used in microbial fuel cells (MFCs) for sustainable energy production. *Bioresour. Technol.*101: 1533–1543.

15 Harnisch, F., Wirth, S., and Schro¨der, U. (2009). Effects of substrate and metabolite crossover on the cathodic oxygen reduction reaction in microbial fuel cells: Platinum vs. iron(II) phthalocyanine based electrodes. *Electrochem. Commun.* 11: 2253–2256.

16 Chaplin, B.P., Roundy, E., Guy, K.A., Shapley, J.R., and Werth, C.J. (2006). Effects of Natural Water Ions and Humic Acid on Catalytic Nitrate Reduction Kinetics Using an Alumina Supported Pd—Cu Catalyst. *Environ. Sci. Technol.* 40: 3075–3081.

17 Kopinke, F.-D., Angeles-Wedler, D., Fritsch, D., and Mackenzie, K. (2010). *Appl. Catal., B* 96: 323–328.

18 Knitt, L.E., Shapley, J.R., and Strathmann, T.J. (2007). Rapid Metal-Catalyzed Hydrodehalogenation of Iodinated X-Ray Contrast Media. *Environ. Sci. Technol.* 42: 577–583.

19 Hildebrand, H., Mackenzie, K., and Kopinke, F.-D. (2009). Highly Active Pd-on-Magnetite Nanocatalysts for Aqueous Phase Hydrodechlorination Reactions. *Appl. Catal., B* 91: 389–396.

20 Schmidt, T.J., Paulus, U.A., Gasteiger, H.A., and Behm, R.J. (2001). The oxygen reduction reaction on a Pt/carbon fuel cell catalyst in the presence of chloride anions. *J. Electroanal. Chem.* 508: 41–47.

21 Munakata, N. and Reinhard, M. (2007). Palladium-catalyzed aqueous hydrodehalogenation in column reactors: Modeling of deactivation kinetics with sulfide and comparison of regenerants. *Appl. Catal., B* 75: 1–10.

22 Kiely, P.D., Rader, G., Regan, J.M., and Logan, B.E. (2011). Long-term cathode performance and the microbial communities that develop in microbial fuel cells fed different fermentation endproducts. *Bioresour. Technol.* 102: 361–366.

23 Martin, E., Tartakovsky, B., and Savadogo, O. (2011). Cathode materials evaluation in microbial fuel cells: A comparison of carbon, Mn2O3, Fe2O3 and platinum materials. *Electrochim. Acta* 58: 58–66.

24 Chaplin, B.P., Reinhard, M., Schneider, W.F., Schu¨th, C., Shapley, J.R., Strathmann, T.J., and Werth, C.J. (2012). Response to Comment on "Critical Review of Pd-Based Catalytic Treatment of Priority Contaminants in Water". *Environ. Sci. Technol.* 46: 3655–3670.

25 Palomares, A.E., Franch, C., and Corma, A. (2010). Nitrates removal from polluted aquifers using (Sn or Cu)/Pd catalysts in a continuous reactor. *Catal. Today* 149: 348–351.

26 Popat, S.C., Ki, D., Rittmann, B.E., and Torres, C.I. (2012). Importance of OH- transport from cathodes in microbial fuel cells. *ChemSusChem* 5: 1071–1079.

27 Logan, B.E., Hamelers, B., Rozendal, R.A., Schrorder, U., Keller, J., Freguia, S., Aelterman, P., Verstraete, W., and Rabaey, K. (2006). Microbial fuel cells: methodology and technology. *Environ. Sci. Technol.* 40: 5181–5192.

28 Hamelers, H., Ter Heijne, A., Sleutels, T., Jeremiasse, A., Strik, D., and Buisman, C. (2010). New applications and performance of bioelectrochemical systems. *Appl. Microbiol. Biotechnol.* 85: 1673–1685.

29 Zhao, F., Slade, R.C.T., and Varcoe, J.R. (2009). Techniques for the study and development of microbial fuel cells: an electrochemical perspective. *Chem. Soc. Rev.* 38: 1926–1939.

30 Qian, F., Wang, G., and Li, Y. (2010). Solar-Driven Microbial Photoelectrochemical Cells with a Nanowire Photocathode. *Nano Lett.* 10: 4686–4691.

31 Lewerenz, H.J., Heine, C., Skorupska, K., Szabo, N., Hannappel, T., Vo-Dinh, T., Campbell, S.A., Klemm, H.W., and Munoz, A.G. (2010). Photoelectrocatalysis: principles, nanoemitter applications and routes to bio-inspired systems. *Energy Environ. Sci.*3: 748–760.

32 Venkata Mohan, S., Srikanth, S., and Sarma, P.N. (2009). Noncatalyzed microbial fuel cell (MFC) with open air cathode for bioelectricity generation during acidogenic wastewater treatment. *Bioelectrochemistry*75: 130–135.

33 Venkata Mohan, S., Mohanakrishna, G., Velvizhi, G., Babu, V.L., and Sarma, P.N. (2010). Bio-catalyzed electrochemical treatment of real field dairy wastewater with simultaneous power generation. *Biochem Eng J*51 (1): 32–39.

34 ElMekawy, A., Diels, L., De Wever, H., and Pant, D. (2013). Valorization of cereal based biorefinery byproducts: Reality and expectations. *Environ Sci Technol*47 (16): 9014–9027.

35 Wang, X., Cheng, S., Feng, Y., Merrill, M.D., Saito, T., and Logan, B.E. (2009). The use of carbon mesh anodes and the effect of different pretreatment methods on power production in microbial fuel cells. *Environ Sci Technol*43: 6870–6874.

36 Sun, M., Sheng, G.P., Mu, Z.X., Liu, X.W., Chen, Y.Z., Wang, H.L., and Yu, H.Q. (2009). Manipulating the hydrogen production from acetate in a microbial electrolysis cell microbial fuel cell-coupled system. *J Power Sources*191: 338–343.

37 Greenman, J., Gálvez, A., Giusti, L., and Ieropoulos, I. (2009). Electricity from landfill leachate using microbial fuel cells: Comparison with a biological aerated filter. *Enzyme Microb Technol*44: 112–119.

38 Yazdi, R.H., Christy, A.D., Dehority, B.A., Morrison, M., Yu, Z., and Tuovinen, O.H. (2007). Electricity generation from cellulose by rumen microorganisms in microbial fuel cells. *Biotechnol Bioeng*97: 1398–1407.

39 Strik, D.P.B.T.B., Terlouw, H., Hubertus, V.M., and Buisman, C.J.N. (2008). Renewable sustainable biocatalyzed electricity production in a photosynthetic algal microbial fuel cell (PAMFC). *Appl Microbiol Biotechnol*81: 659–668.

40 ElMekawy, A., Srikanth, S., Vanbroekhoven, K., De Wever, H., and Pant, D. (2014). Bioelectro-catalytic valorization of dark fermentation effluents by acetate oxidizing bacteria in bioelectrochemical system (BES). *J Power Sour*262: 183–191.

41 Jiang, H., Luo, S., Shi, X., Dai, M., and Guo, R. (2012). A novel microbial fuel cell and photobioreactor system for continuous domestic wastewater treatment and bioelectricity generation. *Biotechnol Lett*34: 1269–1274.

42 Elakkiya, E. and Matheswaran, M. (2013). Comparison of anodic metabolisms in bioelectricity production during treatment of dairy wastewater in Microbial Fuel Cell. *Biores Technol*136: 407–412.

43 Zhao, G., Ma, F., Wei, L., Chua, H., Chang, -C.-C., and Zhang, X.-J. (2012). Electricity generation from cattle dung using microbial fuel cell technology during anaerobic acidogenesis and the development of microbial populations. *Waste Manag*32: 1651–1658.

44 Venkata Mohan, S. and Srikanth, S. (2011). Enhanced wastewater treatment efficiency through microbially catalyzed oxidation and reduction: Synergistic effect of biocathode microenvironment. *Bioresour Technol*102 (22): 10210–10220.

45 Min, B., Kim, J., Oh, S., Regan, J.M., and Logan, B.E. (2005). Electricity generation from swine wastewater using microbial fuel cells. *Water Res*39: 4961–4968.

46 Heilmann, J. and Logan, B.E. (2006). Production of electricity from proteins using a microbial fuel cell. *Water Environ Res*78: 531–537.

47 Zhang, Y., Min, B., Huang, L., and Angelidaki, I. (2009). Generation of electricity and analysis of microbial communities in wheat straw biomass-powered microbial fuel cells. *Appl Environmental Microbiol*75: 3389–3395.

48 Freguia, S., Teh, E.H., Boon, N., Leung, K.M., Keller, J., and Rabaey, K. (2010). Microbial fuel cells operating on mixed fatty acids. *Bioresour Technol*101: 1233–1238.

49 Yang, F., Ren, L., Pu, Y., and Logan, B.E. (2013). Electricity generation from fermented primary sludge using singlechamber air-cathode microbial fuel cells. *Bioresour Technol*128: 784–787.

50 Li, X.M., Cheng, K.Y., Selvam, A., and Wong, J.W.C. (2013). Bioelectricity production from acidic food waste leachate using microbial fuel cells: Effect of microbial inocula. *Process Biochem*48: 283–288.

51 Clauwaert, P., Aelterman, P., Pham, T.H., De Schamphelaire, L., Carballa, M., Rabaey, K., and Verstraete, W. (2008). Minimizing losses in bio-electrochemical systems: the road to applications. *Appl. Microbiol. Biotechnol.*79: 901–913.

52 Logan, B.E. (2010). Scaling up microbial fuel cells and other bioelectrochemical systems. *Appl. Microbiol. Biotechnol.* 85: 16665–21671.

53 Dewan, A., Beyenal, H., and Lewandowski, Z. (2008). Scaling up microbial fuel cells. *Environ. Sci. Technol.*42: 7643–7648.

54 Di Lorenzo, M., Curtis, T.P., Head, I.M., and Scott, K. (2009). A single-chamber microbial fuel cell as a biosensor for wastewaters. *Water Res.*43: 3145–3154.

55 Di Lorenzo, M., Scott, K., Curtis, T.P., and Head, I.M. (2010). Effect of increasing anode surface area on the performance of a single chamber microbial fuel cell. *Chem. Eng. J.*156: 40–48.

56 Fornero, J.J., Rosenbaum, M., and Angenent, L.T. (2010). Meta-analysis of Microbial Fuel Cells Using Waste Substrates. *Electroanalysis*22: 832–843.

57 Pant, D., Singh, A., Satyawali, Y., and Gupta, R.K. (2008). Effect of carbon and nitrogen source amendment on synthetic dyes decolourizing efficiency of white-rot fungus, Phanerochaete chrysosporium. *J. Environ. Biol.* 29: 79–84.

58 Li, W.W., Guo-Ping Sheng, G.P., Liu, X.W., and Yu, H.Q. (2011). Recent advances in the separators for microbial fuel cells. *Bioresour. Technol.*102: 244–252.

59 Wang, H.-Y., Bernarda, A., Huang, C.-Y., Lee, D.-J., and Chang, J.-S. (2011). Micro-sized microbial fuel cell: a mini-review. *Bioresour. Technol.*102: 235–243.

60 Zhou, M., Chi, M., Luo, J., He, H., and Jin, T. (2011). Effect of Epoxy Mixed with Nafion Solution as an Anode Binder on the Performance of Microbial Fuel Cell. *J. Power Sources*196: 427–4435.

61 Yu, E.H., Cheng, S., Scott, K., and Logan, B. (2007). Microbial fuel cell performance with non-Pt cathode catalysts. *J. Power Sources*171: 275–281.

62 Rozendal, R.A., Hamelers, H.V.M., Rabaey, K., Keller, J., and Buisman, C.J.N. (2008). *Trends Biotechnol.*26: 450–459.

63 Foley, J.M., Rozendal, R.A., Hertle, C.K., Lant, P.A., and Rabaey, K. (2010). *Environ. Sci. Technol.*44: 3629–3637.

64 Liu, H. (2008). *Anaerobic Biotechnology for Bioenergy Production: Principles and Applications*. (ed.S.K.Khanal), Oxford, UK: Wiley-Blackwell.221–246.

65 Freguia, S., Rabaey, K., Yuan, Z., and Keller, J. (2007). Non-catalyzed cathodic oxygen reduction at graphite granules in microbial fuel cells. *Electrochim. Acta*53: 598–603.

8

Nanotechnology

Environmental Sustainable Solutions for Wastewater Treatment

CHAPTER MENU

8.1 Introduction

Oceans constitute 71% of the earth's surface while islands and continents cover the remaining portion. With increasing population and human needs the demand increases for freshwater in order to fulfill day-to-day life activities per capita, and also industrial sectors like the food industry, pharmaceuticals, medicine, electronics and other plants that need huge quantities of water to run their processes. Freshwater sources are under immense pressures due to excessive usage and global issues like climate change and population growth. Due to water-borne diseases and lack of proper sanitation around 2 million people die every year in developing countries. A continuous availability of clean water is also important for our use in everyday life. Moreover, the availability of clean water to industry is contingent upon many industry firms processing paper, plastics and metal. Consequently, the use of water can be preserved by recognizing the use and reuse

Aquatic Environmental Bioengineering: Monitoring and Remediation of Contamination, First Edition.
Rouf Ahmad Bhat, Mohammad Yaseen Mir, Gowhar Hamid Dar and Moonisa Aslam Dervash.

of materials used to generate safe water and also drinking water. The need for clean water is rising alongside the need for the reuse of wastewater and freshwater for farming and aquaculture. More new technologies and water management frameworks are required to satisfy this demand as the latter enter the final stage of sufficient water treatment for the everyday needs of humans and the environment. Untreated industrial effluents and the absence of an improved sanitation system pollute the water bodies. Human activity leads to water contamination by using sewage in water supplies such as the river or the sea. As it depends heavily on centralized systems, the conventional method of water storage, water delivery and disposal is no longer feasible. One of the most basic humanitarian essentials, secure, safe water availability, remains a major challenge to satisfy the global demand of the twenty-first century. Increased drinking water prices due to expanding communities affected by a wide range of environmental and climate challenges [1] contribute to people's inability to provide ample drinking water. According to the WHO, in 2015 the supply of adequate drinking water was reportedly insufficient for 1.1 billion people. With rapid industrial development, urbanization, population growth, increased living conditions and agricultural growth, a substantial quantity of waste is generated and released into water sources, and contaminates water which poses a hazard to humans and the aquatic environment. The major causes of pollution, of which heavy metals are the main pollutants, are wastewater and the effluent from numerous small- and medium-sized enterprises (SMEs) or process industries, such as metal boards, battery producers and mining companies. These non-biodegradable heavy metals pose a significant threat to plants, animals and human beings when this polluted water is drunk [2]. Through the construction of multiple hydraulic systems such as dams, rivers and reservoirs, efforts have been made to harness water quantity, while water management, recycling and desalination have been at the forefront of technical options, particularly in countries stressed by water [3]. These technical solutions are often costly and not cost-effective; when there is an increasing awareness that water is finite, attempts are therefore made on wastewater treatment [4]. A move forward in ensuring water protection is the need for technical advancement to allow safe and integrated water resources. According to Davies [5], the term nano means one billionth of a meter (10^{29}) and even smaller materials. Nanomaterials have been categorized specifically on the basis of physical, chemical and biological aspects. Carbon compounds, polymers, metal oxides and metals have a greater proportion of the surface area of the nano-scale particle dimension, hence these nano materials show various mechanical, electrical, magnetic and optical characteristics, typically distinct from the macroscopic scale characteristics, and have a larger number of leads to particle size reduction [6]. Various reports have been put forward to reveal that nano-biotechnology is the flexible option over obtainable water treatment technologies, due to its being an efficient and multifunctional method [7]. Therefore, with other water treatment strategies such as adsorption, coagulation, membrane technology and photocatalysis, this impressive nanotechnology can be used to create a higher performance, lower-cost and environmentally appropriate water treatment system [8]. This chapter offers a brief overview of the common use of nanomaterials based on nano-biotechnology for its use in the treatment of water and in collaboration with other treatment techniques. The primary purpose of this paper will be to address the possibilities and challenges of using nanomaterials for a particular kind of water treatment method in combination with biotechnology.

8.2 Water Nanotechnology

With the rise of nanotechnology, its use in the handling of water and wastewater is becoming inevitable. There have been several studies available on such a topic over the past few decades. We have given a brief analysis in this section of some standard applications of nanotechnology in the handling of water and wastewater, i.e. adsorption and isolation, disinfection and sensing catalytic oxidation.

8.2.1 Adsorption and Separation

The two most frequently used methods for polishing the handling of water and wastewater are adsorbents or membrane-based separation systems. Conventional adsorbents also face problems such as low ability and selectivity as well as the short duration of adsorption-regeneration, which has greatly limited the adsorbents' cost-effectiveness. Adsorbents based on nanomaterials, i.e. nanosized metal or metal oxides, carbon nanotubes (CNTs), graphene and nanocomposites, also have a wide specific region, high reactivity, rapid kinetics and specific affinity with different pollutants. Often their adsorption ability for such toxins is many times better than typical adsorbents [9]. Membrane segregation is also a crucial module in the polishing treatment stage in addition to adsorption, allowing water reclamation from unusual sources of water such as industrial wastewater. The elimination of contaminants by isolation of membranes is primarily based on the exclusion of size. However, many hurdles to more membrane technology exist, i.e. the inherent trade-off between membrane selectivity and permeability, high energy consumption, fouling and operational complexity. Advanced nanocomposite membranes were created to deal with these problems by incorporating functional nanoparticles into the membrane. These membranes exhibit physiochemical properties such as improved mechanical or thermal stability, porosity and hydrophilicity, though they possess specific properties like enhanced permeability, or anti-fouling, antimicrobial, adsorptive or photocatalysis capabilities [10].

8.2.2 Catalysis

An advanced oxidation method for removing trace contaminants and microbial pathogens from water is catalytic or photocatalytic oxidation. In order to improve biodegradability for dangerous and non-biodegradable pollutants, it is a valuable pretreatment process [11]. In order to treat recalcitrant organic compounds, photocatalysis can also be used as a polishing step [8]. High surface-to-volume ratio nanocatalysts have demonstrated dramatically improved catalysis efficiency over their bulky counterparts. In addition, the nanosized semiconductors' band gap and crystalline structure displayed size-dependent activity. With differing sizes, their electron hole redox potential and photo-generated charge distribution varied [12]. In addition, the stability of the nanocatalyst was enhanced by immobilizing nanoparticles on different substrates, and the resulting nanocomposites were consistent with the current photo-reactor [13].

8.2.3 Disinfection

High surface-to – volume ratio nanocatalysts have demonstrated dramatically improved catalysis efficiency over their bulky counterparts. In addition, the nanosized semiconductors' band gap and crystalline structure displayed size-dependent activity. With differing sizes, their electron hole redox potential and photo-generated charge distribution varied [14]. In addition, the stability of the nanocatalyst was enhanced by immobilising nanoparticles on different substrates, and the resulting nanocomposites were consistent with the current photo-reactor [13]. Disinfection is the last but crucial stage of water treatment in order to deter the transmission of waterborne illnesses. The optimal disinfectant should have the following properties: (1) a broad antimicrobial range within a limited period of time; (2) no harmful by-product generation; (3) low toxicity to health and ecosystems; (4) low energy expense and fast operation; (5) fast to store and not corrosive; and (6) have a healthy disposal. Several nanomaterials, including chitosan nanoparticles [15], nano silver (nAg) [16], photocatalytic TiO_2[17] and carbon-dependent nanomaterials [18], have recently been shown to have powerful antimicrobial properties. By emitting poisonous metal ions (e.g. Ag^+), destroying the cell membrane by direct interaction (e.g., chitosan NPs) or producing reactive oxygen species (ROS, e.g., TiO_2), these nanomaterials destroy microorganisms. Unlike traditional chemical disinfectants, microorganisms are inactivated by these antimicrobial nanomaterials by more natural methods, which are intended to reduce the formation of dangerous by-products of disinfection (DBPs) [19]. Furthermore, certain nano-disinfectants may be worked continuously with high performance and low energy consumption when integrated with proper separation techniques, which is highly desirable for decentralized water and wastewater treatment.

8.2.4 Sensing

Traditional sensing and tracking techniques in diverse water sources are incapable of identifying exceedingly low concentrations of micro-pollutants. In emergencies (e.g., water-related accidents), immediate and *in situ* identification of contaminants and highly toxic substances is of considerable importance. There are special electrochemical, optical or magnetic properties of nanomaterials, such as carbon nanotubes, graphene, quantum dots and noble metals (e.g. Ag or Au). Trace contaminants could be selectively preconcentrated for identification by inserting such nanomaterials into sensors or electrodes. Many nanomaterials could increase spectroscopic reaction by several magnitudes (e.g. Raman change or surface plasmon resonances) [20]. Furthermore, nanocomposite sensors have been extensively explored for their environmental tracking and sensing abilities [21].

8.2.5 Carbon-Based Nanoadsorbents

In the handling of wastewater, the elimination of both organic and inorganic contaminants is primarily achieved by adsorption. Through the development of carbonaceous nanomaterials (CNMs), which involve carbon nanoparticles, carbon nano sheets and carbon nanotubes (CNTs), carbon-based nanoabsorbents are emerging. Due to greater adsorption efficiency of various organic chemicals, CNTs are more favored relative to activated carbons [22] (Pan and Xing, 2008). Different CNT applications have excellent chemical resistance, high specific surface area, excellent adsorption power, and mechanical strength [22]. In separating heavy metals such as nickel ions (Ni21) from water, CNTs are particularly useful [23]. By grafting, the surface properties of CNTs can be further changed. The grafting requires functional groups by diverse processes such as microwave, plasma procedure and chemical alteration [24].

Due to the hydrophobic effect of graphitic surfaces, CNT transforms into aqueous aggregates. For trapping organic molecules, interstitial spaces and grooves within the aggregates form high absorption energy. Notably, due to the high adsorption capacity of CNTs, bulky organic molecules become stuck because of their large pores. Other reasons for trapping organic components present in wastewater are: covalent bonding and electrostatic interactions, hydrogen bonding (for compounds with OH, NH_2, COOH and functional groups) and pipi interactions (for polar aromatic compounds, polycyclic aromatic hydrocarbons). Due to surface functional groups (carboxyl, hydroxyl and phenol), chemical bonding and electrostatic attraction produce CNTs to adsorb metal ions [25]. A composite filter based on CNT can very effectively eliminate heavy metal ions from water [26]. The efficient removal of Zn21 by continuous regeneration and reuse of CNT nanoabsorbents is one of the significant advantages of CNT nanoabsorbents. In addition, multiwall carbon nanotubes (MWCNT) [27] have also been used to eliminate heavy metals such as Cu (II) [28], Pb (II) and Mn (II) [27], thus overcoming the barriers of low ability dispersing ability, separation complexity and small particle size.

8.2.6 Metal-Based Nanoadsorbents

In recent years, the use of metal and metal oxide-based nanoadsorbents for wastewater treatment has gained a lot of interest [29]. For example, for the purification of wastewater, metallic iron and its various oxides, titanium dioxide, magnesium oxide, alumina and zinc oxide have been extensively studied. These metal oxides exhibit many beneficial features due to their inherent propensities for harmful metal ions, magnetic activity and tunable size or morphology, such as quicker adsorption kinetics, high performance and quick recovery after adsorption [30].

The detoxification of wastewater containing toxins, polychlorinated biphenyls and other organic compounds has been shown in pure metallic iron [31]. In the adsorption of various arsenic species from water, iron oxides, hydroxides and oxy-hydroxides have been reported to be highly efficient [32]. Via electrostatic interaction, the higher surface area of the nanoscale iron oxide particles and their intrinsic ability to bind As (III) and As(V) species make these magnetically separable nanoadsorbents very common among researchers working on arsenic removal. Furthermore, they have also been used to extract water from cobalt, chromium, copper, lead and nickel [22]. For the unique binding of chromium from water, surface functionalized iron oxides have been researched in the literature [33]. For the binding of iron, nickel, lead, copper, zinc and cadmium, γ-Fe2O3 nanotubes and nanorods are found to be highly efficient. For the unique binding of harmful Hg(II) ions from water, hematite nanoparticles deposited on graphene oxide have been used successfully [34]. Nanostructured ZnO, TiO_2 and MgO are also well researched and reported to be highly successful in extracting various toxic elements from water, apart from both of these iron-based materials [29]. In the removal of arsenic, copper, lead, cadmium, manganese and iron, hydrated amorphous titania is found to be highly successful [35]. Effective adsorbents for arsenic, lead and cadmium are stated to be nano-flowers of magnesium oxide having a high surface area [36] (Srivastava et al., 2015). Some low-cost nanoadsorbents that have been widely used to adsorb and isolate arsenic, cadmium and chromium ions from water include zinc oxide and alumina [37].

8.2.7 Polymer-Based Nanoadsorbents

These nanoabsorbents are capable of removing both organic materials and heavy metals by complex cavities for pollutant removal. Owing to the presence of hydroxyl on the exterior surface, heavy metals are absorbed, while chemical compounds from wastewater are eliminated by hydrophobic activity on the internal surface owing to sorption. The efficient separation of metal ions from aqueous solutions by a polymeric nanoabsorbent such as the dendrimer-ultrafiltration method is demonstrated by Diallo et al. [38]. By mimicking natural detection and recognition, synthetic polymers such as molecularly imprinted polymers (MIPs) are acquiring a reputation for wastewater treatment, showing high affinity and selectivity for their structural analogues [39]. The ability for wastewater treatment using composite materials based on MIP is outstanding. With the use of composite materials such as quantum points, magnetic and nanoparticles, MIPs may reserve their prominent selectivity and high adsorption power. Arsenic (V) (B40 mg/g) and fluoride (B100 mg/g) ions have been successfully isolated from effluent and wastewater by polymeric nanoparticles obtained by synthesis of Fe- and Al-doped, activated micron (B0.8 mm) and nano (B100 nm) sized porous adsorbents [40]. Adhesive biocatalytic coatings with nanoporous microstructure formed by partiallycombined particles of latex polymer trap highly condensed microbial cells (harmful pathogenic bacteria) during wastewater treatment in a dry state stabilized by carbohydrate osmoprotectants [41].

8.3 Zeolites

The metal ion adsorption used in the case of synthetic zeolites are in general functionalized by organosilane moieties particularly mercaptopropyl, aminopropyl, etc. [42]. The harmful anions and carcinogenic organic pollutants are mainly removed from water by using zeolites functionalized with cationic surfactants [43]. It is reported in certain cases that zeolites loaded with lead and silver significantly remove pathogenic microbes and hazardous anions like cyanide, chromate, arsenate and arsenite by forming insoluble complexes [44]. In terms of thermal stability, structural rigidity, specificity and adsorption efficiency because of engineered channel networks the synthetic zeolites are found more beneficial as compared over natural zeolites.

Zeolites, primarily because of their porous nature, are among the most common natural adsorbents, allowing them to accommodate a wide range of metal ions. For adsorption-based wastewater disposal, both naturally occurring and synthetic zeolites have been commonly used. Zeolites such as analcime, clinoptilolite, chabazite, stilbite, natrolite, heulandites and phillipsite are 3-dimensional complexes of microporous alumina silicate which are loosely bound by cations like magnesium, potassium, calcium, sodium, etc. A rapid exchange of these cations occurs with cations of water-based pollutants [42].

8.4 Magnetic Nanocomposites

Magnetic nanocomposites are a class of unique nanomaterials. They usually feature a core-shell nanostructure that could be easily and swiftly recovered with an external magnetic field. Functional nanoparticles or functional groups could be grafted or immobilized onto Fe_3O_4 nanoparticles through chemical bonding or direct deposition [45]. The materials to be coated include Ag, TiO_2, CNTs, GO, Pd and SiO_2, and so on. An SiO_2 coating could provide high surface area and sound porosity, while protecting the magnetic core from corrosion. After coating, the resultant magnetic nanocomposites often exhibited enhanced adsorption or photocatalytic capability with fast kinetics towards contaminants such as heavy metals, dyes, phenolic compounds or microbial [46].

A sequence of porous Mn-based magnetic nanowires have been synthesized by Cui et al. [47] in recent years to eliminate heavy metals, As (III) and some organic contaminants. In order to generate magnetic carbon hollow spheres, a simple one-step solvothermal approach was also proposed, which was highly successful for heavy metal removal [48]. In addition, a nanocomposite of maghemite/silica has been synthesized and used as a heterogeneous Fenton catalyst. This could degrade contaminants such as PNP and MB efficiently [49]. Long-term use has demonstrated negligible leaching of Fe. Jin et al. [50] indicated that amino acid-modified magnetic nanoparticles could trap 94% of bacterial cells in 20 minutes over a pH range of 4–10. Shan et al. [51] found that nanocomposites grafted with poly(1-vinylimidazole) $Fe_3O_4@SiO_2$ could effectively adsorb Hg(II) at pH 7 with a power of 346 mg/g. With 0.50 M HCl, the depleted adsorbent could be produced in just 10 minutes. Zhang et al. [52] formed another especially important magnetic nanoparticle with multi-functionality, where a corona produced from polyethylenimine was coated around the magnetic core. The resulting nanomaterial could inactivate bacteria by penetration of the cell membrane, thus capturing heavy metal through electrostatic attraction by chelation and adsorbing anions. The nanocomposite could be regenerated by EDTA and NaOH and reused. In addition, to accelerate sewage coagulation, magnetic nanocomposites were also used [53], and certain polymer-coated magnetic nanoparticles were tested for *insitu* oil spill remediation [54]. Generally speaking, different products with considerable potential for real water remediation are magnetic nanocomposites. However, in further research, the cost of synthesis and its consistency with the existing infrastructure should be considered.

8.5 Nano Zero Valent Iron (nZVI)

An emerging nanomaterial which is used to extract different organic and inorganic contaminants from water is nano zero valent iron (nZVI). Therefore, many nZVI improvements have been studied, such as surface-modified nZVI, emulsified nZVI (better miscibility), bimetallic nZVI (higher reactivity) and nZVI on carbon support (better distribution). The highly reactive nZVI has a limited lifespan. At the nZVI surface, heavy metals are either diluted (e.g., Cu^{2+}, Ag^{2+}, or adsorbed directly to the nZVI surface where they are made immobile (e.g., Zn^{2+}, Cd^{2+}).The simple separation of the particles from water can be achieved by nZVI with aid of magnet nature. The nZVI shows notable adsorption and precipitation for As(III) and As(V) ions [55, 56] where it was found that the adsorption occurs through weak electrostatic interaction between the sorbate and the binding sites [57]. Gupta et al. [58] reported the application of nZVI encapsulated chitosan nanospheres for the removal of arsenic from an aqueous medium. Within a wide range of pH (from 2 to 9), the authors have found no significant interference from major anions like sulfate, phosphate and silicate on the adsorption behavior of arsenic on nZVI, and removed total inorganic arsenic from contaminated groundwater. Apart from arsenic, nZVI is being used to successfully treat various metallic ions in aqueous solutions, such as Cr^{6+}, Cu^{2+}, Pb^{2+}, Ba^{2+}, As^{2+}, As^{2+}, Co^{2+} and so on. Additionally, nZVI is capable of removing or recovering dissolved metals from solution (e.g., Cr(VI), U(VI)), and is also effective against several organic and inorganic contaminants [59–61].

8.6 Biosorbents

Biosorbents, which are generally extracted from biological or agricultural materials for the effective removal of certain contaminants, are considered to be promising for the efficiency of the removal of certain contaminants because of the very low levels of such pollutants, normally between picograms and nanograms per liter of water. Biosorbents have many benefits relative to traditional absorbents, such as low cost, high quality, low agricultural and biological sludge, no extra nutrient demand and they are also regenerative.

Liu et al. [62] developed a DNA matrix consisting of salmon milt DNA hydrogel percolations and successfully used the matrix for the selective adsorption of dioxin materials. The DNA beads will easily be regenerated after dioxin is adsorbed with hexane by rinsing the beads. Various trial-embedded biomass biosorbents for organical toxins were observed to be extracted from water [63]. The benefits of triolein include their high build-up potential and their low solubility and permeability owing to their large molecular weight of 885.45 da [64]. (105–107 for trace organic compounds in aquatic). Chitosan sorbents have showed a promising finding of highly effective adsorption ability for metal-ion ion adsors, where the adsorption of metal ion takes place by chelation in amino acid group chitosan [65]. Guo et al. [66] developed another analysis into a biosorbent of black alcohol, paper and paper waste. A biosorbent was used for the treatment of heavy metal ions. The order of the biosurgent sorption affinity to different heavy metal ions has been stated under Pb(II) > Cu(II) > Cd(II) > Zn(II) > Ni(II). Strong metal was also extracted from the water from biosorbents prepared from other agricultural materials [67].

8.7 Treatment of Wastewater by Means of Membrane-Based Techniques

Membranes are semi-permeable protective layers which enables the permeation by means of pores of substances having selective size and shape. Nowadays, membrane-based technology has seen wide uses in wastewater treatment. Basically, if a water filtration membrane is used, it stops the pollutants from passing (which have a scale larger than the pores) and gives better water. The membrane technology in modern times has evolved quite rapidly, starting from desalination to antimicrobial filtration. The various pollutants, particularly particulate matters, organic substances and microorganisms which pollute water bodies are trapped in highly efficient developed membranes. The membranes are classified on the basis of mode of use, structure and porosity. Some of these are thin film compound membranes, nanofibre membranes, nanocomposite membranes, self-assembling membranes, aquaporine-based membranes and also membranes for reverse and forward osmosis may be used as membranes.

The process of nanofiltration is a pressurized process in which impure water flows into the thin, 110 nm porous polymer membrane, separating the contaminants from the water having size larger than that of the pores. This process is highly significant in removal of organic pollutants as well as ions from water. The nano-filtration membranes are mainly used in water softening as they block magnesium and calcium ions and allow only monovalent hydrated ions to pass through the water. Nanofiltration membranes are used as solvent recovery systems in the pharmaceutical and chemical industries. They are also used for the purification of condensates in petrochemical industries. They are used in the life sciences to extract amino acids and lipids from blood and other cell cultures [68]. Many methods have been developed for the manufacture and surface change of filtration membranes. As a case, Xi et al. [69] documented the preparation of an interlayer of polydopamine-polyethicylene on an ultrafiltration substratum for the thin film composite nanofiltration membrane. These membranes indicate that Na_2SO_4 is removed by 97% and that $Na_2SO_4 > MgCl_2 > NaCl$ Salt rejection is removed.

It is very useful to eliminate thin ionic membranes via reverse osmosis, such as chloride, sodium and potassium. Consequently, extensive applications are found in seawater desalination. In recent decades, significant efforts have been made to improve cost-effective and energy-efficient desalination membrane filtration technology. Many current commercial RO membranes have a hybrid thin film structure in which the surface of a microporous support membrane is laid with a polyamide sheet [70]. The penetration selectivity of polyamide films can be effectively regulated. Membrane fouling caused by deposition of organic matter and microbial growth upon pores over time decreases the output. Chlorine-assisted oxidative degradation also often deteriorates membranes. The most common practice to solve these problems is by changing membrane surfaces by means of grafting or by functional group replacement [70]. The integration of nanomaterials into the manufacture of new membrane materials has also been documented. Choi et al. [71] used graphene oxide (GO) as a dual-functional protective layer to improve both membrane and chlorine resistance and preserve their separation efficiency as they were coated on the membrane surface. In the recent past there has been much focus on the use of usable carbon nanotubes in composite RO membranes.

Chan et al. [72] reported that the ratio of salt rejection and water flow can be greatly improved by integrating into the polyamide membrane the efficient zwitterion carbon nanotubes. Kim et al. [73] incorporated in the polyamide RO membranes the carbon nanotubes so as to obtain higher water flux. Furthermore, carbon nanotubes containing membranes displayed a greater robustness and chemistry than bare membranes to sodium chlorides. The hydrophobic nano-channels formed by carbon nanotubes and the uniform interactions between CNT and polyamide in the active layer are attributed with these advantageous properties found in the composite membranes [74]. The RO membranes which are commercially used are mainly of mixed type, in which super-hydrophilic nanoparticles are embedded on a polyamide film to repel microbials, salt and impurities. TiO_2 and silver nanoparticles are classically used in RO membranes to prevent fouling of organic pollutants through oxidative degradation [75]. Recently, selectivity-enhanced membranes modified by biomolecules such as proteins have been developed. Ultra-filtration self-assembly membranes of block copolymers have also been studied and have shown a higher selectivity and permeation efficiency [76]. The use of polymeric nanofibers and electro-spinned ceramics is an important class of membranes. Open-pore, interconnected structures with an integrated polymer membrane thickness offer high water flux into membranes based on polymer nanofiber [77]. Additionally, bio-nanofiber membranes based on polyurethane, polylactic acid and polyethylene oxide have been developed to achieve higher anti-fouling properties [78]. There is a greater selectivity and a higher water flow known for biomimetic membranes. The vesicle-incorporated aquaporin-supported nanofiltration membrane is an example of this membrane [79].

8.8 Nanoparticles for Microbial Control and Disinfection

Water infected with pathogenic microorganisms adversely affects human health. As per reports of the World Health Organization, globally 3.4 million people die each year because of numerous water-borne diseases [80]. Long-term use of antibiotics has permitted the microorganisms to undergo and develop resistance to genetic mutations. The production of antibiotic-resistant microbes poses a significant challenge to all living species. Thus, nanotechnology-based methods have recently begun appearing as an alternative to conventional disinfection methods, using some nanomaterials' biotoxicity [43].

Metal nanoparticles are advantageous over antibiotics usage as metal nanoparticles can concurrently induce different attacking mechanisms, which make resistance production very difficult for the microbes through multiple genetic transformations at the same time [81]. A significant number of nanostructured materials having antimicrobial activity have been synthesized and used for microbial control. The important ones among them are TiO_2, ZnO, carbon-based nanomaterials and platinum [82]. Here we will speak briefly about the latest nanotechnology-based solutions and their mode of action for microbial water contamination. The antibacterial and detective effects of silver have long been established [83]. However, its behavior is increased when used in nano form rather than bulk. In the literature, antimicrobial activities have been investigated for metallic silver nanoparticles, silver oxide, silver phosphate and various silver-doped nanoparticles [84]. The high and broad spectrum antimicrobial action of silver nanoparticles destroys all positive and negative bacterial strains. Nano-silver is the most common material for disinfection of wastewater, for pharmaceutical water and for drinking water use [84]. It was used for several commercial water treatment systems including Tata Swach, Blue Star and Aquapure. CNTs are another type of nanomaterials with very high microbial activity and thus suitable for many antimicrobial applications. Its fibrous composition allows microbes to be eliminated by exclusion in size and depth. CNT's electrical conductivity enables adsorbed microbes to be deactivated within seconds, even with a very low voltage of 23 V [85]. Often used to suppress pathogenic microbes such as *Escherichia Coli, Stephylococcus pneumoniae* and *Klebsiella pneumoniae* were TiO_2 nanoparticles of various sizes [86]. Titanium dioxide nanoparticles can create reactive oxygen species in the cells and thus destroy microbes successfully when illuminated by the right source of light energy [86]. Nano-fibers and thin solgel films with silver-doped titania have also been reported to have strong anti-bacterial activity with *E. Epifani* [87] Coli. A variety of investigations [88, 89] have also examined water disinfection using nanostructured zinc oxide and its silver and TiO_2 composites with prominent antimicrobial activity over a wide bacterial species range. As zinc oxide nanoparticles are not documented to be harmful to human cells, antibacterial drugs and industrial water purifiers are added. Often used as antibacterial agents are many other nonmetal oxide nanoparticles such as copper oxide, magnesium oxide and iron oxides and their silver composites [90].

8.9 Antimicrobial Action of Nanoparticles

A general mechanism of antimicrobial activity is difficult to recommend for various forms of nanoparticles, since their nature and chemistry vary. Oxidative stress induction, metal emission and nonoxidative pathways are three of the commonly known ways. The significant anti-bacterial activity of silver-nanoparticles is well known. They are well studied in literature as regards their mechanisms of operation [91]. It begins with the release of silver ions, which bind certain essential proteins quickly and contribute to damage to the enzymes. Silver inhibits the mechanism of DNA replication and cleaves the membrane of the cell. The reactive oxygen (ROS) and H_2O_2 species inside a cell produce nanostructured silver, ZnO and TiO_2 and help in destroying the microbe via membrane clavation and protein damage. The disinfections and microbial controls of carbon-dependent materials such as carbon nanotubes and graphene are very successful as they can contribute to oxidative stress in the cells and oxidize microbes by the critical cell components [92]. CNTs have been reported to be more powerful in nanotubes with smaller sizes against microbes [93] based on their physical-chemical properties.

8.10 Potential Applications in Wastewater Treatment

Highly promising candidates for treatment of water with pathogenic bacteria and viruses are nanostructured materials with good antimicrobial action. Nano-platinum, TiO_2 and ZnO are used to tolerate bacterial contamination in water filters and RO membranes. They were also already used in food processing, medical devices' surface painting, textiles and dental implants. The use of nanomaterials in microbial control is useful compared with other processes of chemical treatment because they are comparatively inert in water and cause reduced amount of harmful substance degradation.

8.11 Benefits of Nano-Biotechnology-Based Applications for Water Sustainability

Because of its tremendous advantages in solving food, water and biodiversity problems, nano-biotechnology is valuable for many applications. It is eminent in terms of biodiversity because of the exciting potential to improve water supply efficiency for human utilization. Innovative and practical biological nano-systems continue to be created and produced within the fields of nanoscience, nanomaterials, biotechnology and its related sectors and researchers. In particular, the regulated discovery of engineered bio-coatings leads to a cost-efficient and high quality biochemical processing of effluent and wastewater. Division technology has gained popularity in the water treatment process, with various applications recently being found in thin-film composites (TFCs) with manufactured antibacterial content [94]. This TFC membrane increases the membrane's antifungal function. It results in more significantly active surfaces and thus improves membrane efficiency, which is beneficial to the TFC membrane [95]. TiO_2 has been immobilized on the surface and Kim et al. [96] found the self-assembled hybrid membrane TiO_2 had a substantially improved bactericidal photocatalytic ability.

The molecule calculation scatter in the effluent at each outlet was analyzed by Stoller [97] and the beneficial effect produced by an auxiliary floculation of the AS floculant by accumulating derived salts close to the membrane layer was emphasized, which led to the movement of particles through tangential instability and minimized fouling. Stoller and Bravi [98] found related findings when they introduced TiO_2 nanoparticles or aerobic digestion to further UV photocatalysis (PC). Both processes of pre-treatment provided final irrigation-friendly sources, but in particular UV/TiO_2 photocatalysis ensured optimum membrane efficiency and suggested the shortest possible time of residence (24 hours). The main advantages of TiO_2/UV light photocatalysis are the high surface/volume ratio of TiO_2 nanoparticles, which, if used as a photocatalytic material, enables mass transfer, even under ambient temperature conditions, through the doping of those nanoparticles in order to allow the sun's activation and photocorrosion resistance (Ch). The findings revealed that the removal values for color, aromatic compound content, chemical oxygen requirement (COD) and the phenols total were up to 42%, respectively 14%, 38% and 31%, but that no reduction in toxicity was achieved. The integration of biological therapy has, however, helped lower the concentration of COD and overall phenols, and the toxicity. *P. chrysosporium* therapy facilitated the greatest toxicity reduction; however, the *P. sajor-caju* produced the highest COD and overall phenol reduction. In addition, pretreatment with hydrogen peroxide improved the efficacy of the biological treatment process.

The cellulose nanofiber (CNF) composite was used by Dwivedi et al. [99]. Four derivatives of CNF have been developed with the incorporation of muscle-inspired dopamine conjugation (a natural water habitat mustard catecholamine) through the exploitation of stable complexation between catechol, metal ions and iron-based cross-linkings. In combination with the commercially available components, the CNF derivatives showed a strong improving efficiency of the use of carcinogenic arsenic and chromium in the mixed polluted water. The implementation of different treatment methods and the size of the trial and successes for different wastewater sources are shown in Table 8.1.

8.12 Challenges and Future Outlook

Several nanotech products are currently commercially available for water purification and pollutant sensing. However, they are considered to have harmful effects on human health in the longer term, so we are becoming more and more vulnerable to the possibility of toxicity of liquidated nanoparticles due to the growing number of water purification systems containing nanomaterials. These nanoparticles are added in terms of scale, shape, reactivity and antimicrobial activity. Because of these same characteristics, they can become poisonous to the human body and dangerous to other living creatures.

Via skin, inhalation and absorption, nanoparticles will quickly penetrate our bodies and enter the bloodstream. The enzymes and protein production of nanoparticles adsorbed on the human skin is impaired. They can pass by blood into all areas of the body, including the brain, kidney, liver and heart, where they can display high reactivity, which disrupts the normal functioning of the tissues. In biological environments, nanomaterials such as carbon nanotubes and metal oxides demonstrate high reactivity by creating multiple reagent oxygen species (ROS). Such ROS can induce oxidative stress and cell inflammation [107]. While the situation and the response to human health of these nanoparticles is still uncertain, processes such as the treatment of freshwater with antimicrobials or nanoadsorbents to remove organic dyes or metal ions from wastewater are suspected to have unwanted effects on the water environment. Their treatment processes are also suspect. Nanoparticles are steadily undergoing bioconcentration until entered into the food chain as the key cause of their toxicity is the aggregation in the body. The researchers, however, are still unaware that nanomaterials may present health risks and exposures, since these are not sufficiently reported. Thus, the real task for science students is to perform proper risk analyses and discuss alternative risk control plans prior to the introduction of a large-scale water filtrations project focused on nanotechnology. Another daunting challenge that has not been tackled consistently in most science literature is healthy disposal of waste from the water process. Many wastewater treatment or methods for the purification of drinking water based on nanotechnology are costly and not energy efficient. These limitations are also not free of the common membrane filtration systems for water purification. Furthermore, during the reverse osmotic filtration process about 75% of the water is lost. Owing to these significant problems, the extensive use of water purification nanotechnology is being hampered.

Table 8.1 Application of various treatment processes for diverse wastewater sources.

Contaminated water source	Treatment processes	Scale	Standards achieved	References
Olive mill	Aerobic digestion and UV photocatalysis along with nanometric TiO_2	Pilot	Final RO permeate streams with COD equal to 456, 242, and 385 mg/L for AS, AD1AS, and PC, respectively	[98]
Olive mill	PC nano-TiO_2	Lab	90.862.7% removal of the phenolic content	[100]
Olive mill	PC oxidation with TiO_2 and Fe_2O_3 nanocatalysts plus biological process	Lab	Nano-TiO_2/UV/H_2O_2 ensured up to 43%, 38%, 31%, and 14% removal of color, COD, total phenols, and aromatic compounds, respectively.	[101]
Synthetic wastewater contain Zn21 and Cd21	Adsorption: magnetic hydroxyapatite nanoparticles	Lab scale	Maximum adsorption capacities towards Zn21 and Cd21 ions are 2.151 and 1.964 mmol/g, respectively	[102]
Municipal wastewater	Ag-NP incubation in aerated and nonaerated activated sludge	Pilot	Results from batch experiments suggested that Ag-NP transformation to Ag2S occured in the nonaerated tank within less than 2 h.	[103]
Synthetic wastewater containing Pb21 ions	Adsorption: carboxylated cellulose nano fibrils-filled magnetic chitosan hydrogel beads	Lab scale	Proven effective in eliminating Pb (II) ions	[104]
Synthetic wastewater treatment plant (WWTP) simulating biological treatment, received 10 μg/L Ag and 100 μg/L TiO2 nanoparticles (NPs) for five weeks.	Ecotoxicological effects of the transformed NPs in the effluent were assessed using a battery of marine	Lab scale	TiO_2 aggregates were detected in the effluent, whereas Ag NPs (0.1–0.22 μg/L) were associated with S, Cu, Zn. Fractionation showed that > 80% of Ag and Ti were associated with the effluent solids	[105]

(*Continued*)

Table 8.1 (Continued)

Contaminated water source	Treatment processes	Scale	Standards achieved	References
Synthetic wastewater containing As (V) and Cr(VI) ions	Adsorption: CNF composite	Lab scale	The CNF derivatives demonstrated efficient uptake of carcinogenic arsenic and chromium from the mixed contaminated waters, a distinct improvement compared to commercially available materials	[99]
Coking wastewater	Adsorption and biodegradation with the help of a modified biocarrier	Lab scale	Simultaneous nitrification/denitrification were reached with the removal of up to 322 mg/L (98%) NH4, 311 mg/L (99%) NO2, and 633 mg/L (97%) total nitrogen	[106]

References

1 Adeleye, A.S., Conway, J.R., Garner, K., Huang, Y., Su, Y., and Keller, A.A. (2016). Engineered nanomaterials for water treatment and remediation: Costs, benefits, and applicability. *Chem. Eng. J.* 286: 640–662.

2 El-Latif, M.M.A., Ibrahim, A.M., Showman, M.S., and Hamide, R.R.A. (2013). Alumina/iron oxide nano composite for cadmium ions removal from aqueous solutions. *Int. J. Nonferrous Metall.* 2: 47.

3 Sa, R.M. and Premalatha, M. (2016). Applications of nanotechnology in waste water treatment: A review. *Imp. J. Interdiscip. Res* 2 (11), 1500–1511.

4 Lingamdinne, L.P., Koduru, J.R., Chang, -Y.-Y., and Karri, R.R. (2018). Process optimization and adsorption modeling of Pb(II) on nickel ferrite-reduced graphene oxide nano-composite. *J. Mol. Liq.* 250: 202–211.

5 Davies, J.C. (2006). *Managing the Effects of Nanotechnology*. Washington, DC: Woodrow Wilson International Centre for Scholars on Emerging Nanotechnology.

6 Theodore, L. and Kunz, R.G. (2005). Nanotechnology: Turning basic science into reality. *Nanotechnol.: Environ. Implic. Solutions.* 61–107.

7 Qu, X., Brame, J., Li, Q., and Alvarez, P.J.J. (2012). Nanotechnology for a safe and sustainable water supply: Enabling integrated water treatment and reuse. *Acc. Chem. Res.* 46: 834–843.

8 Qu, X., Alvarez, P.J.J., and Li, Q. (2013). Applications of nanotechnology in water and wastewater treatment. *Water Res.* 47: 3931–3946.

9 Ali, I. (2012). New generation adsorbents for water treatment. *Chem. Rev.* 112 (10): 5073–5091.

10 Yin, J. and Deng, B.L. (2015). Polymer-matrix nanocomposite membranes for water treatment. *J. Membr. Sci.* 479: 256–275.

11 Reddy, P.A.K., Reddy, P.V.L., Kwon, E., Kim, K.H., Akter, T., and Kalagara, S. (2016). Recent advances in photocatalytic treatment of pollutants in aqueous media. *Environ. Int.* 91: 94–103.

12 Turki, A., Guillard, C., Dappozze, F., Ksibi, Z., Berhault, G., and Kochkar, H. (2015). Phenol photocatalytic degradation over anisotropic TiO2 nanomaterials: Kinetic study, adsorption isotherms and formal mechanisms. *Appl. Catal. B Environ.* 163: 404–414.

13 Petronella, F., Truppi, A., Ingrosso, C., Placido, T., Striccoli, M., Curri, M.L., Agostiano, A., and Comparelli, R. (2016). Nanocomposite materials for photocatalytic degradation of pollutants. *Catal. Today*. doi: http://dx.doi.org/10.1016/j.cattod.2016.05.048.

14 Song, S., Liu, Z.W., He, Z.Q., Zhang, A.L., and Chen, J.M. (2010). Impacts of morphology and crystallite phases of titanium oxide on the catalytic ozonation of phenol. *Environ. Sci. Technol.* 44 (10): 3913–3918.

15 Higazy, A., Hashem, M., ElShafei, A., Shaker, N., and Hady, M.A. (2010). Development of antimicrobial jute packaging using chitosan and chitosan-metal complex. *Carbohydr. Polym.* 79 (4): 867–874.

16 Rai, M., Yadav, A., and Gade, A. (2009). Silver nanoparticles as a new generation of antimicrobials. *Biotechnol. Adv.* 27 (1): 76–83.

17 Hebeish, A.A., Abdelhady, M.M., and Youssef, A.M. (2013). TiO2 nanowire and TiO2 nanowire doped Ag-PVP nanocomposite for antimicrobial and self-cleaning cotton textile. *Carbohydr. Polym.* 91 (2): 549–559.

18 Martynkova, G.S. and Valaskova, M. (2014). Antimicrobial nanocomposites based on natural modified materials: Areview of carbons and clays. *J. Nanosci. Nanotechnol.* 14 (1): 673–693.

19 Li, Q.L., Mahendra, S., Lyon, D.Y., Brunet, L., Liga, M.V., Li, D., and Alvarez, P.J.J. (2008). Antimicrobial nanomaterials for water disinfection and microbial control: Potential applications and implications. *Water Res* 42 (18): 4591–4602.

20 Das, S., Sen, B., and Debnath, N. (2015). Recent trends in nanomaterials applications in environmental monitoring and remediation. *Environ. Sci. Pollut. Res.* 22 (23): 18333–18344.

21 Shrivastava, S., Jadon, N., and Jain, R. (2016). Next-generation polymer nanocomposite-based electrochemical sensors and biosensors: A review. *TrAC Trends Anal. Chem.* 82: 55–67.

22 Lee, C.G., Lee, S., Park, J.A., Park, C., Lee, S.J., Kim, S.B., et al. (2017). Removal of copper, nickel and chromium mixtures from metal plating wastewater by adsorption with modified carbon foam. *Chemosphere* 166: 203–211.

23 Ren, X., Chen, C., Nagatsu, M., and Wang, X. (2011). Carbon nanotubes as adsorbents in environmental pollution management: A review. *Chem. Eng. J.* 170: 395–410.

24 Zhang, C., Sui, J., Li, J., Tang, Y., and Cai, W. (2012). Efficient removal of heavy metal ions by thiol-functionalized superparamagnetic carbon nanotubes. *Chem. Eng. J.* 210: 45–52.

25 Rao, G., Lu, C., and Su, F. (2007). Sorption of divalent metal ions from aqueous solution by carbon nanotubes: A review. *Sep. Purif. Technol.* 58: 224–23.

26 Parham, H., Bates, S., Xia, Y., and Zhu, Y. (2013). A highly efficient and versatile carbon nanotube/ceramic composite filter. *Carbon. N. Y.* 54: 215–223.

27 Tarigh, G.D. and Shemirani, F. (2013). Magnetic multi-wall carbon nanotube nanocomposite as an adsorbent for preconcentration and determination of lead (II) and manganese (II) in various matrices. *Talanta* 115: 744–750.

28 Tang, -W.-W., Zeng, G.-M., Gong, J.-L., Liu, Y., Wang, X.-Y., Liu, -Y.-Y., et al. (2012). Simultaneous adsorption of atrazine and Cu (II) from wastewater by magnetic multi-walled carbon nanotube. *Chem. Eng. J.* 211–212: 470–478.

29 Hu, J., Chen, G., and Lo, I.M.C. (2005). Removal and recovery of Cr(VI) from wastewater by maghemite nanoparticles. *Water Res.* 39: 4528–4536.

30. Xu, P., Zeng, G.M., Huang, D.L., Feng, C.L., Hu, S., Zhao, M.H., et al. (2012). Use of iron oxide nanomaterials in wastewater treatment: A review. *Sci. Total Environ.* 424: 1–10.

31 Wang, C.B. and Zhang, W.X. (1997). Synthesizing nanoscale iron particles for rapid and complete dechlorination of TCE and PCBs. *Environ. Sci. Technol.* 31: 2154–2156.

32 Andjelkovic, I., Tran, D.N., Kabiri, S., Azari, S., Markovic, M., and Losic, D. (2015). Graphene aerogels decorated with α-FeOOH nanoparticles for efficient adsorption of arsenic from contaminated waters. *ACS Appl. Mater. Interfaces* 7: 9758–9766.

33 Burks, T., Avila, M., Akhtar, F., Göthelid, M., Lansåker, P.C., Toprak, M.S., et al. (2014). Studies on the adsorption of chromium (VI) onto 3-mercaptopropionic acid coated superparamagnetic iron oxide nanoparticles. *J. Colloid Interface Sci.* 425: 36–43.

34 Diagboya, P.N., Olu-Owolabi, B.I., and Adebowale, K.O. (2015). Synthesis of covalently bonded graphene oxide _iron magnetic nanoparticles and the kinetics of mercury removal. *R. Sci. Adv.* 5: 2536–2542.

35 Hu, S., Shi, Q., and Jing, C. (2015). Groundwater arsenic adsorption on granular TiO2: Integrating atomic structure, filtration, and health impact. *Environ. Sci. Technol.* 49: 9707–1.

36 Srivastava, V., Sharma, Y.C., and Sillanpää, M. (2015). Green synthesis of magnesium oxide nanoflower and its application for the removal of divalent metallic species from synthetic wastewater. *Ceram. Int.* 41: 6702–6709.

37 Ray, P.Z. and Shipley, H.J. (2015). Inorganic nano-adsorbents for the removal of heavy metals and arsenic: A review. *RSC Adv.* 5: 29885–29907.

38 Diallo, M.S., Balogh, L., Shafagati, A., Johnson, J.H., Goddard, W.A., and Tomalia, D.A. (1999). Poly(amidoamine) dendrimers: A new class of high capacity chelating agents for Cu(II) ions. *Environ. Sci. Technol.* 33: 820–824.

39 Advincula, R.C. (2011). Engineering molecularly imprinted polymer (MIP) materials: Developments and challenges for sensing and separation technologies. *Kor. J. Chem. Eng.* 28: 1313–1321.

40 Kumar, V., Talreja, N., Deva, D., Sankararamakrishnan, N., Sharma, A., and Verma, N. (2011). Development of bi-metal doped micro- and nano multi-functional polymeric adsorbents for the removal of fluoride and arsenic(V) from wastewater. *Desalination* 282: 27–38.

41 Cortez, S., Nicolau, A., Flickinger, M.C., and Mota, M. (2017). Biocoatings: A new challenge for environmental biotechnology. *Biochem. Eng. J.* 121: 25–37.

42 Dąbrowski, A., Hubicki, Z., Podkościelny, P., and Robens, E. (2004). Selective removal of the heavy metal ions from waters and industrial wastewaters by ion-exchange method. *Chemosphere* 56: 91–106.

43 Li, Z. and Bowman, R.S. (1998). Sorption of chromate and PCE by surfactant-modified clay minerals. *Environ. Eng. Sci.* 15: 237–245.

44 Kwakye, A., Bright, C.W., Kenward, M.A., and Radecka, I. (2008). Antimicrobial action and efficiency of silverloaded zeolite X. *J. Appl. Microbiol.* 104: 1516–1524.

45 Gómez-Pastora, J., Dominguez, S., Bringas, E., Rivero, M.J., Ortiz, I., and Dionysiou, D.D. (2016). Review and perspectives on the use of magnetic nanophotocatalysts (MNPCs) in water treatment. *Chem. Eng. J* 310 (2): 407–427. http://dx.doi.org/10.1016/j.cej.2016.04.140.

46. Sharma, V.K., McDonald, T.J., Kim, H., and Garg, V.K. (2015). Magnetic graphene-carbon nanotube iron nanocomposites as adsorbents and antibacterial agents for water purification. *Adv. Colloid Interf. Sci.* 225: 229–240.

47 Cui, H.J., Shi, J.W., Yuan, B.L., and Fu, M.L. (2013). Synthesis of porous magnetic ferrite nanowires containing Mn and their application in water treatment. *J. Mater. Chem. A* 1 (19): 5902–5907.

48 Cui, H.J., Cai, J.K., Zhao, H., Yuan, B.L., Ai, C.L., and Fu, M.L. (2014b). One step solvothermal synthesis of functional hybrid gamma-Fe2O3/carbon hollow spheres with superior capacities for heavy metal removal. *J. Colloid Interface Sci.* 425: 131–135.

49 Ferroudj, N., Nzimoto, J., Davidson, A., Talbot, D., Briot, E., Dupuis, V., Bee, A., Medjram, M.S., and Abramson, S. (2013). Maghemite nanoparticles and maghemite/silica nanocompositemicrospheres as magnetic Fenton catalysts for the removal ofwater pollutants. *Appl. Catal. B Environ.* 136: 9–18.

50 Jin, Y.J., Liu, F., Shan, C., Tong, M.P., and Hou, Y.L. (2014). Efficient bacterial capture with amino acid modified magnetic nanoparticles. *Water Res* 50: 124–134.

51 Shan, C., Ma, Z.Y., Tong, M.P., and Ni, J.R. (2015). Removal of Hg(II) by poly(1-vinylimidazole)-grafted Fe3O4@SiO2 magnetic nanoparticles. *Water Res* 69: 252–260.

52 Zhang, Y.Y., Pan, B.C., Shan, C., and Gao, X. (2016c). Enhanced phosphate removal by nanosized hydrated La(III) oxide confined in cross-linked polystyrene networks. *Environ. Sci. Technol.* 50 (3): 1447–1454.

53 Booker, N., Keir, D., Priestley, A., Ritchie, C., Sudarmana, D., and Woods, M. (1991). Sewage clarification with magnetite particles. *Water Sci. Technol.* 23 (7–9): 1703–1712.

54 Gu, J.J., Jiang, W., Wang, F.H., Chen, M.D., Mao, J.Y., and Xie, T. (2014). Facile removal of oils from water surfaces through highly hydrophobic and magnetic polymer nanocomposites. *Appl. Surf. Sci.* 301: 492–499.

55 Machado, L.C.R., Lima, F.W.J., Paniago, R., Ardisson, J.D., Sapag, K., and Lago, R.M. (2006). Polymer coated vermiculite–iron composites: Novel floatable magnetic adsorbents for water spilled contaminants. *Appl. Clay Sci.* 31: 207.

56 Kanel, S.R., Greneche, J.-M., and Choi, H. (2006). Arsenic(V) removal from groundwater using nano scale zero-valent iron as a colloidal reactive barrier material. *Environ. Sci. Technol.* 40: 2045.

57 Li, X.-Q., Elliott, D., and Zhang, W.-X. (2006). Zero-Valent Iron Nanoparticles for Abatement of Environmental Pollutants: Materials and Engineering Aspects. *Crit. Rev. Solid State Mater.Sci.* 31: 111.

58 Gupta, A., Yunus, M., and Sankararamakrishnan, N. (2012). Zerovalent iron encapsulated chitosan nanospheres – A novel adsorbent for the removal of total inorganic Arsenic from aqueous systems. *Chemosphere*86: 150.

59 Mantha, R., Biswas, N., Taylor, K.E., and Bewtra, J.K. (2002). Removal of Nitroaromatics from Synthetic Wastewater Using Two-Step Zero-Valent Iron Reduction and Peroxidase-Catalyzed Oxidative Polymerization. *Water Environ. Res.*74: 280–287.

60 Babuponnusami, A. and Muthukumar, K. (2013). Treatment of Phenol-Containing Wastewater by Photoelectro-Fenton Method Using Supported Nanoscale Zero-Valent Iron, Environ. *Sci. Pollut. Res.*20: 1596–1605.

61 Ling, X., Li, J., Zhu, W., Zhu, Y., Sun, X., Shen, J., Han, W., and Wang, L. (2012). Synthesis of nanoscale zero-valent iron/ordered mesoporous carbon for adsorption and synergistic reduction of nitrobenzene. *Chemosphere*87: 655.

62 Liu, X.D., Murayama, Y., Matsunaga, M., Nomizu, M., and Nishi, N. (2005). Preparation and characterization of DNA hydrogel bead as selective adsorbent of dioxins. *Int. J. Biol. Macromol.* 35 (3–4):193–199.

63 Huo, J., Liu, H., Qu, J., Wang, Z., Ge, J., and Liu, H. (2005). Preparation and characteristic of triolein-embedded composite sorbents for water purification. *Sep. Purif. Technol.* 44: 37.

64 Huckins, J.N., Tubergen, M.W., and Manuweera, G.K. (1990). Semipermeable membrane devices containing model lipid: A new approach to monitoring the availability of lipophilic contaminants and estimating their bioconcentration potential. *Chemosphere* 20: 533–552.

65 Miretzky, P. and Cirelli, A.F. (2009). Hg(II) removal from water by chitosan and chitosan derivatives: A review. *J. Hazard Mater.* 167: 10.

66 Guo, X., Zhang, S., and Shan, X.Q. (2008). Adsorption of metal ions on lignin. *J. Hazard Mater.*151: 134.

67 Aydın, H., Bulut, Y., and Yerlikaya, C. (2008). Removal of copper (II) from aqueous solution by adsorption onto low-cost adsorbents. *J. Environ. Manage.* 87 (1):37–45.

68 Mohammad, A.W., Teow, Y.H., Ang, W.L., Chung, Y.T., Oatley-Radcliffe, D.L., and Hilal, N. (2015). Nanofiltration membranes review: Recent advances and future prospects. *Desalination* 356: 226–254.

69 Xi, Y., Du, Y., Zhang, X., He, A., and Xu, Z. (2017). Nanofiltration membrane with a mussel-inspired interlayer for improved permeation performance. *Langmuir*33: 2318–2324.

70 Jiang, S., Li, Y., and Bradley, P.L. (2017). A review of reverse osmosis membrane fouling and control strategies. *Sci. Total Environ.*595: 567–583.

71 Choi, W., Choi, J., Bang, J., and Lee, J.H. (2013). Layer-by-layer assembly of graphene oxide nanosheets on polyamide membranes for durable reverse-osmosis applications. *ACS Appl. Mater. Interfaces.*5: 12510–12519.

72 Chan, W.F., Chen, H., Surapathi, A., Taylor, M.G., Shao, X., Marand, E., et al. (2013). Zwitterion functionalized carbon nanotube/polyamide nanocomposite membranes for water desalination. *ACS Nano*7: 5308–5319.

73 Kim, H.J., Choi, K., Baek, Y., Kim, D.G., Shim, J., Yoon, J.et al. (2014). High-performance reverse osmosis CNT/polyamide nanocomposite membrane by controlled interfacial interactions. *ACS Appl. Mater. Interfaces.*6: 2819–2829.

74 Ma, L., Dong, X., Chen, M., Zhu, L., Wang, C., Yang, F., et al. (2017). Fabrication and water treatment application of carbon nanotubes (CNTs)-based composite membranes: A review. *Membranes*7: 16.

75 Mohammad, K., Peyravi, M., and Jahanshahi, M. (2017). The potential of nanoparticles for upgrading thin film nanocomposite membranes-a review. *J. Memb. Sci. Res.*3: 2–12.

76 Miller, D.J., Dreyer, D.R., Christopher, W.B., Paul, D.R., and Freeman, B.D. (2017). Surface modification of water purification membranes. *Angew. Chem. Int. Ed*56: 4662–4711.

77 Gopakumar, D.A., Pasquini, D., Henrique, M.A., Morais, L.C., Grohens, Y., and Thomas, S. (2017). Meldrum's acid modified cellulose nanofiber-based polyvinylidene fluoride microfiltration membrane for dye water treatment and nanoparticle removal. *ACS Sustain. Chem. Eng.*5: 2026–2033.

78. Rana, D. and Matsuura, T. (2010). Surface modifications for antifouling membranes. *Chem. Rev.*110: 2448–2471.

79 Tang, C.Y., Zhao, Y., Wang, R., Hélix-Nielsen, C., and Fane, A.G. (2013). Desalination by biomimetic aquaporin membranes: Review of status and prospects. *Desalination*308: 34–40.

80. Ashbolt, N.J. (2004). Microbial contamination of drinking water and disease outcomes in developing regions. *Toxicology*198: 229–238.

81 Rai, M., Yadav, A., and Gade, A. (2009). Silver nanoparticles as a new generation of antimicrobials. *Biotechnol. Adv.*27: 76–83.

82 Malik, R., Tomer, V.K., Kienle, L., Chaudhary, V., Nehra, S.P., and Duhan, S. (2018). Ordered mesoporous Ag-ZnO@g-CN nanohybrid as highly efficient bifunctional sensing material. *Adv. Mater. Interfaces*5: 1701357.

83 Prabhu, S. and Poulose, E.K. (2012). Silver nanoparticles: Mechanism of antimicrobial action, synthesis, medical applications, and toxicity effects. *Int. Nano Lett.*2: 32.

84 Morones, J.R., Elechiguerra, J.L., Camacho, A., Holt, K., Kouri, J.B., Ramírez, J.T., et al. (2005). The bactericidal effect of silver nanoparticles. *Nanotechnology*16: 2346.

85 Upadhyayula, V.K.K., Deng, S., Mitchell, M.C., and Smith, G.B. (2009). Application of carbon nanotube technology for removal of contaminants in drinking water: A review. *Sci. Total Environ.*408: 1–13.

86 Dizaj, S.M., Lotfipour, F., Barzegar-Jalali, M., Mohammad, H.Z., and Adibkia, K. (2014). Antimicrobial activity of the metals and metal oxide nanoparticles. *Mater. Sci. Eng.*44: 278–284.

87 Kudhier, M. A., Sabry, R. S., Al-Haidarie, Y. K. & AL-Marjani, M. F. (2018). Significantly enhanced antibacterial activity of Ag-doped TiO_2 nanofibers synthesized by electrospinning. *Materials Technology* 33 (3): 220–226. DOI: 10.1080/10667857.2017.1396778

88 Epifani, M., Giannini, C., Tapfer, L., and Vasanelli, L. (2000). Sol_gel synthesis and characterization of Ag and Au nanoparticles in SiO2, TiO2, and ZrO2 thin films. *J. Am. Ceram. Soc.*83: 2385–2393.

89 Padmavathy, N. and Vijayaraghavan, R. (2008). Enhanced bioactivity of ZnO nanoparticles-an antimicrobial study. *Sci. Technol. Adv. Mater.*9: 035004.

90 Azam, A., Ahmed, A.S., Oves, M., Mohammad, S.K., Habib, S.S., and Memic, A. (2012). Antimicrobial activity of metal oxide nanoparticles against Gram-positive and Gram-negative bacteria: A comparative study. *Int. J. Nanomed.*7: 6003–6009.

91 Karlsson, H.L., Cronholm, P., Hedberg, Y., Tornberg, M., Battice, L.D., Svedhem, S., et al. (2013). Cell membrane damage and protein interaction induced by copper containing nanoparticles-Importance of the metal release process. *Toxicology*313: 59–69.

92 Perreault, F., Faria, A.F.D., Nejati, S., and Elimelech, M. (2015). Antimicrobial properties of graphene oxide nanosheets: Why size matters. *ACS Nano*9: 7226–7236.

93 Kang, S., Herzberg, M., Rodrigues, D.F., and Elimelech, M. (2008). Antibacterial effects of carbon nanotubes: Size does matter. *Langmuir*24: 6409–6413.

94 Ong, C.S., Goh, P., Lau, W., Misdan, N., and Ismail, A.F. (2016). Nanomaterials for biofouling and scaling mitigation of thin film composite membrane: A review. *Desalination*393: 2–15.

95 Kwak, S.-Y., Kim, S.H., and Kim, S.S. (2001). Hybrid organic/inorganic reverse osmosis (RO) membrane for bactericidal anti-fouling. 1. Preparation and characterization of TiO2 nanoparticle self-assembled aromatic polyamide thin-film-composite (TFC) membrane. *Environ. Sci. Technol.*35: 2388–2394.

96 Kim, S.H., Kwak, S.-Y., Sohn, B.-H., and Park, T.H. (2003). Design of TiO2 nanoparticle self-assembled aromatic polyamide thin-film-composite (TFC) membrane as an approach to solve biofouling problem. *J. Memb. Sci.*211: 157–165.

97 Stoller, M. (2009). On the effect of flocculation as pretreatment process and particle size distribution for membrane fouling reduction. *Desalination*240: 209–217.

98 Stoller, M. and Bravi, M. (2010). Critical flux analyses on differently pretreated olive vegetation waste water streams: Some case studies. *Desalination*250: 578–582.

99 Dwivedi, A.D., Sanandiya, N.D., Singh, J.P., Husnain, S.M., Chae, K.H., Hwang, D.S., et al. (2017). Tuning and characterizing nanocellulose interface for enhanced removal of dual-sorbate (AsV and CrVI) from water matrices. *ACS Sustain. Chem. Eng.*5: 518–528.

100 Costa, J.C. and Alves, M.M. (2013). Posttreatment of olive mill wastewater by immobilized TiO2 photocatalysis. *Photochem. Photobiol.*89: 545–551.

101 Nogueira, V., Lopes, I., Freitas, A.C., Rocha-Santos, T.A.P., Gonçalves, F., Duarte, A.C., et al. (2015). Biological treatment with fungi of olive mill wastewater pre-treated by photocatalytic oxidation with nanomaterials. *Ecotoxicol. Environ. Saf.*115: 234–242.

102 Feng, Y., Gong, J.-L., Zeng, G.-M., Niu, Q.-Y., Zhang, H.-Y., Niu, C.-G., et al. (2010). Adsorption of Cd (II) and Zn (II) from aqueous solutions using magnetic hydroxyapatite nanoparticles as adsorbents. *Chem. Eng. J.*162: 487–494.

103 Kaegi, R., Voegelin, A., Sinnet, B., Zuleeg, S., Hagendorfer, H., Burkhardt, M., and Siegrist, H. (2011). Behavior of metallic silver nanoparticles in a pilot wastewater treatment plant. *Environ. Sci. Technol.*45 (9): 3902–3908.

104 Zhou, Y., Fu, S., Zhang, L., Zhan, H., and Levit, M.V. (2014). Use of carboxylated cellulose nanofibrils-filled magnetic chitosan hydrogel beads as adsorbents for Pb(II). *Carbohydr. Polym.*101: 75–82.

105 Georgantzopoulou, A., Carvalho, P.A., Vogelsang, C., Ndungu, K., Booth, A.M., Thomas, K.V., and Macken, A. (2018). Ecotoxicological effects of transformed silver and titanium dioxide nanoparticles in the effluent from a lab-scale wastewater treatment system. *Environ. Sci. Technol.*52 (16): 9431–9441.

106 Ahmad, M., Liu, S., Mahmood, N., Mahmood, A., Ali, M., Zheng, M., et al. (2017). Synergic adsorption _biodegradation by an advanced carrier for enhanced removal of high-strength nitrogen and refractory organics. *ACS Appl. Mater. Interfaces*9: 13188–13200.

107 Baun, A., Hartmann, N.B., Grieger, K., and Ole, K.K. (2008). Ecotoxicity of engineered nanoparticles to aquatic invertebrates: A brief review and recommendations for future toxicity testing. *Ecotoxicology*17: 387–395.

9

Biotechnology Intercession in Phytoremediation

CHAPTER MENU

9.1 Introduction

Biochemics and genetic engineering has become an important goal because understanding the factors of critical metabolism in plant species can improve knowledge of the molecular genetic model organisms. A fundamental insight into the uptake and translocation mechanisms of crops and metal hyperaccumulators and their regulatory control and tissue-specific promoters is highly promising to enable scientists to implement efficient metal phytoremediation plants through use of the tools of molecular biology.

9.2 Genetically Engineered Plants and Phytoremediation

Plants such as "*Nicotiana tabacum*", "*Populus agustifoli*" or "*Silene cucubalis*" produced deep metals compared to the corresponding recessive homozygous plant species. They are genetically modified to over-express "glutamy cysteine syntlietase" [1]. Increased nutrient plants that have endogenous phytoremediating potential should be considered for genetic manipulation in short- and long-term plants (trees). The Indian mustard ("*B. juncea*"), sunflowers ("*H. anuus*"), tomato (*L.esculentum*) and poplar (*L.telipifera*) are

Aquatic Environmental Bioengineering: Monitoring and Remediation of Contamination, First Edition.
Rouf Ahmad Bhat, Mohammad Yaseen Mir, Gowhar Hamid Dar and Moonisa Aslam Dervash.

some biomass accumulators that have already been regenerated by the use of a hyper-bushing system [1].

A number of genes inclined to participate in transitting metal oxides can be detected with innovative molecular and genetic techniques. Recent advances in the detection of the plant genes of hyperaccumulator plants representing shuttles of cations of metal and their homology sequence have been significant. Therefore, it is hoped genetic engineering will provide an effective new way to improve plant capacity to repair contaminants [1]. "*B. juncea*" is genetically engineered in order to study limiting "glutathione" and "phytochelatin" development factors. This was accomplished by the *E.coli* "gshl" gene. Transgenic cadmium tolerance of the "β-CES" soils is higher than in wild seedlings with higher levels of "phytochelatin", "α-GluCys", "glutathione" and totally "non-protein-Thiol" [2].

The study revealed that "L-buthionine-sulfoximine (BSO)", an inhibitor of "glutamyl-cysteine synthetase", increases resistance significantly in both unadapted and arsenic-hypertolerant plants, and shows that "phytochelatine sequestration" is important both for usual constituent tolerances and adaptive metalloids [3, 2]. Some genes were isolated from, and incorporated into, improved heavy metal (Cd) susceptibility and penetration of plants, including "AtNramps", "AtPkrs", and "CAD1" from "*A. thaliana*", a C-induced "*Datura innoxia*" CD-induced library, "gshI" from "*B. juncea*" [4–6].

Multifaceted examples of transgenic metal tolerant plants are available, such as tobacco with cadmium accumulation through "Zn/Fe–regulated transporters" (vacuolar transporter) "CAX gene" as shown in Figure 9.1 [7]; "copper" and "sodium" tolerant with tobacco "GST gene" (Ezaki et al., 2000); "tobacco with nickel tolerance" and lead accumulation with tobacco Nt "CBP4 gene" [8], and rice [9] with increased iron accumulation with soybean "Ferretin gene"; "mercury resistant" and tobacco with bacterial

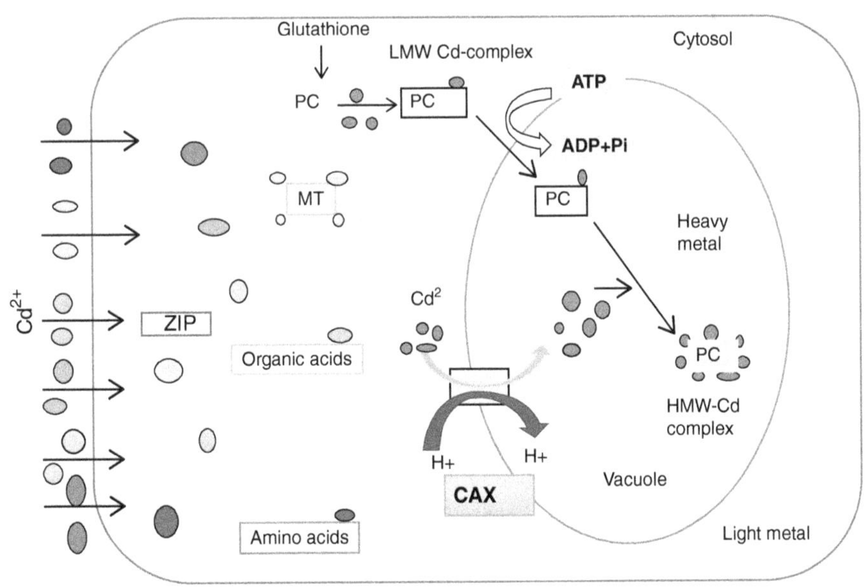

Figure 9.1 Vacuolar sequestration of HMs in plants.

"merA" gene from *A.thaliana* [10], selenium-tolerant indian mustard transformed with bacterial "glutathione reductase" in both cytoplasm and chloroplasm [11] and transgenic plants expressing "SRSIp/ArsC". The tolerance of arsenium for "ACT 2p/Ţ-ECS" was 4 to 17 times greater than that of "α-ECS"or "ArsC" alone, with the fresh shoot being 4 to 17 times larger and 2- to 3-fold hyperarsenic accumulation per gram of tissue. Most, if not all, previously produced transgenic plants use over-expression of metal binding protein and peptide biosynthesis traces, while there is a range of different metal tolerance mechanisms and several transgenic plants that have trace metal tolerance and accretion [1].

At least three distinct engineering protocols for improved metal retention including an increase in the number of reception sites, changes of the reception mechanism in order to minimize unnecessary cation competition and the diversification of intracellular binding facilities can be envisaged. Each metal has its own mechanism for molecular absorption, transport and sequestration [12]. In hyperaccumulation or plant phytovolation plants, new metabolic pathways may be entered in plants with multiple mercury fold-tolerant plates and with transgenic transgenic "*Arabidopsis*" plants containing volatile elementary mercury, which can carry oxygen arsenic to the earth, reduce it to arsenite and secrete it by the transfer of *E.coli*. Altering the enzyme associated with oxidative stress can also result in an improved metal tolerance due to the gene expression of "glutathione-S-transferase" and "peroxidase" [13].

"Aminocyclopropane-1-carboxyl-acide (ACC) deaminase" overextualization resulted in increased metal accumulation [14]. The effective absorption of toxic metals is critical for plants with highly connected root systems on the wide surface [10, 11]. Experiments have found that in certain hyperaccumulatory plants *Agrobacterium rhizogenes* can boost the root biomass. Some hyperaccumulators show high efficiency of the hairy roots for rhizofiltration [15] and heavy metals [16].

The plants' metabolism, mostly related to their enzymes and protein profiles in the field of phytoremediation, is of great importance since the technology can use techniques of genetic engineering to manipulate molecular and cellular plant detoxification mechanisms. Plants also benefit because, over time, they can develop complicated processes to absorb nutrients, detoxify and control toxins locally. They benefit from their presence. The control of the water content of the soil by plants is of major significance in avoiding fluid infiltration, the main input mechanism for pollutants. Soil is enhanced by plant roots and aerated to improve the efficiency of microbial nutrients. Phytoremediation in particular provides more visual impact than some other physical remediation technologies. Phytoremediation, on the other hand, has certain disadvantages that require stringent plant and soil conditions research. One major drawback is that the detoxification approach is too slow or seasonally efficient.

Regulators are also expecting significant progress in remediation work in just a few years' time, making most phytoremediation unacceptable. Pollutant concentrations such as "trichloroethylene" and "carbon tetrachloride" in some cases are not decreased adequately. In addition, pollutants may exceed phytotoxic levels in certain polluted areas which invalidate the plant. Recent studies have been conducted to improve genetic engineering for plant phytoremediation [17]. Multi-metal contaminated soil and water require unique plant remediation metal hyperaccumulators, and therefore a large number of pre-implementing tests. The efficacy of phytoremediation can also be linked to other factors, including the presence of a biomass contaminant. If the metal is closely attached to the organic soil, the plant would not have access to it. In addition, some contaminants are unique to plants. In order to prevent polluted areas from detoxifying with different contaminants, "cadmium" and "zinc" ("*Thlaspi caerulescens*") may be susceptible to other metals such as copper [15, 16]. Soil structural construction technology is used globally, and many investigators face major challenges.

9.3 Qualitative Phytoremediators

The ideal phytoremediator would consist of the capacity for biomass degradation or concentration of contaminants; the ability to absorb huge soil water volumes; and the ability to grow fast and high levels of biomass [18]. While a number of species in certain contaminated areas can be tolerated and cultivated, they usually develop at a very slow rate, produce extremely low levels of biomass and are adapted to very specific environmental conditions. And for trees with large root systems, high biomass and low inputs, pollutants are poorly tolerated and accumulated. There is, therefore, a breach of the requirements of successful phytoremediators [19]. The remediation capacity of plants can be greatly improved by genetic manipulation and plant transformation technologies [20], introducing a new approach to the development of enhanced pest-remediating plants in high-biomass plants for the recovery and accumulation of pollutants [21]. This chapter examines some of the research efforts in this field and highlights future challenges.

9.4 Biotechnology in Plant Mediated Remediation for Contaminants

In plant substrates, the fates of the various pollutants are different so that they have different rates of plant remediation factors, which can be used with success for genetics [22]. The underlying mechanism must be understood in order to use a transgenic approach in raising plants with higher phytoremediation properties. The specific membrane transporters or responsible enzymes can be over-expressed by physiological and biochemical experimentation as a result of possible rate limitation measures. If any organism contains the genes encoding these features, they can be placed in the plant. Transgenic substances can be compared with wild types in the context of pollutant remedies [21].

9.5 Toxic Metals (TMs)

Metals are found in the earth and atmosphere as a natural element and encourage the release and concentration into the environment of these metals, which lead to toxicity including industry, mineral extraction, car-transportation and agriculture [23]. The inorganic pollutants mainly consists of "nitrate" and "phosphate" including metals such as

"As", "Cd", "Cu", "Hg", "Mn", "Se" and "Zn". All ions occur in nature primarily as positive or negative ions loaded by carriers. Inorganically, in some cases, they can be modified, transported to/from plants or, however, volatilized and not degraded. The existing phytoremediation methods in inorganics therefore consist of the stabilization process (phytostabilization), sequestration and phytovolatilization of harvesting plant tissue. According to a report [24], biotechnological approaches that have successfully altered plant capacity in inorganic phytoremedies have focused on tolerance and accumulation. The target genes comprise both genes for the metal conveyor and genes for chelator production. In the case of items that can be volatilized [22], genes that facilitate the conversion into volatile forms were also over-expressed. The following section highlights three inorganics: As, Hg and Se.

9.5.1 Arsenic (As)

From rocks and soil, arsenic is discharged into subterranean seas. Drinking water is contaminated and can result in cancer of the skin, kidneys and also gangrene. In addition, high concentrations of arsenic in agricultural land degrades soil, reduces crops and introduces contaminants into the alimentary chain. Bangladesh is the world's most highly populated poisoned country, described by the World Health Organization (WHO) as a problem that threatens up to 40 million people with arsenic contamination [24]. Arsenic species remain on the surfaces of agricultural soils. Many studies have demonstrated that AsV is an analog phosphate in plants by phosphate uptake systems [25]. The transport of phosphates in *A.thaliana* was shown to be involved in "AsV" uptake. "AsV" represses the hunger reaction genes of phosphate and suggests that AsV interferes with phosphate sensing, altering the phosphate signaling mechanism [26]. *A. thaliana* has nine specific protein phosphate transporters, and different PHTs may differ in the affinity of arsenic.

Various transgenes have been developed in order to improve tolerance and aggregation [27]. The synthesis of phytochelatins or GSH precursors considerably boost tolerance [28]. These studies have shown that vastly increased GSH and PC synthesis on its own is not helpful to enhance the As structure. Both "β-ECS" and co-expression "PCS" in *Arabidopsis* had a larger effect on As tolerance and accretion than the expression of the two genes solely. Changing the GSH and PC levels in plants therefore provides an efficient strategy for increasing plant tolerance and can be used in new plants with high fittings. Shooting build-up was conducted with two bacterial genes [29] co-expressing strongly as tolerant transgenic plants and as enhanced. In the leaves of the small soybean promoter unit "1 (SRS1)"; "RuBisCo"; the *E. coli* reductase arsenate gene, "ars C", was expressed. In addition, μ-ECS was expressed in the form of a powerful actin 2 promoter, both in roots and in shoots [29]. Any double transgenic plants were relatively susceptible compared to plants with μ-ECS alone. Moreover, biomass and three times hyperaccumulative than over ground biomass, were approximately seventy times higher in dual transgenic plants compared with wild plants with 125 microns of sodium arsenate. This work represents an excellent confirmation of the notion for transgenic plant phytoremediation in groundwater sources. The interpretation of arsC in the leaf increases slightly the arsenate significant decrease, while overexpression of μ-ECS tends to increase thiol-rich peptides with AsIII biosynthetic pathway.

9.5.2 Mercury (Hg)

Mercury is an extremely toxic substance found naturally as well as an imported contaminant and poses a major environmental threat. Mercury poisoning is deleterious to foetal development, to the immune system and the nervous system [30]. Terrestrial plantings are normally expected to the harmful effects of synthetic chemicals, but mercury does influence chloroplasts and mitochondria photosynthesis and oxidative metabolism through interference in electron transport. Mercury also inhibits the activity of aquaporins, and reduces the absorption of plant water [31]. The plants require no mercury and are simply inadequate in their biogeochemistry. To date there seems to be no recognized organic plant species with considerable Hg absorption, depletion or reduction. Numerous electrode potential enzymes such as catalase and peroxidase convert numerous species of plants to moderate concentrations of "Hg(II)" to "Hg(0)." "Hg (0)" is released either through the roots or by atmospheric shooting. "Hg (II)" is chemically toxic, and tends to fasten sulfa hydride groups that contain enzymes and ensures quality chemicals with decreased thiols. Thiol-based "Hg (II)" enzyme and associated proteins are generally nonvacuolized, but their statements to specific enzymes and proteins can demolish their interaction.

The tested plants could already effectively inactivate or integrate extremely poisonous methyl mercury into less environmentally hazardous forms. As discussed earlier, the genes that embed BMT were well-characterized [32] and laid the molecular genetic foundations for increased mercury tolerance in plants. In the early 1990s, Richard et al. initiated a framework for the development of mercury deactivation and disinfection plants. In the plant engineering of the Hg process and remediation system [33], the above two bacterial types of the well-characterized marine operon, "merA" and "merB" were used [33]. The adapted "MerA" was, in fact, compared with non-transgenic controls of "*A.thaliana*" [33], "Tobacco" [34], "Yellow poplar"; "cottonwood" and "rice" [34], at least ten times greater in the "Hg (II)" concentrations. The volatilization levels of these transgenic plants were significant in comparison with the controls of "Hg(0)". Chloroplasm and endoplasmic reticulum (ERs) have been shown to be important goals of Hg poisoning [35]. Thus the Hg toleration and chloroplast detoxifying systems used in detoxifying Hg may be high by including "merA" and "merB" genes in the chloroplast genomes. The organomercurial compound "phenylmercuric acetates (PMA)" in the presence of PMA or $HgCl_2$ had an increased pain threshold of transgenic tobacco plants, with mercury accruing 100–104 times more than in unprocessed plants [36] in the shooting. Therefore, chloroplast technology can also prove a cost-effective means of repairing mercury plants.

9.5.3 Organic Pollutants (OPs)

OPs in the ecosystem are typically xerobiological. Many of them are toxic and/or carcinogenic. Organic sources of pollutants in the environment include accidents (fuel consumption, solvents), intensive agriculture and industrial processes (e.g. chemical, petrochemical) and detonators as well as chemical weapons. Moreover, in contaminated sites there is often a mixture of toxic pollutants [37]. The potency of comprehensive pollutant degradation is offered by organic pollutant phytoremediation if a plant can use that chemical and when all of the genes necessary for biodegradation are present. Most organic pollutants must be overcome with phytotoxic effects to be effective. There are various organic contaminant categories: solvents (e.g. "trichloroethylene"); polycyclic, aromatic hydrocarbons (e.g. naphthalene; or "pyrene"); petroleum products including "benzene", "toluene", "ethylbenzene" and "xylene" (BTEX); "polychlorinated byphenyls", "pyramide", "trichloroethylene") and "explosives".

9.5.4 Pesticides

Because severe human deformities have resulted from the use of pesticides and because their environmental pollution is generally reduced, multiple methods have been used to remove these types of pollutants, including incineration and landfilling; however, these physical processes are both costly and inefficient. Some viable methods are arising in bioremediation using microorganisms that are able to decrease pollutant pesticide use and greatly improve pesticide phytoextraction using transgenic plants [38]. Pesticides are being improved through genetic transformation technology. A study [34] has expressed a six-step path in transgenic plants; the "atzA" gene which codes the first enzyme, "atrazine chlorohydrolase". The active expression of "atzA" is due to transgenic tobacco, *arabidopsis* and lubricants, resulting in greater tolerance to various atrazine concentration levels. In all the vascular plants, the pesticide has been hydroxyatrazine dechlorinated. Another way of doing things is to express "CYP1A1" and "CYP1A2" mammalian cytochrome P450 genes, which results in enhanced metabolism of atrazine [39]. In genetically modified rice plants co-expriming "CYP1A1", "CYP2B6" and "CYP2C19" [40], profound improvements were made to the metabolic processes of a wide variety of herbicides including atrazine and metolachlor. Eight different herbicides had a strong tolerance of transgenic plants. While atrazine has killed control plants, inhibiting photosynthesis, transgenic plant growth has not been affected. Genetically modified rice plants expelled two times more herbicides than the control plants after a week for the remediation of contaminated surface water.

9.5.5 Oil Spills (OSs)

A multifactor problem currently ravaging oil-producing communities worldwide, involves environmental pollution caused by the spillage of oil. This causes loss of diversity of species and habitat, destroys aquaculture and sometimes kill organisms and human beings. This problem has a major impact. Environmental degradation due to oil spill has a negative effect on the lives of the oil-producing communities as they can no longer maintain water in their rivers or other water bodies. They are also unable to drink or to swim in their rivers after the pollution by oil and therefore their social life is affected [41]. Petroleum particulates including the hydroxyl groups, and the "aromatic benzene" can be remediated by plants; low level "toluene", low level "ethylbenzenene" and "xylene" (BTEX). Petroleum-degrading bacteria are usually observed in the rhizose seedlings nurtured on sites contaminated with these pollutants [42], and a few hundred pollutant demeaning endocrine-pollutants may focus on improving pest control [43].The "BTEX" degradation component genes, "toluene", were transferred to the endophytic strain and inoculated on euphorbia for the aim of expanding biosorption of formaldehyde contaminants [44]. The toluene concentrations in the inoculated plants were tolerated by ten times the usual phytotoxicity. Poplar triggered plasmid to be somewhat tolerant to toluene by indoctrinating the original toluene degrading strain into the most suitable restaurant in plants [45]. The existence of endophytic fungi also reduced the phytotranspiration of the chemical. In transgenic plants expressing mammalian "cytochrome P450 2E1", the removal rate for "toluene" and "benzene" was also increased significantly [46]. Toluene was excluded from the hydroponic solution inside of two days ten times faster than in vector control systems. Tobacco "Transgenic CYP2E1" almost completely deleted benzene within three days, whereas the vector-control plants deleted unharvested benzene [47–50].

9.6 Conclusion and Future Prospects

Phytoremediation is an expensive technology that mostly deteriorates, reduces, incorporates or prevents water and soil toxins from plant species. The whole research has rapidly garnered praise, as an inventive and premium substitute toward the more sophisticated therapeutic approaches used throughout dumpsites. Therefore, in the recuperation of impacted habitats the most potent and expense exfoliating innovations are crucial. There have been huge efforts to properly assess and detoxify species of plants in those regions. The pathways for the uptake, accretion and denigration of pollutants, among many other issues, diverge per the species of plants but are very crucial factors in determining their unique function in plant restorative actions. Plants can dramatically increase disinfection potentials by exploring and translating specific genes for certain enticing microbes to many other life forms using gene editing technology. It can play a key role when working to improve expenses to make saturated sites extra profitable. The early phase of PHI is still in progress, and genetic modification phytoremediation screening at the concession stands is very limited. Bio-safety misgivings must be adequately addressed and interventions to manage gene flows of wildlife formulated. The technology is particularly only obtainable for a few particulate matter and several areas are infected. In order to fulfill the demands of individual sites, multiple stacked genes should be designed. Of course, in plant biotechnology methodologies, phytoremediation plays an essential role. Genetic engineering is very convincing for establishing transgenic plants with high capacity for cost effective, smooth, affordable and viable bioremediation. The molecular basis of pathways to degrade organic material must be better understood. It is important to even further evaluate and detect genes suitable for phytoremediation. In order to be successfully accepted by the water treatment industry, it is essential that the correct transgenesis components proceed to be evaluated in the area. The use of genetic modification products to make use of phytoremediation is much less burdensome when regulatory restrictions can indeed be frequently renegotiated.

References

1 Mello-Farias, P. and Chaves, A.L. (2008). Advances in Agrobacterium-mediated plant transformation with enphasys on soybean. *Scientia Agricola*, 65: 95–106.

2 Fulekar, M., Singh, A. and Bhaduri, A.M. (2009). Genetic engineering strategies for enhancing phytoremediation of heavy metals. *Afr. J. Biotechnol.* 8: 529–535.

3 Schat, H., Llugany, M. and Bernhard, R. (2000). Metal-specific patterns of tolerance, uptake, and transport of heavy metals in hyperaccumulating and nonhyperaccumulating metallophytes. In: Terry, N., Bañuelos, G (eds). *Phytoremediation of contaminated soil and water*. Boca Raton, FL, USA: Lewis Publishers, 171–188.

4 Thomine, S., Wang, R., Ward, J.M., Crawford, N.M. and Schroeder, J.I. (2000). Cadmium and iron transport by members of a plant metal transporter family in *Arabidopsis* with homology to *Nramp* genes. *Proceedings of the National Academy of Sciences* 97: 4991–4996.

5 Song, L., Singleton, E., Hill, J. and Koh, M. (2004). Improving online learning: Student perceptions of useful and challenging characteristics. *The Internet and Higher Education*, 7(1): 59–70.

6 Ha, S.B., Smith, A.P., Howden, R., Dietrich, W.M. and Bugg, S. (1999). Phytochelatin synthase genes from *Arabidopsis* and the yeast, *Schizosaccharomyces pombe*. *Plant Cell*, 11: 115364.

7 Hirschi, K.D., Korenkov,V.D., Wilganowski, N.L. and Wagner, G.J. (2000). Expression of *Arabidopsis* CAX2 in Tobacco. Altered Metal Accumulation and Increased Manganese Tolerance. *Plant Physiology*, 124: 125–133.

8 Arazi, T., Sunkar, R., Kaplan, B. and Fromm, H. (1999). A tobacco plasma membrane calmodulin-binding transporter confers Ni^{2+} tolerance and Pb^{2+} hypersensitivity in transgenic plants. *Plant J.* 20: 171–182.

9 Goto,H., Kosako, H., Tanabe, K., Yanagida, M., Sakurai, M., Amanoi,M., Kaibuchii,K. and Inagaki, M. (1999). Phosphorylation of Vimentin by Rho-associated Kinase at a Unique Amino-terminal Site That Is Specifically Phosphorylated during Cytokinesis. *The Journal of Biological Chemistry* 273: 11728–11736.

10 Eapen,S. and D'Souza, S. (2005) .Prospects of genetic engineering of plants for phytoremediation of toxic metals. *Biotechnol Adv* 23: 97–114.

11 Eapen, S., Singh, S. and D'Souza, S. (2007). Advances in development of transgenic plants for remediation of xenobiotic pollutants. *Biotechnol Adv* 25: 442–51.

12 Clemens, S., Palmgreen, M.G., Kramer, U. (2002). A long way ahead: understanding and engineering plant metal accumulation. *Trends Plant Sci.* 7: 309–315.

13 Ezaki, B., Gardner, R.C., Ezaki, Y. and Matsumoto, H. (2000). Expression of aluminum-induced genes in transgenic *Arabidopsis* plants can ameliorate aluminum stress and/or oxidative stress. *Plant Physiol*, 122: 657–655.

14 Grichko, V.P. and Glick, B.R (2001). Flooding tolerance of transgenic tomato plants expressing the bacterial enzyme ACC deaminase controlled by the 35S, rolD or PRB-1b promoter. *Plant Physiol Biochem* 39: 19–25.

15 Eapen, D., Barroso, M.L., Campos, M.E., Ponce, G., Corkidi, G., Dubrovsky, J.G. and Cassab, G.I. (2003). A no hydrotropic response root mutant that responds positively to gravitropism in Arabidopsis, *Plant Physiology*, 131: 536–546.

16 Nedelkoska, T.V., Doran, P.M. (2000). Hyper accumulation of cadmium by hairy roots of Thlaspi caerulescens. *Biotechnol. Bioengineer*, 67: 607–615.

17 Danh, L.T., Paul, T., Raffaella, M., Tam, T. and Neil, F. (2009). Vetiver grass, *Vetiveria zizanioides*: a choice plant for phytoremediation of heavy metals and organic wastes', *International Journal of Phytoremediation*, 11: 8, 664–691.

18 Cherian, S. and Oliveira, M.M. (2005). Transgenic plants in phytoremediation: Recent advances and new possibilities. *Environ Sci Technol* 39: 9377–9390.

19 Gratão, L.P. and Braz, J. (2005). Phytoremediation: Green technology for the clean-up of toxic metals in the environment. *Plant Physiol* 17: 53–64.

20 Kraomer, U. (2005). Phytoremediation: Novel approaches to cleaning up polluted soils. *Curr Opin Biotechnol* 16: 133–141.

21 Martanez, M., Bernal, P., Almela, C., Vaolez, D., Garca-Agustan, P., et al. (2006). An engineered plant that accumulates higher levels of heavy metals than *Thlaspi caerulescens*, with yields of 100 times more biomass in mine soils. *Chemosphere* 64: 478–485.

22 Dhankher, O.P., Doty, S.L., Meagher, R.B., Pilon-Smits, E. (2011). Biotechnological approaches for phytoremediation. In: Altman, A., Hasegawa, P.M. (eds.), *Plant Biotechnology and Agriculture*. Oxford: Academic Press, 2011, pp. 309–328.

23 Elizabeth, A.H. and Pilon-Smith, F.J.L. (2006). Environmental cleanup using plants Biotechnolgical advances and ecological consideration. *Front Ecol Environ* 4: 203–210.

24 Pilon-Smits, E.A.H., Hwang, S., Lytle, C.M., Zhu, Y., Tai, J.C., et al. (1999). Overexpression of ATP sulfurylase in Indian mustard leads to increased selenate uptake, reduction, and tolerance. *Plant Physiology* 119: 123–132.

25 Meharg, A.A. and Acnair, M.R. (1992). Suppression of the high-affinity phosphateuptake system: A mechanism of arsenate tolerance in Holcus lanatus L. *Journal of Experimental Botany* 43: 519–524.

26 Catarecha, P., Segura, M.D., Franco-Zorrilla, J.M., Garcia-Ponce, B., and Lanza, M. (2007). A mutant of the Arabidopsis phosphate transporter PHT1;1 displays enhanced arsenic accumulation. *Plant Cell* 19: 1123–1133.

27 Li, Y., Dhankher, O., Carreira, L., Balish, R., and Meagher, R. (2005). Engineered overexpression of γ- glutamylcysteine synthetase in plants confers high levels arsenic and mercury tolerance. *Environmental Toxicology & Chemistry* 24: 1376–1386.

28 Gasic, K. and Korban, S.S. (2007). Transgenic Indian mustard (*Brassica juncea*) plants expressing an Arabidopsis phytochelatin synthase (AtPCS1) exhibit enhanced As and Cd tolerance. *Plant Mol Biol* 64: 361–369.

29 Dhankher, O.P., Li, Y., Rosen, B.P., Shi, J., Salt, D., et al. (2002). Engineered tolerance and hyperaccumulation of arsenic in plants by combining arsenate reductase and γ -glutamylcysteine synthetase expression. *Nature Biotechnology* 20: 1140–1145.

30 Nriagu, J.O. and Pacyna, J.M. (1988). Quantitative assessment of worldwide contamination of air, water and soils by trace metals. *Nature* 333: 134–139.

31 Sas-Nowosielska, A., Galimska-Stypa, R., Kucharski, Z.U., Malkowski, E., et al. (2008). Remediation aspect of mibrobial changes of plants R. rhizosphere in mercury contaminated rhizosphere in mercury contaminated soil. *Envirnomental Monitorin and Asesement* 137: 101–109.

32 Salt, D.E., Smith, R.D., and Raskin, I. (1998). Phytoremediation. *Annu Rev Plant Physiol Plant Mol Biol* 49: 643–668.

33 Rugh, C.L., Wilde, H.D., Stack, N.M., Thompson, D.M., Summers, A.O., et al. (1996). Mercuric ion reduction and resistance in transgenic Arabidopsis thaliana plants expressing a modified bacterial merA gene. *Proc Natl Acad Sci USA* 93: 3182–3187.

34 Heaton, A.C.P., Rugh, C.L., Wang, N.J., and Meagher, R.B. (2005). Physiological responses of transgenic merA tobacco (Nicotiana tabacum) to foliar and root mercury exposure. *Water, Air, & Soil Pollution* 161: 137–155.

35 Bizily, S.P., Kim, T., Kandasamy, M.K., and Meagher, R.B. (2003). Subcellular targeting of methylmercury lyase enhances its specific activity for organic mercury detoxification in plants. *Plant Physiol* 131: 463–471.

36 Ruiz, O.N., Hussein, H.S., Terry, N., and Daniell, H. (2003). Phytoremediation of organomercurial compounds via chloroplast genetic engineering. *Plant Physiol* 132: 1344–1352.

37 Bizily, S.P., Meagher, R.B. (2000). Phytoreduction of environmental mercury pollution. Pp 151–170. In: Raskin, I., Ensley, B.D. (eds). *Phytoremediation of toxic metals: using plants to cleanup the environment*. New York: Wiley.

38 Hussein, H., Ruiz, O.N., Terry, N., and Daniell, H. (2007). Phytoremediation of mercury and organomercurials in chloroplast transgenic plants: Enhanced root uptake, translocation to shoots and volatilization. *Environmental Science & Technology* 41: 8439–8446.

39 Bode, M., Stajbe, P., Thiede, B., Schuphan, I., and Schmidt, B. (2004). Biotransformation of atrazine in transgenic tobacco cell culture expressing human P450. *Pest Manag Sci* 60: 49–58.

40 Kawahigashi, H., Hirose, S., Ohkawa, H., and Ohkawa, Y. (2006). Phytoremediation of the herbicides atrazine and metolachlor by transgenic rice plants expressing human CYP1A1, CYP2B6, and CYP2C19. *J Agric Food Chem* 54: 2985–2991.

41 Ndimele, P.E. (2010). A review on the phytoremediation of petroleum hydrocarbon. *Pak J Biol Sci* 13: 715–722.

42 Siciliano, S.D., Germida, J.J., Banks, K., and Greer, C.W. (2003). Changes in microbial community composition and function during a polyaromatic hydrocarbon phytoremediation field trial. *Appl Environ Microbiol* 69: 483–489.

43 Moore, F.P., Barac, T., Borremans, B., Oeyen, L., Vangronsveld, J., et al. (2006). Endophytic bacterial diversity in poplar trees growing on a BTEX-contaminated site: The characterisation of isolates with potential to enhance phytoremediation. *Systematic and Applied Microbiology* 29: 539–556.

44 Barac, T., Taghavi, S., Borremans, B., Provoost, A., Oeyen, L., et al. (2004). Engineered endophytic bacteria improve phytoremediation of water-soluble, volatile, organic pollutants. *Nat Biotechnol* 22: 583–588.

45 Taghavi, S., Barac, T., Greenberg, B., Borremans, B., Vangronsveld, J., et al. (2005). Horizontal gene transfer to endogenous endophytic bacteria from poplar improves phytoremediation of toluene. *Appl Environ Microbiol* 71: 8500–8505.

46 James, C.A., Xin, G., Doty, S.L., and Strand, S.E. (2008). Degradation of low molecular weight volatile organic compounds by plants genetically modified with mammalian cytochrome P450 2E1. *Environmental Science & Technology* 42: 289–293.

47 Cunningham, S.D. and Berti, W.R. (1993). Remediation of contaminated soils with green plants: An overview. *In Vitro Cell Dev Biol* 29: 207–212.

48 Guo, J., Dai, X., Xu, W., and Ma, M. (2008). Overexpressing GSH1 and AsPCS1 simultaneously increases the tolerance and accumulation of cadmium and arsenic in *Arabidopsis thaliana*. *Chemosphere* 72: 1020–1026.

49 Schnoor, J.L., Licht, L.A., McCutcheon, S.C., Wolfe, N.L., and Carreira, L.H. (1995). Phytoremediation of organic and nutrient contaminants. *Environ Sci Technol* 29: 318A–323A.

50 Suresh, B. and Ravishankar, G.A. (2004). Phytoremediation–a novel and promising approach for environmental clean-up. *Crit sRev Biotechnol* 24: 97–124.

10

Biofilms in Remediation

Current Trends and Future Perspectives

CHAPTER MENU

10.1 Introduction

With the advent of microbiology, microorganisms are mainly characterized as planktonic cells and are categorized in nutritionally rich culture media according to their morphological, physiological and development characteristics. Microbes are usually present in just about all natural habitats. Moreover, they are usually found strongly associated with substrate surfaces and edges, in the shape of a mass of cells combined together with the secretions that are released by them [1]. "Biofilms" are a consortium of microorganisms that emerge at substrate surfaces and are naturally bounded by an "extracellular polymeric substance (EPS)". Their presence traces back to over 3.8 billion years and they exist everywhere on the earth today. "Biofilms" consist mainly of microorganism and exopolysaccharides (EPS) [2]. The EPS is the primary factor of biofilms that signifies resistance and survival and also serves as a barrier to diffusion, molecular sieves and biosorbents. Biofilms are produced on a wide range of substrates, such as medical

Aquatic Environmental Bioengineering: Monitoring and Remediation of Contamination, First Edition.
Rouf Ahmad Bhat, Mohammad Yaseen Mir, Gowhar Hamid Dar and Moonisa Aslam Dervash.

devices, living tissues, indoor environments, industrial or drinking water pipes and aquatic ecosystems [3]. Microbes attached to interfaces and their growth at step boundaries are commonly seen in human, technological and clinical spaces [4].

For a number of factors, biofilm production is helpful to microbes. First, the anionic EPS matrix feeds the biofilm by holding the water and binding cationic solutes [48]. This helps in the growth of biofilm, communication, and exchanging of genetic material horizontally with the microorganisms present in the vicinity [49]. Second, biofilm microorganism can flourish, interact with and exchange genetic information laterally with the microorganisms that are active in the vicinity. Eventually, bacteria in a biofilm under stress become much spongier than planktonic cells. The benefits of biofilm formation can be enhanced to proficiently and cost-effectively filter water contaminated with pollutants.

The species that are present in biofilms are mainly bacteria, algae and protozoa that are produced by a single species of bacteria. Approximately 97% of the biofilm mixture is either water or solvent, whose physical characteristics like viscosity are defined by the dissolved solvent molecules [5]. The diffusion process in the biofilm matrix is determined by its water-holding ability and mobility. In addition, microbial cells also contain secreted polymers, cell metabolites, nutrients, lysis products and debris from the environment [48]. Moreover, it is possible to observe all main classes of macromolecules like proteins, polysaccharides, DNA and RNA within a biofilm environment [4, 5]. The anionic feature of biofilms can be attributedto the presence of ketal-linked pyruvates or uronic acids including “D-glucuronic”, “D-galacturonic” and “mannuronic acids” [48]. The structure of the biofilm affects the process of mass transport which depends on the availability of substrates present in the vicinity. Because of the coexistence of various microenvironments, the biofilm matrix shows a high degree of micro-heterogenity. In comparison to other microorganisms, the variety of bacteria is immense as these form the foundation stone for certain forms of microorganisms to create their own room.

Bacterial biofilms in nearly all dimensions are surprisingly heterogeneous and can be evaluated reliably and reproducibly. These differences: structural, physiocochemical, ecological and electrical, etc., have been implied as the cause of many phenomena characteristic of the attached—as opposed to the planktonic—mode of growth. In the natural environment, pure cultures are practically nonexistent. Microorganisms survive in populations where there is a diversity of interactions between the organisms, for example, pitting corrosion of a stainless steel surface and microbially influenced corrosion (MIC) that occurs due to microbial colonization on metal surfaces. Biofilms are efficiently reactive to the environment. The detachment of bacterial cells from biofilms may be individually or in clumps. When they detach in clumps, their susceptibility to antimicrobials properties of biofilms is reduced. Biofilms can also migrate across surfaces in a variety of ways, under the right conditions over a period of time. Some of the interesting features of life in the bacterial culture are that it creates an atmosphere for bacteria to interact through chemical signals. There are possibilities that some of these chemical signals, generated by cells and transmitted via their outer membranes, can be perceived not only by representatives of the same genus, but also by certain microbial species—and in certain instances even by many more complex organisms.

10.2 Different Methods for Culturing Biofilms In Vitro

10.2.1 Static Microtiter Plate Assays

This is one of the first systematic techniques for evaluating the creation of biofilms and is also referred to as the "microtiter plate static biofilm model". This approach is built for quantification of the degree to which microorganisms bind to an abiotic surface [6]. It was developed as a method to study the coagulase-negative *Staphylococcus aureus* attachment to plastic, but has been since adapted for use in many other species. The initial binding to a surface may be measured using brief incubation periods (1–2 hours), whereas long periods of incubation (20 hours) allow for a measure of the development of biofilms.

10.2.2 Tube Biofilms

Biofilm tube reactors offer a forum for the analysis of biofilms under heat just like flow cells. In the tube reactor there is a key difference in the two approaches: on the inner sheet of silicone tubing biofilms are developed, rather than on a cover slip connected to a container. The tube biofilm allows the accumulation of large mass and can be collected from the tube by scraping; this is the primary advantage of tube biofilm. This method is applicable for quantitative evaluation of the impact of antibacterial agents on biofilms, biochemical and gene expression research on the biomass obtained from tubes and the phenotypical diversification that can occur as a consequence of the development of biofilms [7]. However, tube biofilms are not amenable with microscopy. The use of high-throughput analytical methods shared with flow cells is the main disadvantage of this tube biofilm method.

10.2.3 Colony Biofilms

This biofilm is cultured on a "semi-permeable polycarbonate filter (GE)" mounted on a suitable solid substrate. The filter offers a conveniently survivable flexible layer enabling movement of the "soil-grown biofilm" from one media channel to another. This approach analyzes the production of biofilm at the air-to-surface interface without allowing the biofilm to become immersed in liquid. Colony biofilms takes less time to grow in size. *P. aeruginosa* produces a biofilm that is more or less "150–300 μm" thick and *S. epidermidis* forms a biofilm that is about "100 μm thick" [8–10] after a 48-hour growth period. Colony biofilms display standard stratified profiles identical to anaerobic flow cell biofilms and mainly lysed cells are present in the interior colony biofilm, whereas areas with more oxygen like the colony–air interface contain the exponential phase cells [9].

A colony biofilm is an effective tool to assess the persistence of bacteria in view of antimicrobial therapy. The advantages of using the biofilm filter method is that by using CFU quantification of susceptibility can be carried out, and also assays like penetration of relative antibiotic through the "biofilm", spatial pattern of oxygen content, "antibiotic activity", components of extracellular matrix, and determination of synthesis of protein by using electrodes or microscopy. An added advantage of using these biofilms is that the shifts in viable CFU are as a result of cell death and not because of the system's robust and spatially limited existence. In addition, one concern for evaluating antimicrobial resistance in certain biofilm structures, such as "flow cell biofilm" and PEG biofilm, is that the inoculum would only be comparable when contrasting strains of equivalent capacity to bind and grow into a "mature biofilm". In comparison, "colony biofilm" is easily transferred to nutrient agar with and without antibiotic therapy, and is not chosen for cell attachment since 258 S.B. is not immersed in the tank.

10.3 Biofilm Growth on Peg Lids

The biofilm production methods such as "flow cells", "rotating-disks" or "tube reactors" are somewhat prone to contamination and leakage and are dependable on continuous flow with the ability to produce only a few biofilms. To overcome these drawbacks, one possible experimental design might be using peg lids for growing the biofilms; for example, the Calgary Biofilm Unit, consisting of 96-peg polystyrene covers that can fit into a regular 96-well microtiter tube. Phenotypic screening is the main role of culturing biofilms on pegs [50, 51]. Calgary biofilm methods are not vulnerable to contamination since they are mounted in a "laminar flow hood" and are therefore not susceptible to leaks as they are sealed to a petri plate in an arranged manner. By the use of less frequency sonication to interrupt the cells that are adherent to a surface into a retrieval medium, cells of biofilm are generally retrieved from pegs. Many enhancements to this methodology have included the formation of 384-pin biofilms on lids [52], and even the biofilms' growth on plastic pegs that are chemically engineered, like those used for "solid-surface enzyme-linked immunosorbant assays" [53]. While many of the peg lids in the "Calgary Biofilm System" are made up of polystyrene which is untreated and has a neutral electrostatic charge, the adhesion of particular types of microorganisms can be facilitated by plastic film because those microorganisms would not otherwise adhere to the surface. Surface hydrophobicity may change such type of coatings or these types of coating may also provide special binding sites for human beings or pathogens of plants such as "hydroxyapetite", "human serum", methylcellulose or "metal alloys". For instance, robust biofilms on 260 S.B. polystyrene pegs was developed by "*Candida tropicalis*" coated with "L-lysine" or "bovine serum albumin", but not on the uncoated surface [54]. That suits well-established protocols for increasing *Candida* spp. biofilms in microtiter plates, made from plastics treated with cell culture [55].

10.4 Rotating Disk and Concentric Cylinder Reactors

To assess the efficacy of reproducible biocides rotating disk reactors are designed, and these reactors expand the biofilm under shear stress. Rotating disk reactors are comprised of a spherical disc designed to accommodate interchangeable coupons or chips which are flush with the entire surface of the disc and can be manufactured from different materials on the basis of the matter under investigation [54, 55]. The disc is attached to a star-head magnet and then placed in a "side-arm reactor" vessel containing 1l of glass. On the top of a magnetic stirrer the vessel placed there creates a liquid shear force over the surface of the disc, supplying the disc with adjustable rotational velocity. The biofilm in the concentric cylinder reactors expands on the spinning stainless steel cylinders surface within "concentric chambers", rather than on a rotating disc [56]. In order to generate differing shear strength, it is important to change the rotation speed and diameter of the cylinders. Through the reactor medium flow is maintained as a pump-regulated steady drip in both reactors. Depending on the location of the drain outlet leading to a waste storage tank, from the reservoir the medium is drained and dripped into the vessel of the reactor containing a medium between 180–300 ml [56].

By removing the coupons of CCR from RDR viability can be traced asepticly and plating bacteria at varying growth times for comparatively viable cells. Into the reaction vessels insertion of molecules or chemicals may be done in the measurement of clinical effectiveness or distributed through the constant flow throughout the medium and in a way that mimics the specific criteria correlated with care regimes like periods of dosage and exposure. In addition, coupons removed from the RDR after biofilms have been generated and incubated in potential biocides or therapies followed by listing of live bacteria to examine sufficient amounts for the biofilm abolition or degradation. Using a spinning disk that provides vast quantities of coupons or chips enables the precise assessment of dose–response killing relationships from a single biofilm.

10.5 Methods for Characterization of Biofilms

10.5.1 Confocal Laser Scanning Microscopy (CLSM)

Microscopy plays an important role in studying every biological cycle, since it is the primary technique for evaluating microbial communities. It allows the early identification of issues, like pathogens and diseases, contamination of food products, failure of biotechnological processes in general, and degradation of the efficiency of organic wastes and wastewater treatment. In general, microscopic approaches have been gradually introduced in recent years because of the prospect of combining them with the processing of automated digital online image and interpretation methods of images and their applications for *in situ* studies [11]. CLSM was compared to other forms of microscopy, as it facilitates live biofilm study and the study of biofilms that are completely hydrated [12]. CLSM offers parallel knowledge on the 3D layout of biofilms and the recognition of the various components, either by "autofluorescence (algae)" or through the use of unique fluorescent colors for "bacterial DNA" or "EPS glycoconjugates" [13]. CLSM was used to test "lectin-binding" technology for the *in situ* identification of glycoconjugates in the biofilms of the river that are completely hydrated [14]. In combination with "epifluorescence microscopy" and"confocal laser scanning microscopy", there has been use of "fluorescently labeled lectins" to allow visual representation and in the biofilms of *Pseudomonas aeruginosa* characterization of extracellular polymeric substances (EPS) that contain carbohydrates [15]. Barranguet et al. compared the light attenuation by the biofilm with the signal attenuation by using pulse amplitude modulation (PAM) CLSM and fluorimetry [16].

10.5.2 Scanning Electron Microscopy (SEM)

Combining a traditional optical microscope with a specially built glass flow cell, biofilms developed on opaque thin biomaterials were visualized *in situ* through a clear noninvasive way [17]. Comparisons were rendered with SEM images for the optical microscopy of thin biofilms. Stewart et al. have shown the systemic variability of SEM, CLSM and cryoem bedding to a microbial biofilm, accompanied by sectioning and microscopic analysis [10]. These researchers contrasted the three separate microscopic approaches about the details we offer on the structural heterogeneity of biofilms.

10.5.3 Atomic Force Microscopy (AFM)

Of the "scanning probe microscopy (SPM)" techniques AFM is one of the most significant and commonly used. AFM may be used to visualize atomic-scale dimensional structures up to about 100 mm, so it provides precise depth information on the surface topography. This is also appropriate to check biopolymers. EPS on the biofilm surface of bacteria condensed deoxyribonucleic acid, structures of deoxyribonucleic acid, and associations of DNA and proteins are the basic examples.

10.5.4 Infrared and Raman Spectroscopy

To obtain data from the cells of bacteria that are bound to a substratum there is a method which is widely used named "ATR-IR spectroscopy". By the use of this method one can study the entire cells of bacteria and biofilms, thereby removing the risk of artifacts that could occur during the procedures needed to separate different cellular components. Hence this method is commonly utilized in the study of biofilms [18].

10.5.5 X-ray Spectroscopy

For the study of processes like interfacial geological and geochemical, X-ray spectroscopy and scattering methods were generally used [19]. The broad penetration range and intrinsic flexibility of X-rays to the molecular scale allows them to be suitable instruments to examine the structural features of natural materials under *in situ* conditions. In order to explore the distribution and speciation of heavy elements at interfaces of the environment, including the presentation of a model formalism that allows for quantitative study of fluorescent yield profiles, a study of the implementation of "grazing-angle X-ray fluorescence techniques" has been carried out by Trainor et al. and Al-Bataineh et al. who studied each step of the chain of immobilization that includes examining the surface immobilization of antibacterial furanones using "X-ray photoelectron spectroscopy (XPS)" and AFM taping mode [20].

10.5.6 Nuclear Magnetic Resonance (NMR) Spectroscopy

Knowledge about cell suspensions of living prokaryotes, gel-immobilized cells and extracts, eukaryotic cells and samples of tissue is provided by the techniques of

NMR spectroscopy [21]. 1 Hand 1H-detected 13 C NMR allows the straightforward, time-resolved and noninvasive measurement of metabolite concentrations, metabolic processes and flux levels for live cell suspension studies *in situ* [22]. For the noninvasive analysis of live biofilms NMR techniques are limited to the assessment of the properties of water. Such experiments are imaging of biofilm in a porous medium [23], and the measurement of convective flow of media in and near *in situ* biofilm surfaces [24].

10.6 Biofilm-Based Bioremediation

For the detoxification of contaminated soil, water and air an eco-friendly method is bioremediation [25]. It is a microbial mechanism that relies solely on microbes' potential to treat contaminated environments. Harsh chemicals are not included in this process, and as such, adverse environmental impacts are minimized [26]. Bioremediation is preferable over other approaches as it involves little or negligible ground or habitat disturbance around the region being handled. This is mainly of two types: "*in situ*" and "*ex situ*" based on the pollutant treatment being performed. Throughout the case of *in situ* bioremediation, the polluted sample is handled at the initial location so there is no need to translocate the pollutant to some other treatment facility, while the sample is transferred to another storage site for *ex situ* remediation [27]. *In situ* bioremediation lowers shipping costs with minimal site disturbance. The optimization of physicochemical environments like pH, aeration and moisture is a prerequisite for the bacteria in the polluted site to remove the pollutant faster together with supplemented nutrients. The alternate approach is to use genetically engineered microorganisms that will modify the way chemicals are processed in a competent manner [28]. A community of microorganisms found in biofilm mode is evidently used for the destruction of widespread toxins in both natural and engineered environments. The exponential development of the pharmaceutical industry in recent decades has contributed to environmental pollution due to hazardous waste effluents [29]. The proliferation of industrial contaminants and their issues has taken long-term environmental catastrophes to the center of public consciousness [30]. As a consequence, different approaches are being established and further work is being conducted to devise a method of environmental protection.

For the decontamination of polluted areas remediation using biofilm is an environmentally safe and cost-effective solution. Biofilm use is effective for "bioremediation", because biofilms capture, immobilize and remove different toxins from the environment. Near the highly polluted areas, bacterial biofilms occur inside indigenous communities to thrive, live, maintain and control severe environmental conditions. In the biofilms the gene expression differs considerably from that of free moving planktonic cells. This differential gene expression inside biofilms may be due to varying local nutrient and oxygen concentrations within the biofilm matrix, and also to the division of labor between microbes. These variance gene expressions can also be useful for the oxidation of various toxins through specific metabolic pathways. "Chemotaxis and flagellar motility" are essential factors for the development of biofilms [31]. Various tactic responses like twitching motility, swimming, swarming or chemotaxis sense the presence of pollutants present in soil and increase their efficiency of biodegradation [32]. Typically under natural conditions, most of the bacteria persist in biofilm mode encapsulated in an extracellular polymeric material (EPS) matrix which provides a beneficial structure for biofilm-forming microbes in bioremediation [33]. The extracellular polymeric substance is comprised of polysaccharides, enzymes, lipids, nucleic acids, humic and water [34]. The composition of the EPS differs between various organisms and relies on the conditions of development, and the surface on which biofilms are produced. The environmental conditions in biofilms will cause a structural and functional shift [35]. Biofilms occur in filamentous and champignon-like structures in flowing and stagnant waters respectively [36]. Protection from different environmental stresses such as acid stress, shear stress, antimicrobials, UV exposure, aridity, predation, etc. to microbes is provided by the biofilm matrix [37]. The free-floating planktonic cells, on the other side, have a drawback over biofilms because they decontaminate the toxins by oxidative action, but these cells are flexible and not able to survive under mechanical and environmental tension. Biofilm-forming microbes, on the other hand, are especially skillful in bioremediation, as they are immobilized in extracellular polymeric substances, which also eliminate pollutants throughout degradation [5].

Cyanobacterial EPS may serve as a biosorbent, eliminating heavy metals from the aqueous process [38]. Surfactant-containing EPS can also help to solubilize otherwise unavailable hydrophobic or other refractory substrates for microorganisms [39]. The presence of extracellular polymeric substances also helps nutrients to be exchanged and by-products to be removed in the biofilm culture. Toxins such as toxic metals and organic compounds are decontaminated by extracellular enzymes in EPS biofilms [33]. Extracellular polymeric substances act as metal and metalloid traps because of the presence of functional groups that are negatively charged and involves the eventual elimination of heavy metal complexes and complexes of organic contaminants [40]. The EPS binds a wide range of metals including Pb, Cu, Mn, Mg, Zn, Cd, Fe and Ni [41]. Restriction in the nutrient will also improve the generation of extracellular polymeric substances and can remove toxic elements and contaminants. Extracellular polymeric substance biofilm consists of microorganisms which accumulate phosphorus and serves as a reservoir and helps to extract and recycle phosphorus from polluted water [42]. There are multiple mixed species of biofilms found in nature that can facilitate better production of biofilm and better ability to endure environmental stress [43]. Coexistence inside adjacent biofilms with various microbial organisms facilitates contact with its representatives. In natural and engineered systems, biofilms can be found. Biofilms are generally found in natural environments like soil, aquatic vegetation, sediments, and cover rocks in streams and trees, lakes and waterways and in wetlands. Naturally occurring biofilms have the potential to degrade and eliminate contaminants from soil and water which include bacteria, fungi as well as algae. Naturally formed algal biofilms like Nitzschia present in the organic sediments can facilitate the recovery of sediments by releasing O_2, which facilitates the "biodegradation" of the aerobic bacteria found in the biofilm. Natural biofilms are an essential part of the food chain, in which parasitic protozoa or grazers may use biofilms for their food requirements. In addition, the protozoa can be eaten up by smaller fishes, and eventually smaller fishes by large fishes and land animals. In aquatic ecosystems, under extreme conditions on the surface of the soil, microbial aggregates consisting of phototrophic bacteria and algae prevent predation by grazers and stress. Flocs are fragile structures that are formed during the blooming periods. “Municipal wastewater” treatment is focused on the “activated sludge” system operation of the floc. For the removal of organic compounds and metals from lakes, rivers and pool water, slow-sand filters are commonly used [44]. “Planctomycetes” that are present in biofilms of seaweed in marine water have the ability to remove N from polluted water because of the anammox reactions in which there is anaerobic oxidation of ammonia to dinitrogen [45].

10.7 Nitrogen Fixing Microorganisms in Lakes

Nitrogen is considered as a critical limiting factor for growth and plant production. The main components are chlorophyll which is an essential plant pigment required for photosynthesis, and amino acids are the building blocks of proteins. It is also present in other essential biomolecules including ATP and nucleic acids. While it is present abundantly in the atmosphere of the earth, mostly in the form of nitrogen gas (N_2), a reduced form of this element is also used by plants. Plants obtain these sources of combined nitrogen in the ratio of: 1) ammonia and/or nitrate fertilizer application from manure to soil); 2) through the decomposition of organic matter; 3) the use of natural processes like lightning to convert ambient nitrogen into compounds; 4) “biological nitrogen fixation” [57]. Biological nitrogen fixation is done by a unique community of prokaryotes and was discovered by Beijerinck [46]. These species catalyze the transfer of ambient nitrogen (N_2) to ammonia (NH_3) by the use of nitrogenase enzymes. Plants can assimilate ammonia readily to produce the above-mentioned nitrogen biomolecules.These prokaryotic species include *Cyanobacteria, Azotobacter, Azospirillum*, and *Rhizobium* and *Bradyrhizobium*, which symbiose with legumes and other plants [46, 57]. For nitrogen fixation there is a requirement of 16 ATP moles to each low nitrogen mole [47]. Those molecules must be derived from other species by nonphotosynthetic free-living microorganisms, whereas photosynthetic microorganisms such as cyanobacteria use photosynthesis-produced sugar. These compounds are created from the rhizospheres of their host plants by symbiotic nitrogen-fixing microorganisms [47].

Atmospheric nitrogen is converted into fixed nitrogen by the microorganisms that fix nitrogen. Ninety percent of nitrogen fixation is done by these species of bacteria, and therefore these are very important parts of the nitrogen cycle. Two groups of bacteria that remove N are known. The *Anabaena* and *Nostoc* cyanobacteria (or blue-green algae) and genera including *Clostridium, Beijerinckia* and *Azotobacter* comprise the first community or the “free-living (nonsymbiotic) bacteria”. The second category contains infectious (symbiotic) bacteria; for instance *Rhizobium* associated with leguminous plants (e.g., numerous pea family members); Frankia related with some dicotyledonous organisms (actinorhizal plants); and *Azospirillum* related with cereal grasses. The root hairs of host plants are penetrated by “symbiotic nitrogen-fixing bacteria”, replicating the production of root nodules, plant cell enlargement, and bacteria by promoting intimate interaction.

Free nitrogen is converted into ammonia by bacteria inside the nodules, which is used by the host plant for its growth.

The combination of autotrophic and heterotrophic microorganisms regulates the nitrogen fixation. In aquatic ecosystems, planktonic fixation is mainly done by cyanobacteria, and such fixation rates are higher only when these microbes make up a maximum percentage of the "planktonic biomass". In oligotrophic and mesotrophic lakes the planktonic fixation of nitrogen appears to be lower. Much of the nitrogen on earth comes in the form of N_2, and is naturally inaccessible to living species except for prokaryotic organism fixation. "Nitrogen fixation" may also be a significant input of nitrogen into individual aquatic environments and may enable the continuity of primary development when sources of fixed nitrogen become exhausted. Phosphorus is the resource that normally regulates primary production in wetlands, partially due to the propensity of nitrogen deficits compared to phosphorus by fixing nitrogen [58]. The physiology of such organisms has been extensively studied and it is well established that fixation could even occur in a wide range of "free-living" and "symbiotic prokaryotic organisms", such as "photoautotrophic cyanobacteria (blue-green algae)", "photoautotrophic anaerobic bacteria", "aerobic and anaerobic heterotrophic bacteria" and "chemoautotrophic bacteria". In the nitrogen economy of eutrophic lakes, the fixation of nitrogen by planktonic organisms is often important, but in the nitrogen economies of mesotrophic and oligotrophic lakes in general it is not important.

Unlike planktonic microorganisms fixing nitrogen, there seems to be no significant variations between freshwater and marine environments about fixation by "benthic bacteria". "Nitrogen fixation" levels in the sediments of most lakes and estuaries are small to moderate except in highly organic-rich estuarine sediments, typically co.25 g N m-2 yr-1. Nitrogen oxidation levels in estuarine sediments that are abundant in organic matter vary from 0.4 to 1.6 g N mm2 yr-I. Heterotrophic (and perhaps chemoautotrophic) bacteria mediate most of the benthic fixation in mesotrophic and eutrophic lakes and estuaries, but cyanobacteria often dominate the benthic fixation in oligotrophic ecosystems. Nitrogen fixation levels occur comparably or much greater in freshwater and aquatic wetlands and seagrass beds than in nonvegetated, organic-rich sediments. Fixation rates in cyanobacterial mats are high to very high (1.3–76 g N m-2 yr-I), but normally these mats only cover a small area of the ecosystems in which they reside. The nitrogen-fixing microorganisms in lakes are therefore an essential component of an underwater food chain in which the fresh water environment could be made extinct. These microorganisms are therefore essential to preserving lake fertility, vigor and efficiency.

10.8 Conclusion

In natural ecosystems, most bacteria live in complex assemblages made up of one or more species. Biofilms are characterized by highly structured matrix-enclosed communities separated by a network of open water channels and are examples of such an assemblage. This structure provides an ideal environment for cell–cell interactions, such as the exchange of genetic material, communication signals, and metabolites that allow nutrients to diffuse into the biofilm community. Biofilms are a viable and eco-friendly option that have a tremendous future for use in the remediation of various types of pollution. The presence of extracellular polymeric substances has made this technique unique as it can be used for the removal of byproducts. Toxins such as toxic metals and organic compounds are decontaminated by extracellular enzymes in EPS biofilms. For the decontamination of polluted areas remediation using biofilm is an environmentally safe and cost-effective solution.

References

1 Wimpenny, J. (2000). An overview of biofilms as functional communities. In: *Community Structure and Co-operation in Biofilms. SGM Symposium 59* (ed. D.G. Allison, P. Gilbert, H.M. Lappin-Scott, and M. Wilson), 1. Cambridge, UK: Cambridge University Press.

2 Flemming, H.-C., Szewzyk, U., and Griebe, T. (2000). *Biofilms*. Lancaster, Basel: Investigative Methods and Applications, Technomic Publishing Co. Inc.

3 Donlan, R.M. (2002). Biofilms: Microbial life on surfaces. *Emerg. Infect. Dis.* 8 (9): 881.

4 Hall-Stoodley, L., Costerton, J.W., and Stoodley, P. (2004). Bacterial biofilms: From the natural environment to infectious diseases. *Nature Rev. Microbiol.* 2: 95.

5 Sutherland, I.W. (2001a). Biofilm exopolysaccharides: A strong and sticky framework. *Microbiol* 147: 003.

6 Christensen, G.D., Simpson, W.A., Younger, J.J., et al. (1985). Adherence of coagulase-negative staphylococci to plastic tissue culture plates: A quantitative model for the adherence of staphylococci to medical devices. *J. Clin. Microbiol.* 22: 996–1006.

7 Kirisits, M.J. and Parsek, M.R. (2006). Does *Pseudomonas aeruginosa* use intercellular signalling to build biofilm communities? *Cell Microbiol.*, 8: 1841–1849.

8 Borriello, G., Werner, E., Roe, F., et al. (2004). Oxygen limitation contributes to antibiotic tolerance of *Pseudomonas aeruginosa* in biofilms. *Antimicrob. Agents Chemother* 48: 2659–2664.

9 Werner, E., Roe, F., Bugnicourt, A., et al. (2004). Stratified growth in *Pseudomonas aeruginosa* biofilms. *Appl. Environ. Microbiol.* 70: 6188–6196.

10 Stewart, P.S., Murga, R., Srinivasan, R., and De Beer, D. (1995). Biofilm structural heterogeneity visualized by three microscopic methods. *Wat. Res.* 29 (8): 2006.

11 Wolf, G., Crespo, J.G., and Reis, M.A.M. (2002). Optical and spectroscopic methods for biofilm examination and monitoring. *Environ. Sci. Technol.* 1: 227.

12 Lawrence, J.R. and Neu, T.R. (2003). Microscale analyses of the formation and nature of microbial biofilm communities in river systems. *Rev. Environ. Sci. Bio Tech.* 2: 85.

13 Zhang, T. and Fang, H.H.P. (2001). Quantification of extracellular polymeric substances in biofilms by confocal laser scanning microscopy. *Biotechnol. Lett.* 23: 405.

14 Neu, T.R., Swerhone, G.D., and Lawrence, J.R. (2001). Assessment of lectin-binding analysis for in situ detection of glycoconjugates in biofilm systems. *Microbiol* 147: 299.

15 Strathmann, M., Wingender, J., and Flemming, H.C. (2002). Application of fluorescently labelled lectins for the visualization and biochemical characterisation of polysaccharides in biofilms of Pseudomonas aeruginosa. *J. Microbiol. Meth.* 50: 237.

16 Barranguet, C., Van Beusekom, S.A.M., Veuger, B., Neu, T.R., Manders, E.M.M., Sinke, J.J., and Admiraal, W. (2004). Studying undisturbed autotrophic biofilms: Still a technical challenge. *Aquatic Microbial. Ecol.* 34: 1.

17 Cortizo, M.C. and Lorenzo De Mele, M.F.L. (2003). Microstructural characteristics of thin biofilms through optical and scanning electron microscopy. *World Microbiol. Biotech.* 19: 805.

18 Boualam, M., Quiles, F., Mathieu, L., and Block, J.C. (2002). Monitoring the effect of organic matter on biofilm growth in low nutritive waters by ATR-FT-IR spectroscopy. *Biofouling* 18 (1): 73.

19 Trainor, T.P., Templeton, A.S., and Eng, P.J. (2006). Structure and reactivity of environmental interfaces: Application of grazing angle X-ray spectroscopy and long-period X-ray standing waves. *J. Elec. Spectr. Rel. Phen.* 150: 66.

20 Al-Bataineh, S., Britcher, L.G., and Griesser, H.J. (2006). XPS characterisation of the surface immobilization of antibacterial furanones. *Surf. Sci.* 600: 952.

21 Grivet, J.P., Delort, A.M., and Portais, J.C. (2003). NMR and microbiology: From physiology to metabolomics. *Biochimie* 85: 823.

22 Brecker, L. and Ribbons, D.W. (2000). Biotransformations monitored in situ by proton nuclear magnetic resonance spectroscopy. *Trends Biotechnol.* 18: 197.

23 Paterson-Beedle, M., Nott, K.P., Macaskie, L.E., and Hall, L.D. (2001). Study of biofilm within a packed-bed reactor by three dimensional magnetic resonance imaging. *Microbial Growthin Biofilms Pt B* 337: 285.

24 Manz, B., Volke, F., Goll, D., and Horn, H. (2003). Measuring local flow velocities and biofilm structure in biofilm systems with magnetic resonance imaging (MRI). *Biotechnol. Bioeng.* 84.

25 Prasad, M.N. and Prasad, R. (2012). Nature's cure for cleanup of contaminated environment: a review of bioremediation strategies. *Rev. Environ. Health* 27: 181–189.

26 Vidali, M. (2001). Bioremediation. an overview. *Pure Appl. Chem.* 73: 1163–1172.

27 Vogt, C. and Richnow, H.H. (2014). Bioremediation via in situ microbial degradation of organic pollutants. *Adv. Biochem. Eng. Biotechnol.* 142: 123–146.

28 Singh, J.S., Abhilash, P.C., Singh, H.B., et al. (2011). Genetically engineered bacteria: An emerging tool for environmental remediation and future research perspectives. *Gene* 480: 1–9. 23.

29 Horn, H. and Morgenroth, E. (2006). Transport of oxygen, sodium chloride, and sodiumnitrate in biofilms subsequent environmental. *Chem. Eng. Sci.* 61: 1347–1356.

30 Pandey, G. and Jain, R.K. (2002). Bacterial chemotaxis toward environmental pollutants: Role in bioremediation. *Appl. Environ. Microbiol.* 68: 5789–5795.

31 Pratt, L.A. and Kolter, R. (1999). Genetic analyses of bacterial biofilm formation. *Curr. Opin. Microbiol.* 2: 598–603.

32 Lacal, J., Reyes-Darias, J.A., García-Fontana, C., et al. (2013). Tactic responses to pollutants and their potential to increase biodegradation efficiency. *J. Appl. Microbiol.* 114: 923–933.

33 Flemming, H.C. and Wingender, J. (2010). The biofilm matrix. *Nat. Rev. Microbiol.* 8: 623–633.

34 Branda, S.S., Vik, S., Friedman, L., et al. (2005). Biofilms: The matrix revisited. *Trends Microbiol.* 13: 20–26.

35 Jung, J.H., Choi, N.Y., and Lee, S.Y. (2013). Biofilm formation and exopolysaccharide (EPS) production by Cronobacter sakazakii depending on environmental conditions. *Food Microbiol.* 34: 70–80.

36 Edwards, K.J., Bond, P.L., Gihring, T.M., et al. (2000). An archaeal iron-oxidizing extreme acidophile important in acid mine drainage. *Science* 287: 1796–1799.

37 Davey, M.E. and O'Toole, G.A. (2000). Microbial biofilms: From ecology to molecular genetics. *Microbiol. Mol. Biol. Rev.* 64: 847–867.

38 De Philippis, R., Colica, G., and Micheletti, E. (2011). Exopolysaccharide-producing cyanobacteria in heavy metal removal from water: Molecular basis and practical applicability of the biosorption process. *Appl. Microbiol. Biotechnol.* 92: 697–708.

39 Iwabuchi, N., Sunairi, M., Urai, M., et al. (2002). Extracellular polysaccharides of Rhodococcus rhodochrous S-2 can stimulate the degradation of aromatic components in crude oil by indigenous marine bacteria. *Appl. Environ. Microbiol.* 68: 2337–2343.

40 Li, -W.-W. and Yu, H.-Q. (2014). Insight into the roles of microbial extracellular polymer substances in metal biosorption. *Bioresour. Technol.* 160: 15–23.

41 Ferris, F.G., Schultze, S., Witten, T.C., et al. (1989). Metal interactions with microbial biofilms in acidic and neutral pH environments. *Appl. Environ. Microbiol.* 55: 1249–1257.

42 Zhang, H.L., Fang, W., Wang, Y.P., et al. (2013). Phosphorus removal in an enhanced biological phosphorus removal process: Roles of extracellular polymeric substances. *Environ. Sci. Technol.* 47: 11482–11489.

43 Burmolle, M., Webb, J.S., Rao, D., et al. (2006). Enhanced biofilm formation and increased resistance to antimicrobial agents and bacterial invasion are caused by synergistic interactions in multispecies biofilms. *Appl. Environ. Microbiol.* 72: 3916–3923.

44 Logsdon, G.S., Kohne, R., Abel, S., et al. (2002). Slow sand filtration for small water systems. *J. Environ. Eng. Sci.* 1: 339–348.

45 Kartal, B., Kuenen, J.G., and Van Loosdrecht, M.C. (2010). Engineering Sewage treatment with anammox. *Science* 328: 702–703.

46 Beijerinck, M.W. (1901). Über oligonitrophile Mikroben. *Zbl. Backt* 7: 561–582. biofilm communities? Cell Microbiol 8:1841–1849.

47 Hubbell, D.H. and Kidder, G. (2009). *Biological Nitrogen Fixation*, 1–4. University of Florida IFAS Extension Publication. SL16.

48 Flemming, H. (1995). Sorption sites in biofilms. *Water Science and Technology*, 32: 27–33.

49 Edwards, S.J. and Kjellerup, B.V. (2013). Applications of biofilms in bioremediation and biotransformation of persistent organic pollutants, pharmaceuticals/personal care products, and heavy metals. *Appl Microbiol Biotechnol.*, 97: 9909–9921.

50 Ceri, H., Olson, M. E., Stremick, C., Read, R. R., Morck, D. and Buret, A. (1999). The Calgary Biofilm Device: a new technology for rapid determination of antibiotic susceptibilities of acterial biofilms. *J Clin Microbiol* 37: 1771–1776.

51 O'Toole, G. A. and Kolter, R. (1998). Initiation of biofilm formation in Pseudomonas fluorescens WCS365 proceeds via multiple, convergent signaling pathways: a genetic analysis. *Mol Microbiol* 28, 449–461.

52 Musk, D. J., Banko, D. A., and Hergenrother, P. J. (2005). Iron salts perturb biofilm formation and disrupt existing biofilms of *Pseudomonas aeruginosa*. *Chem. Biol.* 12, 789–796.

53 Spoering, A.L. and Lewis, K. (2001). Biofilms and planktonic cells of *Pseudomonas aeruginosa* have similar resistance to killing by antimicrobials. *J Bacteriol.*, 183(23): 6746–51.

54 Harrison, J.J., Turner, R.J. and Ceri, H. (2005). Persister cells, the biofilm matrix and tolerance to metal cations in biofilm and planktonic *Pseudomonas aeruginosa*. *Environ Microbiol.*, 7(7): 981–94.

55 Pierce, C.G., Uppuluri, P., Tristan, A.R., Wormley, F.L., Jr, Mowat, E., Ramage, G., Lopez-Ribot, J.L. (2008). A simple and reproducible 96-well plate-based method for the formation of fungal biofilms and its application to antifungal susceptibility testing. *Nat Protoc.*, 3: 1494–1500.

56 Meeuwse, M., Lempers, S., van der Schaaf, J. and Schouten, J. C., 2010. Liquid-solid mass transferand reaction in a rotor-stator spinning disc reactor. *Industrial & Engineering Chemistry Research* 49: 10751–10757.

57 Vance, C.P. (2001). Symbiotic Nitrogen Fixation and Phosphorus Acquisition. Plant Nutrition in a World of Declining Renewable Resources, *Plant Physiology*, 127: 390–397.

58 Schindler, D.W., Hecky, R.E., Findlay, D.L., Stainton, M.P., Parker, B.R. Paterson, M.J., Beaty, K.G., Lyng, M. and Kasian, S.E.M. (2008). Eutrophication of lakes cannot be controlled by reducing nitrogen input: Results of a 37-year whole-ecosystem experiment. *Proceedings of the National Academy of Sciences*, 105: 11254–11258.

11

Graphene-Based Absorbents for Wastewater Treatment

CHAPTER MENU

11.1 Introduction

Water contamination and a shortage of clean water sources have become significant problems around the world. Due to population development, industrial and agricultural activities, as well as other ecological and environmental phenomena, the quality of water supplies is currently declining [1]. The continuous release of various forms of contaminants into clean water sources, such as heavy metal ions and organic dyes, has caused increasing concern in many countries [2]. Chemical precipitation, coagulation, reverse osmosis, membrane filtration, solvent extraction, flotation, ion exchange, electrochemical treatments, evaporation, oxidation, adsorption and biosorption are only a few of the methods that have been developed to date for the efficient removal of heavy metal ions from wastewaters [3]. Precipitation and coagulation, for example, require a range of chemicals and result in large quantities of sludge; electrochemical and filtration approaches, on the other hand, require electricity equipment and large capital expenditures [4]. Because of its high removal efficiency, quick and easy operation, cost-effectiveness, and wide range of adsorbent materials, adsorption/ion exchange has been considered as the most important method for removing heavy metal ions among the aforementioned techniques.

Nanotechnology has evolved in the last two decades, with applications in almost every area of science and technology [5]. Nanomaterials have been produced in a wide range

Aquatic Environmental Bioengineering: Monitoring and Remediation of Contamination, First Edition.
Rouf Ahmad Bhat, Mohammad Yaseen Mir, Gowhar Hamid Dar and Moonisa Aslam Dervash.

of scientific and technological fields [6]. Nanomaterials are good adsorbents and catalysts. The use of nanomaterials in the environment has piqued people's interest. Due to their extremely high specific surface area, multiple associated sorption sites, low-temperature alteration, short intraparticle diffusion distance, tunable surface chemistry and size-tunable surface pores, nanomaterials outperformed other materials significantly [7]. As a result, extensive research has been conducted using a variety of nanomaterials to extract heavy metal ions from wastewater. The adsorption potential and the desorption property are two important parameters to consider when evaluating an adsorbent. The absorbent loading, ionic strength, pH, adsorption contact time, temperature, metal ion concentration and competition among metal ions are all important factors in the adsorption process that are frequently studied and optimized.

11.2 Graphene-Based Materials

In 1986, the word "graphene" was coined to describe single atomic layers of graphite [8]. The International Union of Pure and Applied Chemistry (IUPAC) first adopted it in 1995 [9]. Katsnelson [10] developed the basic definition and terminology used by scientists today in the field of graphene.

Pristine graphene is a hydrophobic substance that does not dissolve in most solvents. Graphene composite manufacturing, on the other hand, is primarily concerned with graphene solubilization. Different functional groups have been added to the carbon backbone through chemical modification [11], covalent [12] or noncovalent functionalization to boost graphene's solubility [13].

The rGO sheets have also been modified with metal ions and functional groups. The sequential chemical reaction of GO sheets in water with KOH, followed by hydrazine, produces K-modified graphene (hKMG) sheets, which can lead to a homogeneous aqueous suspension [14]. In addition, p-phenyl-SO_3H groups were added to GO, and the lightly sulfonated rGO can be distributed in water at acceptable concentrations (2 mg/mL) in the pH range of 3 to 10 [15].

The reaction of functional molecules with oxygenated groups on the GO or rGO surface, such as carboxyl groups at the periphery and epoxy, hydroxyl and –C = C– groups in the basal plane, results in covalent modification of grapheme [16]. Converting the edge carboxylic acid groups of rGO into octadecylamines, which sterically stabilized the graphene sheets, yielded rGO dispersions in THF, CCl_4, and 1,2-dichloroethane (EDC) [17] The formation of amides or carbamate esters (RnCO) from the treatment of GO with organic isocyanates could result in the derivatization of both the edge carboxyl and surface hydroxyl functional groups. In the synthesis of GO–polymer composite sheets, the carboxylic groups in GO were esterified with the hydroxyl groups in poly(vinyl alcohol) (PVA) [18] The nucleophilic ring-opening reaction between the epoxy groups of GO and the amine groups of an amine-terminated organic molecular, such as 1-(3-aminopropyl)-3-methylimidazolium bromide (an ionic liquid) and 3-aminopropyltriethoxysilane, will, on the other hand, functionalize the basal surface of GO (APTS) [19]. Also, diazonium salts (e.g., sodium dodecylbenzenesulfonate, SDBS) were used to covalently functionalize rGO platelets, and the resulting rGO was easily dispersed in a variety of polar organic solvents [20]. A variety of different organic compounds, such as azido-phenylalanine and azidotrimethylsilane, have been bound to the graphene surfaces by the [2 + 1] cycloaddition of nitrenes onto the –C = C– double bonds [21]. Graphene's surface has recently been covalently functionalized by grafting polystyrene-polyacrylamide (PS-PAM) onto it through *in situ* free-radical polymerization [22].

Compared to covalent functionalization, noncovalent functionalization based on the vander Waals force or the π—π interaction between rGO and stabilizers has a lower negative effect on the structure of graphene and its derivatives than covalent functionalization, and it also allows for tuning of their solubility and electronic properties. By reducing an aqueous GO dispersion with hydrazine in the presence of poly(sodium 4-styrenesulfonate), the first noncovalent functionalized stable graphene dispersion was developed (PSS). The rGO sheets were stabilized in this experiment by interacting with the hydrophobic backbone of PSS, while the hydrophilic sulfonate side groups held the graphene–PSS complex afloat. Biomolecules have also been used in the van der Waals functionalization of graphene, such as single-strand DNA (ssDNA) to stabilize aqueous suspensions of single-layer graphene sheets [23].

Using the water-soluble aromatic organic molecule 1-pyrenebyturate (PB) as a stabilizer, noncovalent functionalization of graphene sheets through $\pi - \pi$ interaction was recorded, and large-area flexible PB–graphene films with layered structures could be prepared through subsequent vacuum filtration. Through the π—π interaction, this method offers a general route for the preparation of functionalized graphene. Other studies using this technique have shown that the functionalized graphene sheets have substantially improved solubility and conductivity [24].

11.3 Graphene–Polymer Composites

Graphene and its derivatives are important filler materials for polymer composites because of their good conductivity, thermal stability and mechanical strength. The properties and performances of graphene–polymer composites are determined not only by the consistency of the graphene filler and polymer matrix, but also by the filler dispersity, the filler–matrix bonding and the filler–matrix ratio. Fabrication processes are the principal determinants of these variables. Solution mixing [25], melt blending [26], and *in situ* polymerization are the methods used to fabricate graphene–polymer composites, which are close to the methods used to fabricate traditional polymer composites.

Solution mixing is one of the most commonly used methods for preparation of polymer composites, since it is straightforward, requires no special instruments and allows for large-scale production. The solubility of graphene sheets in the polymer solution is one of the major concerns in solution mixing. Water-soluble polymers, such as poly (vinyl alcohol) (PVA), for example, are simple to combine with aqueous GO dispersions at different concentrations [27].

The low solubility of GO and rGO in organic solvents, on the other hand, makes synthesis of graphene–polymer composites in organic solutions extremely difficult (Figure 11.1). Ultrasonication can be used to create short-term metastable dispersions of GO or other graphene derivatives, which are then combined with polymer solutions such as

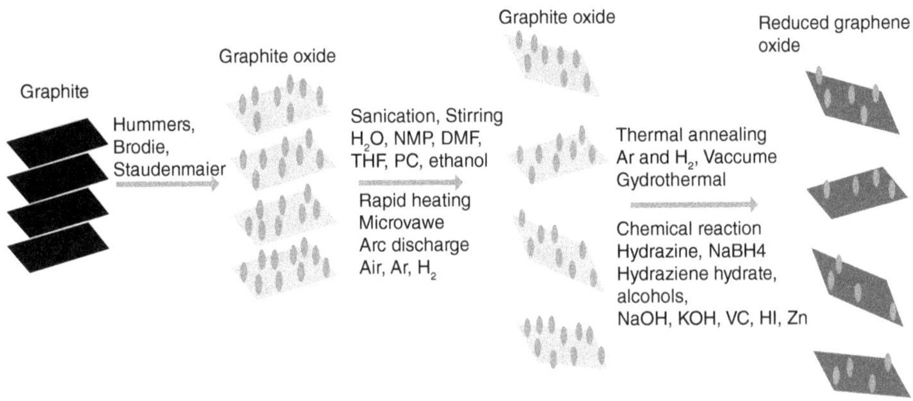

Figure 11.1 Routes to the synthesis of GO and rGO (Peng et al., 2017). *Source*: Peng, W., Li, H., Liu, Y., Song, S. 2017. A review on heavy metal ions adsorption from water by graphene oxide and its composites, Journal of Molecular Liquids 230: 496–504.

poly (methyl methacrylate) (PMMA) [28], polycaprolactone (PCL) [29], polyurethane (PU) and polyaniline (PANI) [30]. High-speed shearing was also used to combine graphene-based fillers with polymer matrices, while ice baths were used to prevent the shearing process from overheating [31]. However, during the slow process of solvent evaporation, the rGO sheets can re-aggregate. As a consequence, it's essential to add functional molecules to graphene sheets or graphene derivatives to improve their solubility in a variety of solvents. For example, GO sheets were reduced to rGO after being functionalized with phenyl isocyanate and mixed with polystyrene (PS) solution in DMF [32]. During the experiment, the polymer matrix stopped rGO sheets from reaggregating, and the dispersed suspension remained stable. However, these studies only use stabilizers with poor electronic delocalization systems (such as small aromatic molecules), which limits the solubility of the rGO-based materials obtained. $PFVSO_3$, a specially formulated conjugated polyelectrolyte (CPE) with a planar backbone and charged sulfonate and oligo(ethylene glycol) (OEG) side chains, was recently used to change rGO sheets in our community. The resulting CPE-functionalized rGO ($PFVSO_3$ -rGO) has excellent solubility and stability in a variety of polar solvents, including water, ethanol, methanol, DMSO and DMF, due to the heavy π—π interaction between $PFVSO_3$ and rGO. An amphiphilic coil-rod-coil conjugated triblock copolymer (PEG-OPE), chemical structure [33] composed of one lipophilic-conjugated oligomer and two hydrophilic PEG coils has been developed to improve the solubility of graphene–polymer composites in both high polarity (such as water and ethanol) and low polarity (such as toluene and chloroform) The heavy π—π interaction allows the conjugated rigid-rod backbone of PEG-OPE to bind to the basal plane of rGO, while the lipophilic side chains and two hydrophilic coils of the backbone fly away from the rGO surface to form an amphiphilic outerlayer. As a result, the graphene–polymer composite obtained is soluble in both low-polarity organic and high-polarity water-miscible solvents.

Melt compounding, which involves the blending of filler materials and polymer matrices using high-shear forces at elevated temperatures, is another common fabrication process in industry for high-yield fabrication of polymer composites. By combining PLA chips and EG in a mechanical mixer at 175–200° C, polylactide (PLA)-exfoliated graphite (EG) composites have been successfully prepared [34]. In another experiment, graphene platelets were produced by thermal reduction of GO and then melt-blended with polyethylene terephthalate (PET) at 285° C to create a PET–graphene composite [26]. Melt compounding, on the other hand, is less efficient than solution mixing in terms of achieving good dispersity and distribution of the filler materials in the matrix. Furthermore, high shear forces will sometimes crack filler materials including carbon nanotubes and graphene sheets [26].

In situ polymerization entails combining a monomer solution with graphene-based material suspensions in the presence of catalysts under controlled reaction conditions. Epoxy is a popular polymer that can be polymerized *in situ* [35]. For example, graphene nanoplatelet suspension was mixed with epoxy resins and placed through a high-shear mixing phase. To complete the polymerization process, epoxy curing agent was added while stirring and heating the mixture to extract the solvent [35]. Another form of polymer is polyaniline (PANI), which is typically made by *in situ* polymerization [36]. Since the polymerization of PANI is an oxidative process, an oxidative agent, such as ammonium persulfate, is added to the reaction solution to facilitate the polymerization [37] In addition to solution-based oxidation, graphene–PANI composite can be prepared by the *in situ* anodic electropolymerization [38], where graphene paper was obtained using H_2 reduction of thermally expanded GO, followed by filtration and vacuum drying. PANI was then electrochemically deposited with the graphene paper as the working electrode in a three-electrode cell containing the electrolyte with aniline monomers. Other polymers have been hybridized with graphene through *in situ* polymerization, including silicone [39], polystyrene [40] and poly(vinyl choloride/vinyl acetate) copolymer [41]. Graphene and graphene derivatives are often uniformly stacked in the polymer matrix using the aforementioned methods. Layer-by-layer (LbL) assembly has been used to prepare freely standing composite membranes to increase the directional mechanical strength of graphene–polymer composites. For example, the Langmuir–Blodgett (LB) technique was used to move GO sheets onto multilayer polyelectrolytes (poly [allylamine hydrochloride] [PAH] and poly[sodium 4-styrene sulfonate] [PSS]) films [42].

Electrical conductivity, electrochemical power, mechanical strength and thermal stability and conductivity have all been improved in graphene–polymer composites to date. PANI is one of the most commonly used host polymers, and several studies have shown that graphene–PANI composites have improved electrochemical efficiency [38]. The electrochemical capacitance of the flexible graphene–PANI composite paper prepared by *in situ* electropolymerization, for example, was 233 F/g and 135 F/cm^3 for gravimetric and volumetric capacitance, respectively. It outperforms several carbon-based flexible electrodes as a supercapacitor candidate [40]. In another study, a high capacitance of 1046 F/g was achieved by coating graphene nanosheets (GNS) with 2 nm PANI nanoparticles, compared to 115 F/g for pure PANI [39]. In addition to PANI nanoparticles, PANI nanofibers have been paired with GO sheets, resulting in a specific capacitance of 531 F/g at a PANI:GO mass ratio of 100 : 1 (323). Insulating polymers have been used to host graphene-based fillers in addition to conducting polymers.

At various composition ratios, the electrical conductivity of polystyrene–graphene composites was investigated [32]. Percolation of functionalized graphene sheets (FGS) occurred at just 0.1 vol % loading, which was lower than many other 2D filler materials. The high aspect ratio of the graphene sheets, their homogeneous dispersion in the polymer matrix, and the strong filler–matrix interfacial interaction caused by the surface functional groups are all factors that contribute to the low percolation threshold. With the other polymer matrices, poly(vinyl chloride/vinyl acetate) copolymer/graphene composites had a percolation threshold of 0.15 vol % [43], while PET–graphene composites had a percolation threshold of 0.47 vol %, and a conductivity of 2.11 S/m was achieved with only a loading of 3.0 vol % [34]. The single-layer graphene sheet, as stated in the introduction, is one of the strongest materials ever measured, with a breaking strength of 42 N m 1 and a Young's modulus of 1.0 TPa [44]. A number of studies have shown that graphene–polymer composites with graphene derivatives as filler materials have improved mechanical properties. LbL assembly was used to embed functionalized GO layers in the polyelectrolyte multilayers of PAH and PSS. The elastic modulus of the resulting composite membrane increased dramatically, from 1.5 GPa for a pure polymer membrane to 20 GPa for an 8 vol % graphene loading. These polyelectrolyte-based membrane materials are conductive, versatile and have strong mechanical properties, making them suitable for capacitive pressure sensors and as a replacement for rigid silicon in microelectromechanical systems (MEMS) [43]. Vacuum filtration was used to create solid and ductile poly(vinyl alcohol) (PVA)–graphene composites in another analysis. A Young's modulus of 4.8 GPa and tensile yield strength of 110 MPa were achieved at 3 wt % GO loading. Polyurethane (PU) was covalently bonded with GO platelets to strengthen the interaction of the graphene filler and the polymer matrix, enabling efficient load transfer. The Young's modulus and hardness were increased by ~ 900% and ~ 327%, respectively. The intrinsic features of graphene, as well as the filler's high dispersion and the filler's effective interfacial interaction with the host polymer, all contribute to the effective increase of mechanical properties of these polymer–graphene composites. Higher hardness implies to increase scratch resistance, which is important in surface coating applications [45]. Because the intrinsic properties of graphene, as well as the strong dispersion of filler and the efficient interfacial contact between the filler and host polymer, prevent polymer chains from moving under external stress, these polymer–graphene composites have improved mechanical properties significantly.

The glass transition temperature (Tg), above which the molecular chains begin to slip past each other when an external force is applied, is used to describe the thermal stability of polymers. The Tg of polymer composites can be increased by inserting stiff filler materials that interfere with the chain's flow phase. When poly(amido amine) (PAMAM) was mixed with graphitic nanoplatelets (FNPs) in THF using ultrasonication and high-speed shearing, the resulting composite had a Tg increase of up to 30° C with 1–5 wt% filler [28]. The composites' efficiency was improved even further by replacing the FNPs with functionalized graphene sheets (FGSs), which were made by thermal exfoliation of highly oxidized graphite oxide. FGSs have a higher proportion of single-layer graphene sheets with wrinkled morphologies and functionalized surfaces than FNPs, resulting in stronger filler–matrix interactions. Tg for the shaped

PAMAM–FGS composite increases by 30° C with a loading of 0.05 wt % FGS, while Tg for the poly (acrylonitrile) (PAN)–FGS composite improves by more than 40° C with a loading of 1 wt % FGS [31].

Importantly, the FGS outperforms all other carbon-based nanofillers in terms of improved polymer glass transition temperature at a 1 wt% loading of nanofiller. Since elastomers like silicone foams are commonly used at temperatures above the glass transition, the degradation temperature is used to determine their thermal stability. For example FGSs have been successfully embedded in silicone using an *in situ* foaming technique. The degradation temperature of silicone–FGS composites with 0.25 wt% filler loading increased by up to 55° C, according to thermo gravimetric analysis (TGA) [325, 326]. Another example shows that when PLA, a biodegradable polymer widely used in biological applications, was composited with EG, the thermal degradation temperature of the PLA–EG composite was increased by 10° C with just 0.5 wt% EG loading [40]. At room temperature, graphene has excellent thermal conductivity, with a calculated k value of~ 2000 W/mK for a single-layer sheet [46]. As a result, embedding it in a polymer matrix is supposed to greatly improve the composite's thermal conduction. In an epoxy matrix, thermally exfoliated graphite nanoplatelets were incorporated. With a loading of 25 vol percent graphene, it was discovered that the incorporation of few-layer graphene (layer number: 4) resulted in a thermal conductivity enhancement of more than 30 times at k = 6.44 W/mK [37]. This outperforms traditional fillers like silver, alumina and silica, which require 50–70% loading to produce a comparable performance. Furthermore, when FGSs is used as a filler in a silicon foam matrix, a 0.25 wt% loading resulted in a 6% improvement in ultimate thermal conductivity [47]. The increased thermal conductivity in these reports is due to graphene's broad aspect ratio, 2D geometry, high stiffness and low thermal interface resistance. Such heat-dissipating composite foams are ideal for coatings on electrical wires to remove conductive heat.

11.4 Applications of Graphene as an Adsorbent in Water Remediation

Surface area and pore size distribution are two significant surface properties that affect graphene adsorption. The graphene in its purest form has an extremely high specific surface area but no porosity. By combining other porous materials such as chitosan [48], gelatin [49] and silica, a breakthrough in graphene adsorption efficacy can be achieved (Fan et al., 2019). The implementation of different functional groups is another way to achieve high adsorption efficiency. These functional groups are essential in the formation of bonds with the adsorbate. This is particularly true of GO and rGO, which have many oxygen-containing functionalities capable of bonding with the adsorbate [50]. For efficient application in environmental remediation, fabrication of functionalized graphene-based nanocomposites with wide surface area, high porosity and oxygen-containing functionalities is important. As mentioned in the sections below, graphene-based adsorbents have been used to extract a number of organic contaminants, with various modifications to the pristine graphene structure.

11.4.1 Polycyclic Aromatic Hydrocarbons (PAHs)

Graphene-based materials have been used to remove aromatic polycyclic polar and non-polar compounds in several studies. Between graphene-based adsorbents and polar and non-polar PAHS, the π–π interactions, hydrophobic effect and van der Waals forces facilitate their adsorptive removal. The adsorption of naphthalene (NAP), anthracene (ANT) and pyrene (PYR) on rGO and GO was studied theoretically and experimentally, and it was discovered that at neutral pH, 95 and 60% of NAP was adsorbed on rGO and GO, respectively [51]. As a result, the rGO has a higher surface adsorption affinity for PAHs than the GO. PAH adsorption on rGO was found to be pH independent, while GO adsorption was pH dependent due to the deprotonation of the acidic functional groups of GO (-COOH and -OH). The hydrophobic interactions, on the other hand, controlled the adsorption activity at higher pH. NAP > PYR > ANT were the order of rGO's adsorption capacities. All of the PAHs' adsorption isotherms fit the Polanyi–Dubinin–Ashtakhov (PDA) model, indicating that pore filling and flat surface adsorption were involved. For the adsorptions of NAP, PYR and ANT, the Metropolis Monte Carlo method generated similar results. Wang et al. [52] compared the adsorption potential of GO and graphene nanosheets (GNS) by using NAP, phenanthrene (PNT) and phenanthrene (PYR) as targets. They theorized that low PAH adsorption causes the graphene surfaces to crumple, resulting in grooves with high adsorption energy sites. At low adsorption concentrations, these sites produced favorable adsorption. Because of the different accessibility of adsorption regions on graphene for molecules of different sizes, this sieving effect played an important role in adsorption, while the adsorptive sites of GO changed to carboxyl groups attached to the edges of GO as the groove region disappeared and the polar nanosheets surface surface decreased the π–π interactions. DFT studies by AraujoContreras et al. [53] revealed that London dispersion forces are responsible for favorable adsorption of chrysene, dibenzo[a,h]anthracene and dibenzo[a,h]pyrene onto graphene. The PBE-D functional, rather than the HCTH functional, provided more accurate geometry, electronic properties and energy results. This is due to the fact that the HCTH functional ignores dispersion correction, which affects the values of interaction distances.

Wang et al. [54] studied the adsorptive properties of colloidal graphene oxide nanoparticles (GONPs) for a variety of aromatic compounds, including non-polar (PYR, PNT, NAP, 1,3-dichlorobenzene) and polar (PYR, PNT, NAP, 1,3-dichlorobenzene). PYR and PNT have poor adsorption affinities due to their high hydrophobicity. PYR > PNT > NAP > 1,3-dichlorobenzene was the adsorption affinity order based on normalized adsorption data for the hydrophobic effect. In addition to the hydrophobic effect and van der Waals forces, electron donor-acceptor (EDA +) interactions interplay between electron-rich PAHs and electron-depleted regions of strongly polarized graphitic surfaces. The adsorption abilities of GONPs were improved by—interactions and Lewis acid-base interactions between 1-naphthylamine and H-bonding between 1-naphthol and O-functionalities. Yang et al. [55] used reduced graphene oxide/iron oxide (rGO/FeO.Fe2O3) composites and registered a similar order of adsorption (NAP 1- naphthol 1-naphthylamine).

On graphene and GO, Pei et al. [56] investigated the adsorption of 1,2,4-trichlorobenzene (TCB), 2,4,6-trichlorophenol (TCP), 2-naphthol and NAP. Since TCB and

NAPH are non-ionizable, their adsorption on graphene and GO was almost pH-independent. Because of π—π EDA interactions, there was a substantial increase in adsorption potential for 2-naphthol at higher pH (> 8.0), despite the fact that it occurred in the anionic form. This was also supported by FTIR and DFT research. Since GO has a smaller surface area and electron density than graphene, its adsorption was lower at the same pH values. On graphene, TCB, NAPH, TCP, and 2-naphthol had equal adsorption capacities, while adsorption on GO was in the order NAPH TCB TCP 2-naphthol.

11.4.2 Phenolic Compounds

There are various studies on the ability of graphene-based adsorbents for the removal of phenolic compounds, close to PAHS. The surface area of the adsorbent and the degree of reduction have a significant impact on adsorption. Peng et al. [57] created a series of highly porous rGO by activating rGO with CO2, dubbed A-rGO-25, ArGO-60 and A-rGO-75, with activation times of 25, 60 and 75 minutes respectively. The surface area was significantly improved, going from 200 to 1200 $m^2\,g^{-1}$. To test the removal ability of as-synthesized porous rGO, the authors used phenol and methylene blue dye (MB) as target molecules. A-rGO-60 had a much higher adsorption ability for phenol than activated carbon and carbon nanotubes (CNT). A-rGO-60, as synthesized, could extract more than 65% of MB dye, whereas rGO could only remove 30%. The adsorption potential of both contaminants followed the pattern rGO A-rGO-25 A-rGO-75 A-rGO-60, which was also the order of the particular surface area (SSA). The effect of graphene oxide reduction on the adsorption of phenolic compounds such as 4-hydroxyacetophenone (4-HAP), 4-hydroxybenzaldehyde (4-HBA), 2,4-dichlorophenol (2,4-DCP), 4-nitrophenol and 4-chlorophenol was investigated by Wang et al. (2014 c). They discovered that lowering GO increased the number of π—π interactions between aromatic compounds and rGO.

Ghahghaey et al. [58] used DFT-D3 calculations with the revPBE/def2- TZVP theory model to investigate the interaction of phenol and water molecules with graphene and graphene nanobuds. In this case, DFT-D3 analysis of non-local dispersion forces is critical since these forces influence the binding structure of the interacting organisms. In comparison to water molecules, the presence of electrostatic interactions and—stacking allowed stronger phenol adsorption on both graphene and graphene nanobud. Because of its higher SSA, graphene nanobud proved to be a stronger adsorbent than graphene. DFT experiments using GO as the adsorbent were used to examine the adsorption process and actions of phenol, 4-CP, 2,4-dichlorophenol and 2,4,6-trichlorophenol in another theoretical analysis [59]. They used DFTD3 calculations to account for dispersion corrections to find the most stable geometry since the weak interactions in the sorption method affect the geometry. The authors concluded that the hydrophobic effect, H-bonding, and—interactions drive the adsorption process based on the theoretical results. MD simulations backed up this point even further. It was also discovered that the adsorption affinity increases as the number of hydroxyl groups on GO and the number of chloro groups on phenols decreases. Electrostatic interactions and H-bonding were also aided by the presence of acidic conditions and polar solvents.

11.4.3 Pharmaceutical Compounds

The favorable adsorption of pharmaceutical compounds is allowed by surface area, pore size, π—π interactions, and hydrogen bonding. To investigate the adsorption of ciprofloxacin, Zhuang et al. [60] synthesized grapheme protein biocomposites with varying protein contents. Higher protein content hydrogels were more soluble, and the protein also avoided graphene agglomeration. The adsorption capacity was stated to be 500 mg g^{-1}. The adsorption was chemisorption, and the favorable adsorption was due to both π—π adsorption and hydrogen bonding interactions. The authors also investigated the effect of copper on ciprofloxacin adsorption, finding that the adsorption potential of ciprofloxacin increased within certain copper concentration ranges due to Cu^{2+} coordination with the amide and hydroxyl groups of GS. Due to competitive adsorption, ciprofloxacin adsorption decreased once the copper concentration surpassed 10 mg L^{-1}.

Akpotu and Moodley [61] synthesized silica nanotubes (SNT) and encapsulated them in graphene oxide (SNTGO) and reduced graphene oxide (SNTG). SNTGO absorbed 125 mg sulfamethoxazole per gram of body weight, while SNTG absorbed up to 248 mg g^{-1}. Because of the stronger hydrophobic interactions and hydrogen bonding between the OH group of SNTG and the oxygen atom of sulfamethoxazole molecules, SNTG has a higher adsorption potential. The Freundlich isotherm model was used, as well as pseudosecond order kinetics. Jauris et al. [62] examined the theoretical and experimental adsorption of sodium diclofenac (an anti-inflammatory drug) on graphene-based adsorbents. Theoretical studies revealed that graphene, vacancy graphene, rGO and functionalized graphene nanoribbons physisorb. Batch adsorption studies showed that rGO is the strongest adsorbent. The same research team investigated physisorption between graphene–based adsorbents and the adsorbate in the case of another anti-inflammatory medication, nimesulide [63]. With a maximum adsorption capacity of 82.4 mg g^{-1}, rGo was found to be an excellent candidate.

11.4.4 Pesticides

Various research groups have also looked at using graphene-based materials as adsorbents to remove pesticides from aqueous solutions. Pesticide adsorption is reduced by strong π—π and electrostatic interactions. Xiaotong et al. [64] used graphene coated silica (GCS) to remove eleven organophosphorus pesticides (OPP). The adsorption process was facilitated by the use of silica as a graphene carrier, which made it faster and easier. According to the scientists, silica had almost no impact on OPP adsorption, and graphene covered the majority of the surface. The Langmuir model was found to be the best fit for the experimental results. Profenofos > fenthion > primiphosmethyl > parathion-methyl > phorate > chlorphenvinphos > methidathion > malathion > isocarbophos > phonamiphos > dimnethoate were the OPPs with the highest adsorption power. The adsorption was aided by the electron-donating capacity of N, S and O atoms, as well as π-bonding networks in the phenyl rings.

Yang et al. [65] developed UiO-67/GO hybrid metal-organic framework/graphene oxide nanocomposites for the removal of OPP glyphosate. With 482.69 mg g^{-1} of glyphosate adsorbed at pH 4, the Langmuir model and pseudo-second-order kinetics model were found to be correct. The interaction of the glyphosate phosphate group with the oxygen-containing functional groups on the surface of UiO-67/GO is responsible for the favorable adsorption, according to XPS studies. The adsorption of chlorpyrifos (CP), endosulfon (ES) and malathion (ML) onto GO and rGO was investigated by Maliyekkal et al. [66]. It has been stated that CP, ES and ML have adsorption capacities of 1200, 1100 and 800 mg g-1, respectively. The adsorption was found to be pH insensitive, suggesting that the COO functional group on the rGO surface did not communicate directly with the pesticide. They used first-principles pseudopotential dependent DFT to elucidate the atomistic mechanism of interactions and discovered that only the polar bonds present in the pesticide molecules changed significantly when graphene and water were combined.

11.4.5 Dyes

Electrostatic interactions between positively charged amino groups of dye molecules and the negatively charged surface of the adsorbate, as well as van der Waals forces and interactions, are used to extract dye. Graphene-based products have been used to extract cationic dyes. Sharma et al. [67] used GO and rGO nanosheets to investigate the adsorption of methyl blue (MB) dye. The activation energy for MB adsorption on rGO was also much lower than for GO, suggesting that rGO nanosheets adsorb MB much faster than GO nanosheets. The endothermic and spontaneous nature of the adsorption mechanism, as well as a pseudo-second order kinetics model, were discovered through thermodynamics and kinetics studies. Peng et al. [68] created a magnetic cellulose/GO composite bioadsorbent and discovered that the Langmuir isotherm and pseudo-second order kinetics models were both accurate.

Zhu et al. [69] synthesized zeolite-rGO composites and found that the MB adsorption potential decreased due to ion exchange with H^+ ions released from aqueous solutions. This suggested that dye absorption is dependent on a dual adsorption and ion exchange process. Liu et al. [70] used -cyclodextrin/poly(acrylic acid) grafted onto GO (-CD/PAA/GO) nanocomposites to remove MB and safranine T dyes from aqueous solutions in another experiment. The isotherms were based on the Langmuir adsorption model, and the kinetics model suited the pseudo-second order model the best. The high adsorption capacity values were a result of the nanocomposites' high concentration of functional groups, which offered a large number of adsorption sites. Second, the abundant carboxylic groups increased the availability of adsorption sites for soluble dyes by increasing their hydrophilicity and dispersibility in aqueous solutions. π—π conjugation interactions, electrostatic interactions, and host-guest supramolecular interactions were all used to absorb the dyes.

Pourjavadi et al. [71] created GO-hydrogel porous nanocomposites and investigated the effect of the hydrogel's porosity on dye adsorption. They reached different porosities by adding different quantities of $CaCO_3$ and then extracting it. Because of the ultra high porosities, they discovered an extremely high adsorption power. The Langmuir isotherm model and pseudo-second order kinetics model best represented the adsorption. Mercante et al. [72] used poly(methylmetacrylate) nanofibers wrapped in rGO (PMMArGO) to adsorb MB and found that the spontaneous adsorption was triggered by

π—π stacking interactions. The adsorption of MB dye was better characterized by the Langmuir isotherm model and pseudo-second order kinetics model. Microcrystalline cellulose graphene oxide (MCC/GO) aerogels were used in another study [73], and extremely high adsorption ability was achieved, which was several times higher than pure GO aerogel. Because of the charge attraction between GO and MB, adsorption occurred spontaneously, and pseudo-second order kinetics were used for both GO and MCC/GO aerogels.

For the adsorption of two toxic cationic dyes, basic red 12 (BR12) and basic red 46 (BR46), Moradi et al. [74] used graphene and GO (BR46). The removal of BR12 and BR46 followed a pseudo-second order model, with the maximum removal potential at pH 9. The sum adsorbed was in the following order: GO-BR12 > GO-BR46 > G-BR12 = GBR46. Because of its smaller size and lower molecular mass, BR12 is adsorbed more readily. On both GO and graphene, thermodynamic studies showed that BR12 adsorption is endothermic while BR46 adsorption is exothermic. Sun et al. [75] produced magnetite/reduced graphene oxide (MrGO) nanocomposites and investigated the rhodamine B (RhB) dye adsorption. The Freundlich model and the pseudo-second order model best represented the adsorption behavior.

Kan et al. (2016) studied the interactions of Rhodamine 6 G (R6G) dye with ideal and defective graphene surfaces using DFT calculations. Because of the increase in orbital overlap caused by the dopant between the substrate and R6G, the B/N doping cases have a higher adsorption potential than perfect graphene. Only the graphene substrate with monovacancy defect, trivacancy defect and B-doped graphene layer exhibited chemisorption. This is because, prior to R6G adsorption, their measured electronic structures showed a single peak around the Fermi range. Under-coordinated atoms, which can undergo a chemical reaction, are correlated with such peaks. In other cases, no such peak was observed and hence they displayed physisorption mode.

11.5 Future Scope

The applications of graphene and its derivatives for the removal of organic contaminants are discussed in this study. Because of the abundance of adsorption sites, such as oxygen containing functional groups, electrons and an outer surface adorned with nanoparticles, graphene-based nanocomposites outperform GO and rGO in terms of pollutant removal. In addition, the graphene sheets serve as a support and prevent metal ions from leaching into the solution. The finer properties of graphene-based nanocomposites than their constituents motivate researchers to establish new synthesis routes and use the resulting composites for water remediation. A extensive DFT-based analysis can be an extremely useful tool in this direction, as well as a tool for synthesizing customized "functional" materials. DFT may also be used to assess the efficacy of adsorption when a specific adsorbent is used to target a specific adsorbate. Furthermore, DFT research is critical for comprehending the atomistic process of adsorption. However, the main challenge in using graphene-based adsorbents is the recovery of these materials. One of the ways to tackle this hurdle is by synthesizing magnetic composites which can be easily retrieved from the analyte using a magnet and can thus prevent crosscontamination.

Also, rGO has been proved to be a better adsorbent than graphene or GO. This is because the porosities, dislocations and lattice vacancies in rGO lead to the formation of micropores which create heterogeneous sites with different adsorption energies and make rGO a better adsorbent.

The use of graphene-based adsorbents in the environment should not be limited to water remediation, but should also include soil and air purification. The high adsorption potential and physisorption of the material can be used to extract and isolate pollutants from the air, soil and water without further degradation. The majority of researches have concentrated on common contaminants such as dyes, pesticides, organic solvents and so on. As a result, future research should concentrate on emerging contaminants such as octanoic acid derivatives, oxygen-demanding wastes, tar, grease and so on. Despite numerous evidences demonstrating the effectiveness of graphene-based adsorbents, their widespread use in environmental remediation is not acknowledged. This is due to the fact that the issue of toxicity, which includes both short- and long-term exposure to humans and the environment, has gone unaddressed. Graphene-based materials tend to be excellent candidates for water treatment applications so far, but the study is still in its infancy and requires further input. The applications of graphene-based adsorbents will be revolutionized as physics and chemistry advance. Future research could also concentrate on developing more graphene-based materials using nanomaterial-processing technologies. This review will assist researchers in better understanding the various properties and phenomena associated with graphene-based materials, especially as adsorbents for water treatment, and in further developing the potential of graphene-based adsorbents.

References

1 Rashed, M.N. (2013). Adsorption technique for the removal of organic pollutants from water and wastewater. *Inf. Technol.* 7: 167–194.

2 Ariffin, N., Abdullah, M.M.A.B., Zainol, M.A., Murshed, M.R.R., Zain, M.F.H., Faris, M.A., and Bayuaji, R. (2017). Review on adsorption of heavy metal in wastewater by using geopolymer. *MATEC Web Conf* 97: 01023.

3 Wang, Sh. and Peng, Y. (2010). Natural zeolites as effective adsorbents in water and wastewater treatment. *Chem. Eng. J.* 156: 11–24.

4 Ali, I. and Jain, C.K. (2005). Wastewater treatment and recycling technologies. In: *Water Encyclopedia, Domestic, Municipal, and Industrial Water Supply and Waste Disposal*, (ed. J. Lehr), 968. New York: springer.

5 Ali, I. (2012). New generation adsorbents for water treatment. *Chem. Rev.* 112: 5073–5091.

6 Santhosh, Ch., Velmurugan, V., Jacob, G., Jeong, S.K., Grace, A.N., and Bhatnagar, A. (2016). Role of nanomaterials in water treatment applications, a review. *Chem. Eng. J.* 306: 1116–1137.

7 Ren, X., Chen, C., Nagatsu, M., and Wang, X. (2011a). Carbon nanotubes as adsorbents in environmental pollution management, a review. *Chem. Eng. J.* 170: 395–410.

8 Boehm, H., Setton, R., and Stumpp, E. (1986). Nomenclature and terminology of graphite intercalation compounds. *Carbon* 24: 241–245.

9 Fitzer, E., Kochling, K.H., Boehm, H.P., and Marsh, H. (1995). Recommended terminology for the description of carbon as a solid (IUPAC Recommendations 1995). *Pure Appl. Chem.* 67: 473.

10 Katsnelson, M.I. (2012). *Graphene, Carbon in Two Dimensions*. New York: Wiley–Interscience.

11 Si, Y. and Samulski, E.T. (2008). Synthesis of water soluble graphene. *Nano Lett.* 8: 1679–1682.

12 Liu, L.-H. and Yan, M. (2011). Functionalization of pristine graphene with perfluorophenyl azides. *J. Mater. Chem.* 21: 3273–3276.

13 Choi, E.Y., Han, T.H., Hong, J.H., Kim, J.E., Lee, S.H., Kim, H.W., and Kim, S.O. (2010). Noncovalent functionalization of graphene with end-functional polymers. *J. Mater. Chem.* 20: 1907–1912.

14 Dong, X., Cheng, J., Li, J., and Wang, Y. (2010). Graphene as a novel matrix for the analysis of small molecules by MALDI-TOF MS. *Anal. Chem.* 82: 6208–6214.

15 Si, Y. and Samulski, E.T. (2008). Synthesis of water soluble graphene. *Nano Lett.* 8: 1679–1682.

16 Lerf, A., He, H., Forster, M., and Klinowski, J. (1998). Structure of Graphite Oxide Revisited||. *J. Phys. Chem. B*102: 4477–4482.

17 Niyogi, S., Bekyarova, E., Itkis, M.E., McWilliams, J.L., Hamon, M.A., and Haddon, R.C. (2006). Solution properties of graphite and graphene. *J. Am. Chem. Soc.* 128: 7720–7721.

18 Salavagione, H.J., Gómez, M.n.A., and Martìnez, G. (2009). Polymeric modification of graphene through esterification of Graphite Oxide and Poly(vinyl alcohol. *Macromolecules* 42: 6331.

19 Yang, H.F., Li, F.H., Shan, C.S., Han, D.X., Zhang, Q.X., Niu, L., and Ivaska, A. (2009). Covalent functionalization of chemically converted graphene sheets viasilane and its reinforcement. *J. Mater. Chem.* 19: 4632.

20 Lomeda, J.R., Doyle, C.D., Kosynkin, D.V., Hwang, W.-F., and Tour, J.M. (2008). Diazonium functionalization of surfactant-wrapped chemically converted graphene sheets. *J. Am. Chem. Soc.* 130: 16201–16206.

21 He, H. and Gao, C. (2010). General approach to individually dispersed, Highly soluble, and conductive graphene nanosheets functionalized by Nitrene chemistry. *Chem. Mater.* 22: 5054–5064.

22 Shen, J., Hu, Y., Li, C., Qin, C., and Ye, M. (2009). Synthesis of amphiphilic graphene nanoplatelets. *Small* 5: 82–85. doi: 10.1002/smll.200800988.

23 Patil, A.J., Vickery, J.L., Scott, T.B., and Mann, S. (2009). Aqueous stabilization and self-assembly of graphene sheets into layered bio-nanocomposites using DNA. *Adv. Mater.* 21: 3159–3164.

24 Su, Q., Pang, S., Alijani, V., Li, C., Feng, X., and Müllen, K. (2009). Composites of graphene with large aromatic molecules. *Adv. Mater.* 21: 3191–3195.

25 Xu, Y.X., Hong, W.J., Bai, H., Li, C., and Shi, G.Q. (2009). Strong and ductile poly(vinyl alcohol)/graphene oxide composite films with a layered structure. *Carbon* 47: 3538–3543.

26 Dasari, A., Yu, -Z.-Z., and Mai, Y.-W. (2009). Electrically conductive and super-tough polyamide-based nanocomposites. *Polymer* 50: 4112–4121.

27 Liang, J., Huang, Y., Zhang, L., Wang, Y., Ma, Y., Guo, T., and Chen, Y. (2009). Molecular-level dispersion of Graphene into Poly(vinyl alcohol) and Effective reinforcement of their nanocomposites. *Adv. Funct. Mater.* 19: 2297–2302.

28 Ramanathan, T., Stankovich, S., Dikin, D.A., Liu, H., Shen, H., Nguyen, S.T., and Brinson, L.C. (2007). Graphitic nanofillers in PMMA nanocomposites—An investigation of particle size and dispersion and their influence on nanocomposite properties. *J. Polym. Sci., Part B: Polym. Phys.* 45: 2097–2112.

29 Cai, D.Y. and Song, M. (2009). The mechanical properties and morphology of a graphite oxide nanoplatelet/polyurethane composite. *Nanotech.* 20: 315708. doi: 10.1088/0957-4484/20/8/085712.

30 Liang, J., Xu, Y., Huang, Y., Zhang, L., Wang, Y., Ma, Y., Li, F., Guo, T., and Chen, Y. (2009). *J. Phys. Chem. C* 113: 9921–9927.

31 Ramanathan, T., Abdala, A.A., Stankovich, S., Dikin, D.A., Herrera-Alonso, M., Piner, R.D., Adamson, D.H., Schniepp, H.C., Chen, X., Ruoff, R.S., Nguyen, S.T., Aksay, I.A., Prud'homme, R.K., and Brinson, L.C. (2008). Functionalized graphene sheets for polymer nanocomposites. *Nat. Nanotech.* 3: 327–331.

32 Stankovich, S., Dikin, D.A., Dommett, G.H.B., Kohlhaas, K.M., Zimney, E.J., Stach, E.A., Piner, R.D., Nguyen, S.T., and Ruoff, R.S. (2006). Graphene-based composite materials. *Nature* 442: 282–286.

33 Qi, X.Y., Pu, K-Y., Li, H., Zhou, X.Z., Wu, S.X., Fan, Q-L., Liu, B., Boey, F., Huang, W., and Zhang, H. (2010). Amphiphilic graphene composites. *Angew Chem Int Ed.* 49: 9426–9429, doi: 10.1002/anie.201004497.

34 Zhang, H.B., Zheng, W.G., Yan, Q., Yang, Y., Wang, J.W., Lu, Z.H., Ji, G.Y., and Yu, Z.Z. (2010). Electrically conductive polyethylene terephthalate/graphene nanocomposites prepared by melt compounding, *Polymer*, 51 (5): 1191–1196.

35 Yu, A.P., Ramesh, P., Itkis, M.E., Bekyarova, E., and Haddon, R.C. (2007). Graphite Nanoplatelet–Epoxy composite thermal interface materials. *J. Phys. Chem. C*111: 7565–7569.

36 Zhou, X.S., Wu, T.B., Hu, B.J., Yang, G.Y., and Han, B.X. (2010). Synthesis of graphene/polyaniline composite nanosheets mediated by polymerized ionic liquid. *Chem. Commun.* 46: 3663–3665.

37 Yan, J., Fan, Z.J., Wei, T., Qian, W.Z., Zhang, M.L., and Wei, F. (2010). Fast and reversible surface redox reaction of graphene–MnO_2 composites as supercapacitor electrodes. *Carbon* 48: 3825–3833.

38 Wang, D.W., Li, F., Zhao, J.P., Ren, W.C., Chen, Z.G., Tan, J., Wu, Z.S., Gentle, I., Lu, G.Q., and Cheng, H.M. (2009). Fabrication of Graphene/Polyaniline composite Paper via in Situ Anodic Electropolymerization for High-performance flexible electrode. *ACS Nano* 3: 1745–1752.

39 Verdejo, R., Barroso-Bujans, F., Rodriguez-Perez, M.A., de Saja, J.A., and Lopez-Manchado, M.A. (2008). Functionalized graphene sheet filled silicone foam nanocomposites. *J. Mater. Chem.* 18: 2221–2226.

40 Hu, H., Wang, X., Wang, J., Wan, L., Liu, F., Zheng, H., Chen, R., and Xu, C. (2010). Preparation and properties of graphene nanosheets-polystyrene nanocomposites via in situ emulsion polymerization. *Chem. Phys. Lett.* 484: 247–253.

41 Wei, T., Luo, G.L., Fan, Z.J., Zheng, C., Yan, J., Yao, C.Z., Li, W.F., and Zhang, C. (2009). Preparation of graphene nanosheet/polymer composites using in situ reduction–extractive dispersion. *Carbon* 47: 2296–2299.

42 Kulkarni, D.D., Choi, I., Singamaneni, S.S., and Tsukruk, V.V. (2010). Graphene oxide–polyelectrolyte nanomembranes. *ACS Nano* 4: 4667–4676.

43 Kotov, N.A., Dekany, I., and Fendler, J.H. (1996). Ultrathin graphite oxide–polyelectrolyte composites prepared by self-assembly: transition between conductive and non-conductive states. *Adv. Mater.* 8: 637–641.

44 Lee, C., Wei, X.D., Kysar, J.W., and Hone, J. (2008). Measurement of the elastic properties and intrinsic strength of monolayer grapheme. *Science* 321: 385–388.

45 Cai, D.Y., Yusoh, K., and Song, M. (2009). The mechanical properties and morphology of a graphite oxide nanoplatelet/polyurethane composite. *Nanotech.* 20: 085712.

46 Balandin, A.A., Ghosh, S., Bao, W., Calizo, I., Teweldebrhan, D., Miao, F., and Lau, C.N. (2008). Superior thermal conductivity of Single-layer graphene. *Nano Lett.* 8: 902–907.

47 Kim, Y.-K., Na, H.-K., and Min, D.-H. (2010). The direct growth of gold rods on graphene thin films. *Chem. Commun.* 46: 3185–3187.

48 Liu, J. and Fu, L. (2019). Graphene: Controllable growth of graphene on liquid surfaces (Adv. Mater. 9/2019). *Adv. Mater.* 31: doi: 10.1002/adma.201970060.

49 Pei, S. and Cheng, H.-M. (2012). The reduction of graphene oxide. *Carbon* 50. doi: 10.1016/j.carbon.2011.11.010.

50 Pei, S. and Cheng, H.-M. (2012). The reduction of graphene oxide. *Carbon* 50: 3210–3228. doi: 10.1002/adma.201970060.

51 Wang, F., Haftka, J.J., Sinnige, T.L., Hermens, J.L., and Chen, W. (2014). Adsorption of polar, nonpolar, and substituted aromatics to colloidal graphene oxide nanoparticles. *Environ Pollut* 186: 226–233.

52 Araujo-Contreras, V., Yepez, F., Castellano, O., Urdaneta, J., and Cubillan, N. (2019). Interaction of Chrysene, Dibenzo[a,h]anthracene and Dibenzo[a,h]pyrene with Graphene Models of Different Sizes: Insights from DFT Molecular Electrical Properties. *Polycyclic Aromatic Compounds* 39: 99–110. doi: 10.1080/10406638.2016.1267020.

53 Wang, J., Chen, Z., and Chen, B. (2014b). Adsorption of polycyclic aromatic hydrocarbons by graphene and graphene oxide nanosheets. *Environ Sci Technol* 48: 4817–4825.

54 Yang, X., Chen, C., Li, J., Zhao, G., Ren, X., and Wang, X. (2012). Graphene oxide-iron oxide and reduced graphene oxide-iron oxide hybrid materials for the removal of organic and inorganic pollutants. *Rsc Advances* 2: 8821–8826. doi: 10.1039/c2ra20885g.

55 Pei, Z.-G., Li, L., Sun, L., Zhang, S., Shan, X.-Q., Yang, S., and Wen, B. (2013). Adsorption characteristics of 1,2,4-trichlorobenzene, 2,4,6-trichlorophenol, 2-naphthol and naphthalene on graphene and graphene oxide. *Carbon* 51: 156–163. doi: 10.1016/j.carbon.2012.08.024.

56 Peng, li, Xu, Z., Liu, Z., Wei, Y., Sun, H., Li, Z., Zhao, X., and Gao, C. (2015). An iron-based green approach to 1-h production of single-layer graphene oxide. *Nature Communications* 6: 5716. doi: 10.1038/ncomms6716.

57 Ghahghaey, Z., Hekmati, M., and Ganji, M. (2019). Removing phenol contaminants from wastewater using graphene nanobuds: DFT and reactive MD simulation investigations. *Journal of Molecular Liquids* 286. doi: 10.1016/j.molliq.2019.04.149.

58 Ghahghaey, Z., Hekmati, M., and Ganji, M. (2019). Removing phenol contaminants from wastewater using graphene nanobuds: DFT and reactive MD simulation investigations. *J. Mol. Liq.* 286. 110872. doi: 10.1016/j.molliq.2019.04.149.

59 Zhuang, Y., Yu, F., and Chen, J. (2015). Facile synthesis of three-dimensional graphene–soy protein aerogel composites for tetracycline adsorption. *Desalination and Water Treatment* 57: 1–10. doi: 10.1080/19443994.2015.1029530.

60 Akpotu, S. and Moodley, B. (2018). Encapsulation of Silica Nanotubes from Elephant Grass with Graphene Oxide/Reduced Graphene Oxide and Its Application in Remediation of Sulfamethoxazole from Aqueous Media. *ACS Sustainable Chemistry & Engineering* 6. doi: 10.1021/acssuschemeng.7b02861.

61 Akpotu, S. and Moodley, B. (2018). Encapsulation of silica nanotubes from elephant grass with Graphene Oxide/Reduced Graphene Oxide and its application in remediation of Sulfamethoxazole from aqueous media. *ACS Sustain. Chem. Eng.* 6. 4539–4548. doi: 10.1021/acssuschemeng.7b02861.

62 Jauris, I., Ferreira de Matos Jauris, C., Saucier, C., Lima, E., Zarbin, A., Fagan, S., Machado, F., and Zanella da Silva, I. (2015). Adsorption of sodium diclofenac on graphene: A combined experimental and theoretical study. *Phys. Chem. Chem. Phys.* 18: 1526–1536, doi: 10.1039/C5CP05940B.

63 Jauris, I., Ferreira de Matos Jauris, C., Zarbin, A., Umpierres, C., Saucier, C., Lima, E., Fagan, S., Zanella da Silva, I., and Machado, F. (2017). Adsorption of anti-inflammatory nimesulide by graphene materials: A combined theoretical and experimental study. *Phys. Chem. Chem. Phys.*19: 22099–22110. doi: 10.1039/C7CP04272H.

64 Yang, Q., Wang, J., Chen, X., Yang, W., Pei, H., Hu, N., Li, Z., Suo, Y., Li, T., and Wang, J. (2018). Simultaneously Detection and Removal of Organophosphorus Pesticide by A novel Zr-MOF based Smart Adsorbent. *Journal of Materials Chemistry A* 6. doi: 10.1039/C7TA08399H.

65 Yang, Q., Wang, J., Chen, X., Yang, W., Pei, H., Hu, N., Li, Z., Suo, Y., Li, T., and Wang, J. (2018). Simultaneously detection and removal of organophosphorus pesticide by A novel Zr-MOF based smart adsorbent. *J. Mater. Chem. A* 6: 2184–2192. doi: 10.1039/C7TA08399H.

66 Maliyekkal, S., Sreenivasan, S., Krishnan, D., Kouser, S., Mishra, A.K.U.M.A.R., Waghmare, U., and Thalappil, P. (2013). Graphene: A reusable substrate for unprecedented adsorption of pesticides. *Small* 9: doi: 10.1002/smll.201201125.

67 Peng Zhi, S. and Ye, N. (2014). Magnetite-graphene oxide composites as a magnetic solid-phase extraction adsorbent for the determination of trace sulfonamides in water samples. *Anal. Methods* 6. doi: 10.1039/C4AY02027H.

68 Peng Zhi, S. and Ye, N. (2014). Magnetite-graphene oxide composites as a magnetic solid-phase extraction adsorbent for the determination of trace sulfonamides in water samples. *Anal. Methods* 6: 9725–9730. doi: 10.1039/C4AY02027H.

69 Zhao, L., Dong, P., Xie, J., Li, J., Wu, L., Yang, S., and Luo, J. (2014). Porous graphene oxide-chitosan aerogel for tetracycline removal. *Mater. Res. Express* 1: 015601.

70 Pourjavadi, A., Nazari, M., Kabiri, B., Hosseini, S.H., and Bennett, C. (2016). Preparation of Porous Graphene Oxide/Hydrogel Nanocomposites and Their Ability for Efficient Adsorption of Methylene Blue. *RSC Adv.* 6. doi: 10.1039/C5RA21629J.

71 Pourjavadi, A., Nazari, M., Kabiri, B., Hosseini, S.H., and Bennett, C. (2016). Preparation of porous graphene Oxide/Hydrogel nanocomposites and their ability for efficient adsorption of Methylene blue. *RSC Adv.* 6. 10430–10437. doi: 10.1039/C5RA21629J.

72 Mercante, L., Facure, M., Locilento, D., Sanfelice, R., Migliorini, F., Mattoso, L., and Correa, D. (2017). Solution blow spun PMMA nanofibers wrapped with reduced graphene oxide as efficient dye adsorbent. *New J. Chem.* 41. 9087–9094. doi: 10.1039/C7NJ01703K.

73 Wei, X., Huang, T., Yang, J.-H., Zhang, N., Wang, Y., and Zhou, Z.-W. (2017). Green synthesis of hybrid graphene Oxide/Microcrystalline cellulose aerogels and their use as superabsorbents. *J. Hazard. Mater.* 335: 28–38. doi: 10.1016/j.jhazmat.2017.04.030.

74 Moradi, O., Gupta, V., Agarwal, S., Tyagi, I., Asif, M., Makhlouf, A., Sadegh, H., and Shahryari-ghoshekandi, R. (2015). Characteristics and electrical conductivity of graphene and graphene oxide for adsorption of cationic dyes from liquids: Kinetic and thermodynamic study. *J. Ind. Eng. Chem.*28: 294–301. doi: 10.1016/j.jiec.2015.03.005.

75 Sun, H., Cao, L., and Lu, L. (2011). Magnetite/reduced graphene oxide nanocomposites: One step solvothermal synthesis and use as a novel platform for removal of dye pollutants. *Nano Research* 4: 550–562. doi: 10.1007/s12274-011-0111-3.

12

Sewage Sludge

Use in Agriculture Practices

CHAPTER MENU

12.1 Introduction

Eighty percent of wastewater is released without treatment into the world's water resources around the globe [1]. As a result, untreated toxic contaminants present in the water cause a serious threat to the whole environment as well as to the climate and human health. "Sustainable development" and a circular economy have made a goal that the wastewater shall be "the source of water, then energy, organics, metals, and other resources". As a result, SS (sewage sludge) is documented as a source of "renewable energy" and "material recovery" [2]. Therefore, it is no longer deemed as waste, because now it is recycled into many products such as "nature as energy", "matter" or both. Because of the various properties of sewage sludge its management is difficult. Sewage sludge is a basin of organic material and nutrients, so it encompasses a possible substrate for selection of probable circumstances. Much water, high amounts of organic matter and the presence of high amounts of pathogenic microorganisms in solids are removed during preliminary sedimentation [3, 4]. Sludge is mainly composed of two categories: primary and secondary sludge. Primary sludge is generated by the use of gravity from the physical water treatment process; when there is capture of suspended solids and organics and when the microorganisms that consume organic matter are involved in the water

Aquatic Environmental Bioengineering: Monitoring and Remediation of Contamination, First Edition.
Rouf Ahmad Bhat, Mohammad Yaseen Mir, Gowhar Hamid Dar and Moonisa Aslam Dervash.

treatment process, then secondary sludge is produced. Primary sludge is foul smelling, grayish brown and should be dewatered mechanically [3]. Compared to primary sludge, due to the presence of microorganisms in secondary sludge, it does not produce so much foul smell and it is quite defiant to hydrolysis. Sewage sludge is produced during manufacturing processes in industries and municipal wastewater treatment. In municipal wastewater treatment plants, secondary sludge frequently has chemical remains of "phosphorus precipitation". In treatment plants of industrial wastewater various processes like "neutralization, precipitation of toxic compounds, or coagulation" are there, which are the cause of the generation of chemical sludge. Generation of chemical sludge is due to the quantity of sewage sludge and quality depends upon the characteristics of wastewater and the various process used for wastewater treatment. Generally, the quantity of sludge depends upon the degree of wastewater; if there is an increase in the degree of wastewater, the quantity of sludge also increases. At the same time, high demand for sludge quality elevates the sludge infectivity level [3]. It is yellowish brown in color, active biologically and not easy to dewater [3]. There is a high amount of phosphorus (P) in polyphosphate form in the secondary sludge where "biological nutrient removal systems" are applied to treat wastewater. P complexes collected in secondary sludge are released to the wastewater under anaerobic conditions in "methane digestion chambers" [5]. "Heavy metals", "polycyclic aromatic hydrocarbon" (PAH) and "polychlorinated biphenyls" are the micropollutants which are the causes of sewage sludge pollution. There is a difference in the quality of both the primary sludge and secondary sludge. Methods of stabilization of sludge treatment of sludge and disposal methods are dependent upon its composition. The composition of SS is highly uneven and it depends upon many aspects like "seasons", "technology behind different wastewater treatment plants" and "the specificity of pollutant source area", etc. For example, sources that can be extracted like P are expected to become extinct in the coming 50–100 years; therefore, recovery of these sources from the wastewater is an alternative in front of us [6]. However, less content of cellulose and lignin can be used to easily decompose the organic matter. Therefore, its application without pretreatment causes its fast degradation and sharp peak in the nitrate and concentration of pollutants in the soil. The huge amount of sludge handling involves a major proportion of the large amount of operating expenses of water-treatment workings, and there is a need of suitable strategies that are sustainable from an environmental as well as the economic point of view. "Incineration", landfill for sanitary wastes, use for land-based applications such as "structural soil improvement, soil buffer and soil amendment" are the primary routes of sludge disposal [7]. Sewage sludge application after treatment to the land could be an alternative and fulfill the requirements of nitrogen and phosphorus for most of the crops. In addition to this, the use of sewage sludge on soil, primarily in agricultural fields, as compared to "incineration" or landfills for sanitary disposal has fewer expenses. González-Ubierna et al. [8], reported that in the Mediterranean region of Europe, primarily where there is high temperature in summer joined with intensive and unsuitable cultivation practices, these endorse a steady reduction in the soil organic matter. Milieu [9] used 40% of sewage sludge as an organic alteration of soil because of its high organic matter content. The use of sewage sludge to add nutrients and the other organic substance could be useful for the soil, but its use also signifies the risk because of the presence of organic pollutants and heavy metals in it. The organic compounds which are commonly present in municipal sewage sludge are "absorbable organic halogens (AOX)", "linear alkylbenzenesulfonates (LAS)", "nonylphenols and nonylphenolethoxylates (NP and NPnEOs)", "di-ethylhexylphthalate (DEHP)", "polyaromatic hydrocarbons (PAH)", "polychlorinated biphenyls (PCB)", "polychlorinated dibenzo-pdioxins" and "(PCDD/F)–furans" [10]. The presence of nonylphenols and nonylphenolethoxylates and of di-ethylhexylphthalate is present abundantly because these are used widely. "Nonylphenolethoxylates" are widely used "alkylphenol ethoxylates", their concentrations range from 1 to 20 and these account for "80–85%" of the total use [11]. As they act as a nonionic surfactant these are extensively used for agriculture, industries and domestic use. During sewage treatment, nonylphenolethoxylates continuously undergo a chain of rapid conversions, where the "polyethoxylate group" get broken up into short-chain nonylphenols like "nonylphenol (NP)", "nonylphenolmonoethoxylate (NP1EO)" and "nonylphenol diethoxylate (NP2EO)". Among the phthalic acid esters "Di-ethyl-hexyl phthalate (DEHP)" is studied extensively due to its perseverance. For the production of different products like resins of cellulose and polyurethane, PVC, PVA, etc., phthalates act as industrial chemicals and are used as additives [12]. Pak et al. [13] has proved that these pollutants are responsible for causing various side-effects like "reproductive damage", "carcinogenicity" and "metabolic and obesity diseases". Considering their toxicity and persistent environmental risks, alkylphenol ethoxylates and phthalates have been listed as main contaminants and these are limited in use because of health issues [14].

12.2 Characteristics of Sewage Sludge

In considering the useful properties of sewage sludge in agriculture application, it may act as a fertilizer or soil conditioner, unless it contains noxious substances. Sewage sludge produced at various treatment plants varies in composition, and due to the differences between seasons there is a requirement of information about the chemical composition of SS before applying it to agricultural land. Processes of wastewater treatment and sewage sludge treatment show the uniqueness of sewage sludge. "Usually various organic complexes, broad array of micronutrients, supplementary trace metals, macronutrients, organic micro contaminants and microorganisms are the composition of sewage sludge" [15]. Macronutrients that are present in sewage sludge provide useful soil conditioning properties and acts as a fine plant nutrient [16]. The urban sewerage system hardly transports domestic sludge to wastewater treatment plants. The effluent from industries, stormwater runoff and many types of waste from different areas are frequently released into the sewerage systems. That is the reason why sewage sludge contains various toxic and organic pollutants. The presence of other dangerous toxic materials has also been detected in sewage sludge, such as the presence of detergents, pesticides and chemicals due to waste from industries and municipal waste, and these also act as hormone disruptors.

12.3 Activation of Sewage Sludge

Nowadays micropollutants are the main problem in treatment plants of sewage sludge, particularly "polycyclic aromatic hydrocarbons, polychlorinated biphenyls, pharmaceuticals and pesticides", etc. During the processes of aerobic and anaerobic stabilization, one part from them is removed [17, 18]. "Advanced oxidation methods" are used for the breakdown and the better removal of micropollutants from sewage sludge. Until now, it was confirmed that "Fenton's reagent" is effective for the removal of organic micropollutants [19, 20]. In recent times, new methods for micropollutant removal have been developed such as manufacturing of "activated carbon" from SS by its activation [21, 22]. "Activated carbons" are derived from SS and are used as an adsorbent. Municipal sludge and industrial sludge are both used for the purpose of preparation of activated carbons. Sludge activation is required because direct application of dried sewage sludge is useless as an "adsorbent" [23]. "Pyrolysis" is the chemical activation of the sludge which gets adsorbents with a considerably high surface area [24]. Before the process of "pyrolysis" the material is generally used as a source for activation of sludge, though chemical activation may also be the primary step for the preparation of activated carbon before the process of "pyrolysis" [23]. The change of sewage sludge to activated carbon depends upon the temperature of the pyrolysis process [25]. "KOH, $ZnCl^2$ and $H3PO^4$" are the variety of chemical reagents that are used for sewage sludge activation. Generally, the process of chemical activation and pyrolysis are joined. Other chemicals like $Fe^{2+/}Fe^{3+}$ might also be used for sludge activation [23, 26]. Acids alter the surface of char with additional "oxygen-containing groups" [25, 27].

Textile sewage sludge activated carbon has also been prepared from textile sludge by an activation process using KOH with "pyrolization" and "carbonization" processes [22]. Various sorts of activated sludge can also be manufactured by changing the temperature of activation and ratio of impregnation with different sorption distinctiveness [21]. The sorption abilities of activated sludge found thus depends upon "primary pH", "temperature" and "the class of pollutants" among others. For sludge activation "iron chloride ($FeCl^3$)" has also been used [28]. Also, aerobic grainy sludge released from cosmetic industrial waste water has been the subject of experiment. Distilled water was used to wash the sludge to obtain stable conductivity. The "Electro-Fenton activation" process may also be used for sludge activation. At the same time, oxidative radicals and ferric ions are generated during the process of the "Electro-Fenton" process. Fe is used as an anode and from the dissolved oxygen, hydrogen peroxide is produced. A group of hydrogen peroxide and Fe^{2+} generates strongly "oxidative hydroxyl radicals". "Disintegration and stabilization of sludge flocks" may take place during the "Electro-Fenton" process [26].

12.4 Disposal of Sludge to Land

At present, there are various substitutes for sludge disposal, such as dumping of sludge at the sea area, dewatered landfill, destroyed sludge and application of dewatered SS to agricultural land. About 80% of sewage sludge that was generated during treatment of sewage in the UK was released on the land; roughly half of this has been applied to land as estimated in 1970 [29]. Sterritt and Lester [30] studied these extents and described that disposal of sludge to agricultural land is a primary outlet. A large amount of residue of SS has been released to sea, and only 3% has been reduced to ashes. Most of the sewage sludge is applied to agricultural land and grazing land; a small quantity is used in forestry and horticulture [30]. The elevation in sludge quantity that was discarded at sea was predicted to be less than 10% in two years [30], while the other important passage is the use of landfill, which is probably limited due to the cost and availability of sites. To evaluate the costs of dumping of sludge at sea and for land application is difficult due to the involvement of many factors. Moreover, the cost of land application is usually lower than that of disposal of sludge at sea, but land application of sludge requires many additional treatments for its use in agriculture like digestion, conditioning and dewatering and all these treatment are cost-comparable. Transport of sludge from domestic sewerage treatment to the sea area may elevate the costs above normal; as a result, this method is economically acceptable only for coastal areas. So, it was considered that disposal of sludge to agricultural land is a best practice and it should be opted for if not considered dangerous to the health of human beings [29]. "Sewage sludge" could have useful effects on the physical properties of soil [31] and acts as a possible source of N and P.

The array of sludge application rates to agricultural land has been reported in the previous literature, and it covers a large range. This is due to extreme rates of application used by researchers in search of heavy metal uptake by crops. "450 t ha^{-1}" has been used as the application rate in an experiment [32] though the total metal quantities that were added most likely equal the application rates that were used in an another experiment, i.e. "80 t ha^{-1}" [33]. In some other studies where trace metals were of smaller quantity, poor stable grazing land has been enhanced after receiving about "10 m 3 ha^{-1} $year^{-1}$" over a period of ten years [34]. The application of dry SS compost at the rate of "240 t ha^{-1}" increased the "soil cation-exchange capacity" [34]. It was reported in the previous research articles that the scarcity of information available on the total sludge quantities applied to the land in the United States makes it hard to differentiate the sensible rate from extreme rates. Furthermore, strategies restricting the sludge application to crop land first came out in the 1970s [35, 36] and have since been reorganized [37, 38].

The options for disposal of sewage sludge generated inland depend upon the relative costs and feasibilities if considerations of health are not preventive factors. There are many risks related with the use of SS on agricultural land that may result in limitations to its application. Progress in sludge disposal policies will consequently depend to an important point on a full consideration of the cost of contaminated sludge application along with trace elements to agricultural soil.

12.5 The Effect of Sludge Application on Soil Properties

12.5.1 Physico-Chemical Properties

Application of sewage sludge to land is becoming increasingly accepted because of its opportunities for recycling costly components, including organic matter, nitrogen, phosphorus and other plant nutrients [39]. The use of SS to agricultural soil facilitates the nutrient recycling found in sewage sludge and eradicates the requirement of synthetic fertilizers in agricultural land [40]. Because sludge is an organic fertilizer it increases soil fertility for a long time period [41]. Unwise modifications to sludge amendment can change the properties of soil, particularly when there is a high concentration of heavy metals and toxic substances. The physical properties of soil have been enhanced by the application of SS [42]. It was reported that the application of municipal SS increases the pH of soil [43]. A reduction in the pH of soil has also been reported [34]. These alterations in the pH of soil are due to the connection between the $CaCo_3$ content in sludge and manufacturing of acid during breakdown of sludge [40]. Deliberation of the pH of soil is important because of the presence of abundant toxic elements in sewage sludge. Sewage sludge compost with organic matter added to the agricultural soil enhances the properties of soil including "bulk density", "porosity" and "the water holding capacity" [44]. Sludge-soil chemical properties do not depend only on the soil properties, sludge and rate of application, but also depend upon the interactions between them and the pH of the soil [44]. The effect of 0.5% application of SS to land on "water retention, hydraulic conductivity and aggregate stability" demonstrates that "raw and digested sludge", elevates the entire "soil-water retention capacity" with the utmost elevation in the raw sludge revised soil [42]. The major elevation in soil hydraulic conductivity was caused by the addition of sludge to soil subsequent to 27 days of incubation. Treatment using raw sludge had the high fraction of constant aggregates in the first 118 days of incubation, but the percentage of constant aggregates for the sludge treatments were the same after 175 days, averaging 34% in contrast to untreated soil which had 17% [42]. The concentrations of metal in the SS depend on a number of aspects including "sewage origin, processes of sewage treatment, and processes of sludge treatment" [45]. Properties of soil including "pH, redox potential, sesquioxide content, organic matter and sludge application rate" influence the bioavailability of metals present in the sewage sludge [45]. The recommended upholding of pH above 6.0 is for grassland, and for arable oil pH is 6.5 and pH varies as Zn, Cu and Ni behave equally [46]. The consequences of pH on the discharge of "zinc, copper and nickel" from the sewage sludge demonstrated that concentration of metal discharged to the supernatant liquid became elevated, as pH reduced below the threshold value. The threshold value of zinc is 5.8, for nickel 6.3 and for copper 4.5 [47]. The supernatant metal content was small and moderately stable over the aforementioned values of pH (Table 12.1)

The quantity of soluble copper as copper ions (Cu^{2+}) in copper chloride was noticed to be linked to pH, while the quantity of soluble zinc as zinc ions (Zn^{2+}) was hardly linked to pH [47]. A study was conducted to check the effect of SS application to a nutrient-rich soil on the soil accessibility of macronutrients like "N, P, K" and trace metals like "Fe, Cu, Mn, Zn and Pb" (Hernandez et al., 1991). In the sludge amended soil, the quantity of total nitrogen and extractable nitrogen and phosphorus contents was elevated, while the extractable potassium remained unchanged. The extractability of "Fe, Cu, Mn, Zn and Pb" is elevated because of the application of sludge in contrast to the control group. The comparatively high rate of application of sludge elevated the "cation exchange capacity", which facilitates the holding of important plant nutrients inside the zone of rooting as a result of extra binding sites of cation [48]. Such answers, though, depend on the ratio of "sewage : soil". The high content of organic matter present in sludge leads to a reduction in mass density and elevates cumulative stability [49]. Enhancement in the physical

Table 12.1 Showing the effect of sewage sludge on various properties of soil.

Properties of soil	Parameters	Effect	References
Physical properties	pH of soil	Increases Decreases	[34, 43]
	Soil aggregate stability	Increase	[49]
	Bulk density	Increases	[98]
	Water-holding capacity	Increases	[98]
	Porosity	Increases	[98]
	Humus content	Increases	[15]
Chemical properties	Toxic elements	Increases	[47, 15, 99]
	Soil organic carbon	Increases	[100]
	Electrical conductance	Increases	[39, 98]
	N and P	Increases	[39, 101]
	Cation exchange capacity	Increases	[98]
Microbial properties	Soil microbial activity	Increases	[60, 63–65]
	Soil biomass	Increases	[66, 67]
	Aerobic bacteria	Increase	[15]

properties of soil increases the water holding capacity which further endorsed the water retention quality in sludge-adjusted soil [49].

A study was carried out in southeastern France on a vineyard to check the effects of the composting of "sewage sludge" application at different rates such as "10, 30, and 90 t/ ha fresh wt." and to measure *in situ* nitrogen mineralization and organic matter in soil [50], and also to assess the risks to the environment such as nitrogen and phosphorus leaching and heavy metal accumulation in soil [50]. The results showed that organic matter present in the soil was elevated at all the doses of treatment, but there is no increase in the total or available concentration of heavy metals. Mostly in the extractable and non-exchangeable forms of sewage sludge, low level of heavy metals are present [51]; the accessibility of heavy metals is decreased by composting because of the "adsorption or complexing" by decomposed substances. In the summer season, the content of nitrogen minerals was elevated in the upper layers in soil of adjusted designs. The hazard of nitrogen leaching was much less, but phosphorus turned out to be the limiting aspect at the suggested SS alteration rate. A significant elevation in phosphorus content was found in the top- and sub-soil of all plots that had been given treatment, and the greatest increase was found in the plot that had been given the highest rate of application of sludge. There were no significant variations observed at lower rates. In soil, heavy metals are distributed in different structures such as "solid phases", "free ions in soil solution", "soluble organic mineral complexes" or "adsorbed on colloidal particles". The possible accessibility of heavy metals can be affected by the addition of SS to soils [52]. "The solubility" and as a result, the "mobility of metals" that are added with SS were not less than in part prohibited by the decomposition of "organic matter" and ensuing soluble organic metals carriers [53]. The bioavailability of trace metal was dependent upon soluble or insoluble organic matter [54]. Metal uptake was inhibited by insoluble organic matter that was strongly bonded to organic matter and decreased the bioavailability. The organic complex of soluble metal was formed when there is an increase in the availability of soluble organic matter [54]. With the organic matter stabilization rate of decaying, the level of soluble organic matter decreases, and it leads to a decrease in the "metals bioavailability". From sewage sludge, the bioavailability of zinc is favored by the acidic pH of the lithic haplumbrept (LH) soil, while in the alkaline soil, copper bioavailability was higher. The higher organic matter present in the acidic soil leads to the production of copper complexes and therefore damages its uptake by plants. In acidic and alkaline soil a complementary drift was found concerning the concentration of metal. There is a reduction in the concentration of metals like zinc, copper, lead and nickel in plants, with elevation in doses of sludge for acidic soil, while zinc and copper elevates in alkaline soil. The small changes were found in the concentrations of metals of plants developed in calcixerollic xerochrepts[2] and paralithic xerorthent soil due to elevation in doses of sludge.

12.5.2 Microbial Parameters of Soil

Microorganisms that are found in the soil play a major role in the main processes of soil [55, 56]. Due to the sensitivity, quick responses, integrative nature and biological significance, "microbial parameters" have gained much significance for the estimation of disturbance impacts as well as practices in agricultural land and contamination on the functioning of soil [57–59]. With regard to this, an increase in the microbial activity and biomass of soil is due to its association with the application of organic adjustments [60, 61]. Here, stimulation in the microbial activity of soil is due to the application of SS, because it is replicated by the elevated fractions of enzyme activities of soil. This may be because of organic matter supply and, alongside, substrates liable for hydrolysis of enzymes [62]. Subsequent to the application of SS, there is an elevation in the microbial activity of soil as reported by [60, 63–65]. Concerning the "microbial biomass of soil", there is an elevation in total bacteria due to the sewage sludge application; these might be elucidated by the elevation in existing sources of carbon in the absence of limitation of nitrogen (in the sewage sludge C:N ratio "5.5"). It was also found that microbial biomass of soil was improved subsequent to the application of sewage sludge [66, 67]. Despite the application rate and frequency among the various sewage sludge amended soils, there were nonsignificant disparities observed on the basis of microbial activity of soil and biomass. Actually, only one significant variation was found among sewage sludge amended soil corresponding to the enzyme activities of "β glucosidase" and "β-glucosaminidase"; rates of these were found to be significantly elevated in high amounts of sewage sludge amended soil. The activities of these enzymes are related to the carbon cycle and the high values are possibly described by the high rates of organic manure applied to the soils. The microbial activity and biomass of soil is reduced by the presence of toxic elements in SS (72, 64, 60). On the contrary, it was reported that after six years from its last application, there was an increase in the value of microbial biomass of sewage sludge amended soil regardless of the higher concentration of heavy metals

[68]. These conflicting results might probably be because of the composite nature of the connections between the "microorganisms", "heavy metals", "sewage sludge" and "the matrix of soil". It was specified that the stability of sewage sludge might have a greater affect on its "ecotoxicity" than the toxic metals present in it [60]. Moreover, the elevated rates of total copper and zinc in soil as a result of application of sewage sludge didn't appear to considerably influence the status of soil microorganisms, as shown by the different rates of microbial indicators on soil quality observed here. Regarding microbial diversity, generally the microbial diversity of soil was not affected considerably by the application of SS. Sewage sludge amended soils showed small, but considerably less complex prokaryotes richness, compared to control soil. Lloret et al. [69], have also reported this effect and might be associated with a selection filter given by the introduction of a significant quantity of exact substrates supplied by the modification, or to the unfavorable impact of trace elements [70]. Contrary to the results, Chen et al. [71], reported an elevation in "bacterial diversity" of soil subsequent to the long-term application of SS. Likewise, under controlled laboratory conditions, it was reported that the sewage sludge based modifications improved the prokaryotic diversity soil following 28 days of incubation [65]. Alterations in the microbial community composition of soil are induced by the use of organic modifications [72], and it might direct attention to alterations in the functions of soil [73] and particularly the food web of soil [74]. It was pointed out that sewage sludge applications didn't considerably alter the composition of soil "prokaryotic operational taxonomic units" (OTU), as replicated by the "non-metric multidimensional scaling (NMDS) analysis" [75]. The resistance and elasticity of "prokaryotic communities" of soil can explain this fact [75]. Nonetheless, "eukaryotic operational taxonomic units" composition showed variations between "sewage sludge amended" and "unamended control soil". Moreover, numerous "prokaryotic and eukaryotic taxa" showed significant association with the total quality of SS applied to the soil after 24 years. Concerning "prokaryotic taxa", a great quantity of eight genera has been considerably interrelated with the whole quantity of sludge applied after 24 years. The elevation in "bacterial biomass" that was observed in sewage sludge amended soil could be due to the prokaryotic and eukaryotic taxa. Lysobacter is abundantly present there, and also clostridium. Lysobactor is a genus that belongs to the "Gamma proteobacteria" which has been associated with the deprivation of a number of phenols present in sewage sludge, like "bisphenol A" [76] and "nonylphenols" [77], so the abundant presence of lysobactor might be related to the presence of phenol in large quantity with the high amount sewage sludge amended soil. Clostridium is a genus which comes under the "Clostridia class"; it is the only "prokaryotic taxon" at the class rank that demonstrated a considerable relation with the whole quantity of sludge applied after 24 years. It can be supported by the process of "anaerobic digestion" and is usually found in municipal solid wastes [78], because through the "anaerobic decomposition" of organic manure, it can oxidize acetate [79]. It was reported that due to the application of SS to the soil there was increase in the abundance of a wastewater pathogen, i.e. *Clostridium perfringens* [80]. In contrast, the other genus was Perlucidibaca, which showed an interrelation with the whole quantity of SS applied after 24 years, this genus contains only one species, i.e. "*Perlucidibaca piscinae*" [81]. This species is most commonly present in drinking water wells [82]. Its functions in the environment are not well known [83]. Regarding the communities of

eukaryotes, there was one genus, i.e. Prototheca, which showed the best interrelation with the whole quantity of sewage sludge applied after 24 years. Prototheca is an opportunist, "pathogenic green algae" which can cause "protothecosis" in man [84]. Species of this genus have the potential to degrade "polycyclic aromatic hydrocarbons" (PAH's) as reported by [85]. Hence, these are increasing abundantly due to their presence in the high amount in soil amended with a large quantity of sewage sludge. On the other hand, there is one family "Herpotrichiellaceae", which comes under the order "Chaetothyriales", negatively interrelated with the whole quantity of sewage sludge applied after 24 years. Several species of the family "Herpotrichiellaceae" are recommended because these have potential to control "soil-borne pathogens" [86]. In addition to this, there are some other members of the order "Chaetothyriales" which have the potential to degrade toluene (volatile aromatic hydrocarbon) [87]. Lastly, regarding the difficulty in antibiotic resistance, sewage sludge has been used as a reservoir of "antibiotic resistant bacteria" (ARB) and "antibiotic resistant genes" ARGs [88]. The process of stabilization by "anaerobic digestion" or lime addition may considerable decrease the load of antibiotic resistant bacteria in sewage sludge [89]. However, antibiotics sorption on sewage sludge particles is only due to such stabilization and production of them is less bioavailable [90]. It was reported that in composted sewage sludge there were present 156 various antibiotic resistance genes which encode resistance to almost all well-known groups of antibiotics and "mobile genetic elements" [91]. It was also reported that in sewage sludge itself there was the presence of large quantities of antibiotic-resistant genes as compared to the amended or unamended soil with sewage sludge (Urra et al., 2018). Moreover, the sewage sludge application may elevate the large quality of certain "antibiotic resistance genes" which are previously there in the soil and it also inserts new genes which were not present there earlier. After the application of SS, an elevation in the abundance of "antibiotic resistance genes" and "mobile genetic elements" genes was often identified in soil after sewage sludge application [71, 80, 92]. On the large quantity of genes, the sewage sludge application did have a significant effect; soil treated with sewage sludge showed a significant enrichment of seven types of genes such as "gene shows resistance to aminoglycodise, tetracycline and vancomycin and a gene encoding transposase" in contrast to the unamended soil control. An elevation in the gene abundance in soil amended with sewage sludge may be because of "horizontal gene transfer" from the population there in the sewage sludge itself or because of the stimulation of the "microbial biomass" of soil and activity. Fascinatingly, an increase in the abundance of genes occurred due to the absence of the alterations in the composition of prokaryotic operational taxonomic units. Through the use of horizontal gene transfer "mobile genetic elements" (MGEs) can favor the distribution of antibiotic resistance genes among the populations of bacteria. Moreover, a main role in the evolution of coresistance is played by mobile genetic elements, because these can carry antibiotic resistance and metal resistance simultaneously [93]. Subsequently, trace elements present in the SS can lead to an elevation in "antibiotic resistance" because of "coselection" [94]. It was reported that abundances in ARGs are related to the concentration of metal in soil [95, 96]. Regarding this, it was found that there is a correlation between the total concentration of Cu and Zn in the soil and antibiotic resistance genes and "mobile genetic elements" genes [97]. The presence of metals like copper and zinc in soil amended with sewage sludge may generate processes of co-selection between metals and antibiotic resistance which leads to the high abundance of some "antibiotic resistance genes" identified here [96]. It has been recognized that some "structural and functional resistance systems" are common for both metals and antibiotics [94]. Furthermore, the positive association between the concentration of copper and zinc and the large quantity of "transposase-encoding" mobile genetic element genes shows the spreading risk of elevated antibiotic resistance genes.

12.5.3 Concentration of Nutrients and the Heavy Metals in Sewage Sludge and Soil

N and P contents found in sewage sludge are the main nutrients present in it. It was found that when sewage sludge was evaluated, the "normal concentration of total N was 2.5% of the dry matter of which only 5% was in inorganic form" [102]. The concentration of P was "1.5% to 2%" as phosphorus pentoxide of dry matter. These element concentrations found in sewage sludge generated at a specific sewage sludge treatment plant may also differ as operational conditions differ. It was found in the sewage treatment workings in the United Kingdom that the concentration of N in the heat digested sludge varies from 1.6% to 10.7% [102]. There is a deficiency of N, P and K in most of the agricultural soil. As a result, there is a requirement of N as a widely used fertilizer. A report by the UK Working Party [29] in 1970 stated that application of SS can fulfill the requirements of N and P because it could supply the same 4.5% of the country's N and P fertilizer requirements. This is only possible when the concentration of toxic elements present in the sewage sludge, those not applied in selected amounts, and not currently being applied to agricultural land, were to become reduced. Compared to sewage sludge other bulky manures have low quantities of N and P content, but the sewage sludge usually contains a low amount of K [102]. Therefore, an extra amount of K would always need to be added where SS is applied as a fertilizer. Usually, the concentration of other heavy metals in soil is less than that in sewage sludge. Moreover, chemical and biological processes isolate heavy metals from soil by their insertion in the mineral structures. As a result, there is an increase in the concentration of many toxic elements in soil due to the addition of SS to agricultural land (Figure 12.1). The concentration of toxic elements differs according to the type of sludge, e.g. sludge from industries or domestic sources [103]. It can clearly be seen by comparing the ordinary values for the concentration of trace elements in soil and SS that the concentration of heavy metals like Ag, Bi, Cd, Cu, Pb and Sn are higher in sewage sludge than in soil. Cadmium naturally has less than "$1\mu gg^{-1}$" concentration in the soil [104–106], but the application of sewage sludge at higher rates resulted in "$76\mu gg^{-1}$" in soil concentrations and where topography has made the sludge distribution complex concentrations of soil of up to "$100\ \mu gg^{-1}$" have been reported (Doyle et al., 1973). Some elements which are present in sludge like Ba, Be, Mn and Sr are the same as their concentrations in soil. If there is repetition in the sludge application to agricultural land then it leads to an increase in the concentration of these elements in soil as organic matter in sludge is degraded and mineralization takes place. As compared to other elements, the toxicity and deficiency of Mn are most prevalent in agricultural practices [107, 108]. If the concentration of Mn in soil is less than $50\mu gg^{-1}$, then it is considered as deficient in

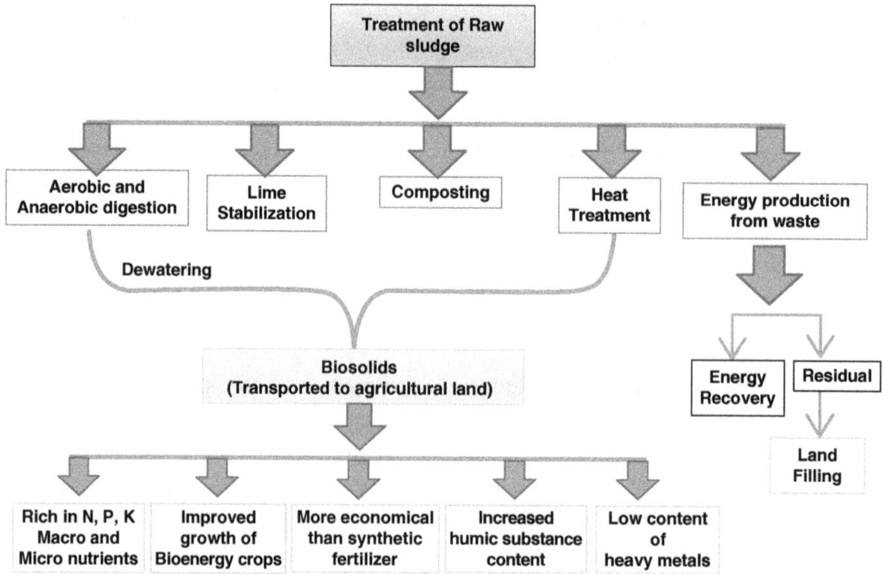

Figure 12.1 Management of Sewage Sludge.

Mn and if the concentration of Mn is about $100\mu gg^{-1}$ then it is fit to produce usual crop yields [109]. It is not likely that sludge application to agricultural land, apart from extreme rates on agricultural soil with maximum Mn concentration, would result in Mn toxicity.

12.6 Outlines of Nutrients and Harmful Metals in Sludge and Soil

Nitrogen is present in two forms in sewage sludge: partially in organic form and partially in inorganic form and these two forms are easily available to plants. The presence of total N in sewage sludge is between 30% and 70% in the form of ammonium and a significant amount of nitrates is found in aerobic sludge [110, 111]. As compared to the dried sludge, there is higher inorganic and total N content present in liquid digested sludge because the soluble form of N can be lost in drainage water during the process of drying. It was suggested that the maximum amount of P present in sludge is moderately in the insoluble form and it is not integrated into the soil as readily as P from other types of bulky manure [112]. However, it seems that if there is an increase in the application of sludge in the soil there is an increase in the amount of available P in the soil [112]. A 70% concentration of the total element in the agricultural soil may be contributed by the organic forms of P. Extraction techniques have been used to investigate the forms of metals present in soil and sewage sludge. To extract those metals “organic, soluble, exchangeable, adsorbed and precipitated forms” different types of reagents have been used. Sequential extraction procedure by the use of "1 M potassium nitrate", "0.5 M potassium fluoride", "0.1 M sodium pyrophosphate", "0.1 M ethylene diamine tetraacetic acid (EDTA)" and "1 M nitric acid" have been used to obtain various forms such as "exchangeable, adsorbed, organic, carbonate and sulphide" correspondingly in sewage sludge [111]. Only arbitrary divisions were represented by this extraction procedure between various forms of trace elements. Reagents that are used in this procedure were not only for one exact form such as only 68% of cadmium carbonate was extracted by EDTA; it extracted more than 90% of model precipitates of Pb, Zn and Cu carbonates. Likewise, sodium pyrophosphate ($Na_2H_2P_2O_7$) extracted 79% of lead carbonates. $Na_2H_2P_2O_7$ is a reagent which was mainly used for extracting fractions of organic metals. Another procedure was implemented by McLaren and Crawford [113]. Cu fractionation in the soil was authenticated by an association with parameters of soil like organic carbon and contents of clay which are known to affect various forms of metals. "Calcium chloride", "acetic acid", "pyrophosphate", "acidic oxalate" and "hydrofluoric acid" are the various reagents that are used in this procedure to signify various fractions such as "soluble-exchangeable, weakly adsorbed, organic, occluded and residual" [113]. Two procedures were compared to find out the mining of metals from SS which showed the quantity of Zn taken out by the use of $Na_2H_2P_2O_7$ was less in the McLaren and Crawford procedure than in that of Stover et al. [111]. It can be recognized by the comparatively higher amount of Zn formerly taken out with acetic acid. Various element forms found in sewage sludge may broadly differ on the basis of the type and characteristics of treated wastewater and the type of treatment used. It was found that the primary outlines of Cd and Pb in sewage sludge were "carbonates", while 56% of Zn was present in the form of organics. Cu and Ni were found as sulphides and redeemable forms correspondingly, and only 10% of them were present there as organic forms [111]. In view of the various researches on water solubility of heavy metals in SS it was found that "less than 1% of Pb" and "2% of Cu" were leached with distilled water or "0.06 M calcium chloride", while "31% of Zn" in sludge was found to be soluble in water [114]. The same results for Cu were found by other scientists but a difference was found in the results of Zn; it was found that less than "2% of zinc in digested sludge" was soluble in water [115]. Cr proves to be comparatively insoluble in sewage sludge, but only "1%" being taken out by "2.5% acetic acid" [103]. In soil, Cr exists mostly in the trivalent form [116], and forms "stable organic complexes" [117]. It is possible that the forms of metals present in soil vary less extensively than those metals present in sludge. It was found that the fraction of total Cu in soil solution didn't vary widely between eleven types of agricultural land of greatly different quality [118]. It was shown in the study of 24 different type of soils that metals form stability in soil and may change to keep the given amount in solution [113].

12.7 The Accumulation of Nutrients by Crops

In most agricultural soil there is a deficiency of required minerals like N, P and K. As a result, these minerals need to be added to the soil, N being the widely used fertilizer. It was reported that SS possibly provides the equivalent of 4.5% of the fertilizer requirement of N and P around the country. The heavy metals that are not currently being applied to soil or not applied at optimal rates have become low due to this reason; their contribution might be slightly increased. In terms of the N and P contents, additional heavy manures were used to compare with sewage sludges, but they usually have low quantities of K [102]. There is always a necessity to add K where the application of sludge has been used as a pure fertilizer. SS combined with manure from the farmyard may be beneficial in fulfilling the requirements of a complete fertilizer. It has been reported that "liquid digested sludge" worked faster than "conventional fertilizer" due to the equipped accessibility of "ammonium nitrogen" [110]. Once "decomposition and mineralization" takes place, then the organic N present in the digested sludge most probably becomes accessible for the uptake of plants [119]. In the dried SS, which has very small amounts of ammonium nitrogen it is just gradually mineralized [31], and it is not unpredicted that early experimentation through "heat dried sludges" has not been hopeful [102]. It is not known how much content of P is accessible to plants, and it is possible that if the application of sludge is enough to satisfy N supplies to agricultural crops, the above amount of P will also need to be added.

12.8 Future Views

As a fertilizer, SS is an assessment in agriculture land due to the presence of N and P in SS. The disposal of "sewage sludge" on the land is considered as a major outlet for its disposal, hence, a significant sludge quantity of the total nutrients present in sludge is used to good results. Disposal of sludge at sea is an extensively used alternative, but due to this route an extensive amount of compatible minerals are lost from the land. It was reported by "the UK Standing Committee" in 1975 that "2100 t of nitrogen and 5000 t of phosphorus" were dumped at sea [30]. The amount of SS used on agricultural land is not likely to reduce in the future. Sewage sludge applied to agricultural crops was suggested as a best practice [29]. An elevation in the amount of sewage sludge certified to be dumped at sea has been predicted to be minimal [30]. Any extensive contribution from incineration as the substitute of sewage disposal would involve the expenses of other plant with its related capital and rates of operation. SS incineration can cause emissions into the atmosphere which will require treatment and control to avoid air pollution. The profits of SS in crop field application as a fertilizer are related with many hazards because of toxic element concentrations. It has been shown in various studies that green vegetables, cereals and grasses may build up these kind of toxic elements from SS, and practices with higher amounts of sewage sludge application holding the maximum concentration of toxic elements showed that in several cases the concentration of plant tissues could attain "phytotoxic" levels. The degree of buildup is simply not a total soil concentration function of toxic elements, other than it is linked to the things which affect the appearance of these toxic elements in sewage sludge and soil amended with sewage sludge. Leafy vegetables are accumulating cadmium very efficiently; it is an element presently of specific interest (Figure 12.1)

References

1 Opec, F.F. I.D. (2018) Wastewater Report 2018: The Reuse Opportunity. International Water Association: IWA. https://www.iwa-network.org/wp-content/uploads/2018/02/OFID-Wastewater-report-2018.pdf

2 Christodoulou, A. and Stamatelatou, K. (2016). Overview of legislation on sewage sludge management in developed countries worldwide. *Water Sci Technol.* 73: 453–462.

3 Sanin, F.D., Clarkson, W.W., and Vesilind, P.A. (2011). *Sludge Engineering: The Treatment and Disposal of Wastewater Sludges*. Lancaster, Pa (USA): DEStech Publications, Inc.

4 Bień, J.B. and Wystalska, K. (2002). Osady ściekowe. Teoria *I Praktyka*, Wyd. Politechniki Częstochowskiej, Częstochowa.

5 Bien, J.B., Malina, G., Bien, J.D., and Wolny, L. (2004). Enhancing anaerobic fermentation of sewage sludge for increasing biogas generation. *Journal of Environmental Science and Health, Part A*. 39: 939–949.

6 Connor, R., Renata, A., Ortigara, C., Koncagül, E., Uhlenbrook, S., Lamizana-Diallo, B.M., and Hendry, S. (2017). The United Nations world water development report 2017. Wastewater: The untapped resource. The United Nations World Water Development Report.

7 Babatunde, A.O. and Zhao, Y.Q. (2007). Constructive approaches toward water treatment works sludge management: An international review of beneficial reuses. *Critical Reviews in Environmental Science and Technology*. 37: 129–164.

8 González-Ubierna, S., Jorge-Mardomingo, I., Carrero-González, B., De La Cruz, M.T., and Casermeiro, M.Á. (2012). Soil organic matter evolution after the application of high doses of organic amendments in a Mediterranean calcareous soil. *Journal of Soils and Sediments*. 12: 1257–1268.

9 Milieu Ltd, W. (2010). Risk and Policy Analysts Ltd (RPA). Environmental, Economic and Social Impacts of the Use of Sewage Sludge on Land. Final Report, Part III: Project Interim Reports.

10 Kapanen, A., Vikman, M., Rajasärkkä, J., Virta, M., and Itävaara, M. (2013). Biotests for environmental quality assessment of composted sewage sludge. *Waste Management*. 33: 1451–1460.

11 Loyo-Rosales, J.E., Rice, C.P., and Torrents, A. (2010). Fate and distribution of the octyl-and nonylphenol ethoxylates and some carboxylated transformation products in the Back River, Maryland. *Journal of Environmental Monitoring*. 12: 614–621.

12 Stasinakis, A.S., Gatidou, G., Mamais, D., Thomaidis, N.S., and Lekkas, T.D. (2008). Occurrence and fate of endocrine disrupters in Greek sewage treatment plants. *Water Research*. 42: 1796–1804.

13 Pak, V.M., McCauley, L.A., and Pinto-Martin, J. (2011). Phthalate exposures and human health concerns: A review and implications for practice. *AAOHN Journal*. 59: 228–235.

14 European Commission. (2012) http://ec.europa.eu/environment/water/waterframework/priority_substances.htm. Accessed 04 Apr 2018.

15 Kulling, D., Stadelmann, F., and Herter, U. (2001). Sewage sludge–fertilizer or waste. In UKWIR Conference, Brussels. 9–11.

16 Logan, T.J. and Harrison, B.J. (1995). Physical characteristics of alkaline stabilized sewage sludge (N-Viro Soil) and their effects on soil physical properties. *Journal of Environmental Quality*. 24: 153–164.

17 Sadecka, Z., Myszograj, S., Sieciechowicz, A., Płuciennik-Koropczuk, E., and Włodarczyk-Makuła, M. (2016). Impact of selected insecticides on the anaerobic stabilization of municipal sewage sludge. *Desalination and Water Treatment*. 57: 1213–1222.

18 Barret, M., Carrère, H., Delgadillo, L., and Patureau, D. (2010). PAH fate during the anaerobic digestion of contaminated sludge: Do bioavailability and/or cometabolism limit their biodegradation? *Water Research*. 44: 3797–3806.

19 Wisniowska, E. (2008). Effect of chemical stabilisation of sewage sludge on the fate of PAHs. *Archives of Environmental Protection*. 34: 249–257.

20 Pignatello, J.J., Oliveros, E., and MacKay, A. (2007). Advanced oxidation processes for organic contaminant destruction based on the Fenton reaction and related chemistry. *Critical Reviews in Environmental Science and Technology*. 37: 273–275.

21 Kacan, E. and Kutahyali, C. (2012). Adsorption of strontium from aqueous solution using activated carbon produced from textile sewage sludges. *The Journal of Analytical and Applied Pyrolysis*. 97: 149–157.

22 Kacan, E. (2016). Optimum BET surface areas for activated carbon produced from textile sewage sludges and its application as dye removal. *Journal of Environmental Management*. 166: 116–123.

23 Monsalvo, V.M., Mohedano, A.F., and Rodriguez, J.J. (2011). Activated carbons from sewage sludge: Application to aqueous-phase adsorption of 4-chlorophenol. *Desalination*. 277: 377–382.

24 Chingombe, P., Saha, B., and Wakeman, R.J. (2005). Surface modification and characterisation of a coal-based activated carbon. *Carbon*. 43: 3132–3143.

25 Wang, X., Gu, L., Zhou, P., Zhu, N., Li, C., Tao, H., and Zhang, D. (2017). Pyrolytic temperature dependent conversion of sewage sludge to carbon catalyst and their performance in persulfate degradation of 2-Naphthol. *Chemical Engineering Journal*. 324: 203–215.

26 Gu, L., Li, C., Wen, H., Zhou, P., Zhang, D., Zhu, N., and Tao, H. (2017). Facile synthesis of magnetic sludge-based carbons by using Electro-Fenton activation and its performance in dye degradation. *Bioresource Technology*. 241: 391–396.

27 Yu, Y., Wei, H., Yu, L., Zhang, T., Wang, S., Li, X., ... Sun, C. (2015). Surface modification of sewage sludge derived carbonaceous catalyst for m-cresol catalytic wet peroxide oxidation and degradation mechanism. *RSC Advances*. 5 (52): 41867–41876.

28 Bedia, J., Monsalvo, V.M., Rodriguez, J.J., and Mohedano, A.F. (2017). Iron catalysts by chemical activation of sewage sludge with FeCl3 for CWPO. *Chemical Engineering Journal*. 318: 224–230.

29 Ministry of Housing and Local Government. (1970). Taken for granted. Report of the Working Party on Sewage Disposal.

30 Sterritt, R.M. and Lester, J.N. (1980). The value of sewage sludge to agriculture and effects of the agricultural use of sludges contaminated with toxic elements: A review. *Science of the Total Environment*. 16: 55–90.

31 Lunt, H.A. (1959). *Digested Sewage Sludge for Soil Improvement.* Connecticut Agricultural Experiment Station.

32 Dowdy, R.H. and Larson, W.E. (1975). The availability of sludge-borne metals to various vegetable crops. *Journal of Environmental Quality*. 4: 278–282.

33 Baerug, R. and Martinsen, J.H. (1977). The influence of sewage sludge on the content of heavy metals in potatoes and on tuber yield. *Plant and Soil*. 47: 407–418.

34 Epstein, E.L., Taylor, J.M., and Chaney, R.L. (1976). Effects of sewage sludge and sludge compost applied to soil on some soil physical and chemical properties. *Journal of Environmental Quality*. 5: 422–426.

35 Emmelin, L. (1973). Heavy metal content in sewage sludge behind Swedish recommendations for limited use. *Ambio*. 2: 126–127.

36 Garrigan, G.A. (1977). Land application guidelines for sludges contaminated with toxic elements. *Journal (Water Pollution Control Federation)*, 49: 2380–2389.

37 US Environmental Protection Agency. (1977) Municipal Sludge Management: Environmental Factors. *EPA 430/9-77-004*.

38 Department of the Environment and National Water Council. (1977) Report of the Working Party on the Disposal of Sewage Sludge to Land.

39 Fernando, M., Gabriela, C., Teresa, I., and Ingrid, W. (2002). Urban organic wastes effects on soil chemical properties in a degraded semiarid ecosystem. In: *17. World Congress of Soil Science Bangkok (Thailand) 14–21 Aug 2002*.

40 Sommers, L.E. (1977). Chemical composition of sewage sludges and analysis of their potential use as fertilizers. *Journal of Environmental Quality*. 6: 225–232.

41 Archie, S.G. and Smith, M.I.L.T.O.N. (1981). Survival and growth of plantations in sewage sludge treated soil and older forest growth study. municipal sludge application to Pacific North-West forest lands. In: *College of Forest Resources*, (ed. C.B. Bledose). 105–113. Washington, DC: Univ of Washington.

42 Epstein, E. (1975). Effect of sewage sludge on some soil physical properties. *Journal of Environmental Quality*. 4: 139–142.

43 Tsadilas, C.D., Matsi, T., Barbayiannis, N., and Dimoyiannis, D. (1995). Influence of sewage sludge application on soil properties and on the distribution and availability of heavy metal fractions. *Communications in Soil Science and Plant Analysis*. 26: 2603–2619.

44 Parkpain, P., Sirisukhodom, S., and Carbonell-Barrachina, A.A. (1998). Heavy metals and nutrients chemistry in sewage sludge amended Thai soils. *Journal of Environmental Science & Health Part A*. 33: 573–597.

45 Hue, N.V. and Ranjith, S.A. (1994). Sewage sludges in Hawaii: Chemical composition and reactions with soils and plants. *Water, Air, and Soil Pollution*. 72: 265–283.

46 Department of the Environment and National Water Council. (1981) Report of the Sub-Committee on the Disposal of Sewage Sludge to Land.

47 Adams, T.M. and Sanders, J.R. (1984). The effects of pH on the release to solution of zinc, copper and nickel from metal-loaded sewage sludges. *Environmental Pollution Series B, Chemical and Physical*. 8: 85–99.

48 Soon, Y.K. (1981). Solubility and sorption of cadmium in soils amended with sewage sludge. *Journal of Soil Science*. 32: 85–95.

49 Ojeda, G., Alcañiz, J.M., and Ortiz, O. (2003). Runoff and losses by erosion in soils amended with sewage sludge. *Land Degradation & Development*. 14: 563–573.

50 Korboulewsky, N., Dupouyet, S., and Bonin, G. (2002). Environmental risks of applying sewage sludge compost to vineyards: Carbon, heavy metals, nitrogen, and phosphorus accumulation. *Journal of Environmental Quality*. 31: 1522–1527.

51 Breslin, V.T. (1999). Retention of metals in agricultural soils after amending with MSW and MSW-biosolids compost. *Water, Air, and Soil Pollution*. 109: 163–178.

52 Wang, P., Qu, E., Li, Z., and Shuman, L.M. (1997). Fractions and availability of nickel in loessial soil amended with sewage or sewage sludge. *Journal of Environmental Quality*. 26: 795–801.

53 Chaney, R.L. and Ryan, J.A. (1994). Heavy Metals and Toxic Organic Pollutants in MSW-composts: Research Results on Phytoavailability, Bioavailability, Fate, Etc US Environmental Protection Agency.

54 McBride, M.B. (1995). Toxic metal accumulation from agricultural use of sludge: Are USEPA regulations protective? *Journal of Environmental Quality*. 24: 5–18.

55 Bastida, F., Kandeler, E., Moreno, J.L., Ros, M., García, C., and Hernández, T. (2008). Application of fresh and composted organic wastes modifies structure, size and activity of soil microbial community under semiarid climate. *Applied Soil Ecology*. 40: 318–329.

56 Fernández, J.M., Plaza, C., García-Gil, J.C., and Polo, A. (2009). Biochemical properties and barley yield in a semiarid Mediterranean soil amended with two kinds of sewage sludge. *Applied Soil Ecology*. 42: 18–24.

57 Mijangos, I., Pérez, R., Albizu, I., and Garbisu, C. (2006). Effects of fertilization and tillage on soil biological parameters. *Enzyme and Microbial Technology*. 40: 100–106.

58 Muñoz-Leoz, B., Garbisu, C., Antigüedad, I., and Ruiz-Romera, E. (2012). Fertilization can modify the non-target effects of pesticides on soil microbial communities. *Soil Biology and Biochemistry*. 48: 125–134.

59 Burges, A., Epelde, L., and Garbisu, C. (2015). Impact of repeated single-metal and multi-metal pollution events on soil quality. *Chemosphere*. 120: 8–15.

60 Roig, N., Sierra, J., Martí, E., Nadal, M., Schuhmacher, M., and Domingo, J.L. (2012). Long-term amendment of Spanish soils with sewage sludge: Effects on soil functioning. *Agriculture, Ecosystems & Environment*. 158: 41–48.

61 Reardon, C.L. and Wuest, S.B. (2016). Soil amendments yield persisting effects on the microbial communities—a 7-year study. *Applied Soil Ecology*. 101: 107–116.

62 Singh, R.P., Singh, P., Ibrahim, M.H., and Hashim, R. (2012). Land application of sewage sludge: Physicochemical and microbial response. *Rev Environ Contam Toxicol.*, 214:41-6163.

63 Kızılkaya, R. and Bayraklı, B. (2005). Effects of N-enriched sewage sludge on soil enzyme activities. *Applied Soil Ecology*. 30: 192–202.

64 Carbonell, G., Pro, J., Gómez, N., Babín, M.M., Fernández, C., Alonso, E., and Tarazona, J.V. (2009). Sewage sludge applied to agricultural soil: Ecotoxicological effects on representative soil organisms. *Ecotoxicology and Environmental Safety*. 72: 1309–1319.

65 Mattana, S., Petrovičová, B., Landi, L., Gelsomino, A., Cortés, P., Ortiz, O., and Renella, G. (2014). Sewage sludge processing determines its impact on soil microbial community structure and function. *Applied Soil Ecology*. 75: 150–161.

66 Antolín, M.C., Pascual, I., García, C., Polo, A., and Sánchez-Díaz, M. (2005). Growth, yield and solute content of barley in soils treated with sewage sludge under semiarid Mediterranean conditions. *Field Crops Research*. 94: 224–237.

67 Xue, D. and Huang, X. (2013). The impact of sewage sludge compost on tree peony growth and soil microbiological, and biochemical properties. *Chemosphere*. 93: 583–589.

68 Parat, C., Denaix, L., Lévêque, J., Chaussod, R., and Andreux, F. (2007). The organic carbon derived from sewage sludge as a key parameter determining the fate of trace metals. *Chemosphere*. 69: 636–643.

69 Lloret, E., Pascual, J.A., Brodie, E.L., Bouskill, N.J., Insam, H., Juárez, M.F.D., and Goberna, M. (2016). Sewage sludge addition modifies soil microbial communities and plant performance depending on the sludge stabilization process. *Applied Soil Ecology*. 101: 37–46.

70 Martin-Laurent, F., Philippot, L., Hallet, S., Chaussod, R., Germon, J.C., Soulas, G., and Catroux, G. (2001). DNA extraction from soils: Old bias for new microbial diversity analysis methods. *Applied and Environmental Microbiology*. 67: 2354–2359.

71 Chen, Q., An, X., Li, H., Su, J., Ma, Y., and Zhu, Y.G. (2016). Long-term field application of sewage sludge increases the abundance of antibiotic resistance genes in soil. *Environment International*. 92: 1–10.

72 Marschner, P., Kandeler, E., and Marschner, B. (2003). Structure and function of the soil microbial community in a long-term fertilizer experiment. *Soil Biology and Biochemistry*. 35: 453–461.

73 Fierer, N., Lauber, C.L., Ramirez, K.S., Zaneveld, J., Bradford, M.A., and Knight, R. (2012). Comparative metagenomic, phylogenetic and physiological analyses of soil microbial communities across nitrogen gradients. *The ISME Journal*. 6: 1007–1017.

74 Morriën, E. (2016). Understanding soil food web dynamics, how close do we get? *Soil Biology and Biochemistry*. 102: 10–13.

75 Allison, S.D. and Martiny, J.B. (2008). Resistance, resilience, and redundancy in microbial communities. *Proceedings of the National Academy of Sciences*. 105: 11512–11519.

76 Yang, J., Li, H., Ran, Y., and Chan, K. (2014). Distribution and bioconcentration of endocrine disrupting chemicals in surface water and fish bile of the Pearl River Delta, South China. *Chemosphere*. 107: 439–446.

77 Zheng, G., Wang, T., Niu, M., Chen, X., Liu, C., Wang, Y., and Chen, T. (2018). Biodegradation of nonylphenol during aerobic composting of sewage sludge under two intermittent aeration treatments in a full-scale plant. *Environmental Pollution*. 238: 783–791.

78 Goberna, M., Simón, P., Hernández, M.T., and García, C. (2018). Prokaryotic communities and potential pathogens in sewage sludge: Response to wastewaster origin, loading rate and treatment technology. *Science of the Total Environment*. 615: 360–368.

79 Karakashev, D., Batstone, D.J., Trably, E., and Angelidaki, I. (2006). Acetate oxidation is the dominant methanogenic pathway from acetate in the absence of Methanosaetaceae. *Applied and Environmental Microbiology*. 72: 5138–5141.

80 Rahube, T.O., Marti, R., Scott, A., Tien, Y.C., Murray, R., Sabourin, L., and Topp, E. (2014). Impact of fertilizing with raw or anaerobically digested sewage sludge on the abundance of antibiotic-resistant coliforms, antibiotic resistance genes, and pathogenic bacteria in soil and on vegetables at harvest. *Applied and Environmental Microbiology*. 80: 6898–6907.

81 Song, J., Choo, Y.J., and Cho, J.C. (2008). Perlucidibaca piscinae gen. nov., sp. nov., a freshwater bacterium belonging to the family Moraxellaceae. *International Journal of Systematic and Evolutionary Microbiology*. 58: 97–102.

82 Navarro-Noya, Y.E., Suárez-Arriaga, M.C., Rojas-Valdes, A., Montoya-Ciriaco, N.M., Gómez-Acata, S., Fernández-Luqueño, F., and Dendooven, L. (2013). Pyrosequencing analysis of the bacterial community in drinking water wells. *Microbial Ecology*. 66: 19–29.

83 Vandermaesen, J., Lievens, B., and Springael, D. (2017). Isolation and identification of culturable bacteria, capable of heterotrophic growth, from rapid sand filters of drinking water treatment plants. *Research in Microbiology*. 168: 594–607.

84 Velez-Mejia, C. and Velez-Londoño, J. (2017). Prototheca: A Danger Underwater. *IDCases*. 9: 106–108.

85 Ueno, R., Wada, S., and Urano, N. (2008). Repeated batch cultivation of the hydrocarbon-degrading, micro-algal strain Prototheca zopfii RND16 immobilized in polyurethane foam. *Canadian Journal of Microbiology*. 54: 66–70.

86 Narisawaa, K., Ohkib, K.T., and Hashibab, T. (2000). fungus, Heteroconium chaetospira. *Plant Pathology*. 49: 141–146.

87 Badali, H., Prenafeta-Boldu, F.X., Guarro, J., Klaassen, C.H., Meis, J.F., and De Hoog, G.S. (2011). Cladophialophora psammophila, a novel species of Chaetothyriales with a potential use in the bioremediation of volatile aromatic hydrocarbons. *Fungal Biology*. 115: 1019–1029.

88 Rizzo, L., Manaia, C., Merlin, C., Schwartz, T., Dagot, C., Ploy, M.C., and Fatta-Kassinos, D. (2013). Urban wastewater treatment plants as hotspots for antibiotic resistant bacteria and genes spread into the environment: A review. *Science of the Total Environment*. 447: 345–360.

89 Munir, M., Wong, K., and Xagoraraki, I. (2011). Release of antibiotic resistant bacteria and genes in the effluent and biosolids of five wastewater utilities in Michigan. *Water Research*. 45: 681–693.

90 Li, W., Shi, Y., Gao, L., Liu, J., and Cai, Y. (2013). Occurrence, distribution and potential affecting factors of antibiotics in sewage sludge of wastewater treatment plants in China. *Science of the Total Environment*. 445: 306–313.

91 Su, J.Q., Wei, B., Ou-Yang, W.Y., Huang, F.Y., Zhao, Y., Xu, H.J., and Zhu, Y.G. (2015). Antibiotic resistome and its association with bacterial communities during sewage sludge composting. *Environmental Science & Technology*. 49: 7356–7363.

92 Xie, W.Y., Yang, X.P., Li, Q., Wu, L.H., Shen, Q.R., and Zhao, F.J. (2016). Changes in antibiotic concentrations and antibiotic resistome during commercial composting of animal manures. *Environmental Pollution*. 219: 182–190.

93 Allen, H.K., Donato, J., Wang, H.H., Cloud-Hansen, K.A., Davies, J., and Handelsman, J. (2010). Call of the wild: Antibiotic resistance genes in natural environments. *Nature Reviews Microbiology*. 8: 251–259.

94 Baker-Austin, C., Wright, M.S., Stepanauskas, R., and McArthur, J.V. (2006). Co-selection of antibiotic and metal resistance. *Trends in Microbiology*. 14: 176–182.

95 Knapp, C.W., McCluskey, S.M., Singh, B.K., Campbell, C.D., Hudson, G., and Graham, D.W. (2011). Antibiotic resistance gene abundances correlate with metal and geochemical conditions in archived Scottish soils. *PloS One*. 6: e27300.

96 Hu, H.W., Wang, J.T., Li, J., Li, J.J., Ma, Y.B., Chen, D., and He, J.Z. (2016). Field-based evidence for copper contamination induced changes of antibiotic resistance in agricultural soils. *Environmental Microbiology*. 18: 3896–3909.

97 Urra, J., Alkorta, I., Mijangos, I., Epelde, L., and Garbisu, C. (2019). Application of sewage sludge to agricultural soil increases the abundance of antibiotic resistance genes without altering the composition of prokaryotic communities. *Science of the Total Environment*. 647: 1410–1420.

98 Sree Ramulu, U.S. (2001). *Reuse of Municipal Sewage and Sludge in Agriculture*. Scientific Publishers, Jodhpur India.

99 Lopez-Mosquera, M.E., Moiron, C., and Carral, E. (2000). Use of dairy-industry sludge as fertiliser for grasslands in northwest Spain: Heavy metal levels in the soil and plants. *Resources, Conservation and Recycling*. 30: 95–109.

100 Kladivko, E.J. and Nelson, D.W. (1979). Changes in soil properties from application of anaerobic sludge. *Journal of Water Pollution Control Federation* 51: 325–332.

101 Häni, H., Siegenthaler, A., and Candinas, T. (1996). Soil effects due to sewage sludge application in agriculture. In Fertilizers and Environment. 267–274.

102 Bunting, A.H. (1963). Experiments on organic manures, 1942–49. *The Journal of Agricultural Science*. 60: 121–140.

103 Berrow, M.L. and Webber, J. (1972). Trace elements in sewage sludges. *Journal of the Science of Food and Agriculture*. 23: 93–100.

104 Klein, D.H. (1972). Mercury and other metals in urban soils. *Environmental Science & Technology*. 6: 560–562.

105 Dudas, M.J. and Pawluk, S. (1977). Heavy metals in cultivated soils and in cereal crops in Alberta. *Canadian Journal of Soil Science*. 57: 329–339.

106 Frank, R., Ishida, K., and Suda, P. (1976). Metals in agricultural soils of Ontario. *Canadian Journal of Soil Science*. 56: 181–196.

107 Batey, T. (1971). Manganese and boron deficiency. In: *Trace elements in soils and crops*, 21 (ed. J. Webber), 137–149. London: United Kingdom Ministry Agric., Fish., and Food Tech. Bull.

108 RUSSEL, E. (1973). *Soil Conditions and Plant Growth*, 7th e. London: Longmans, Green and Co.

109 Sherman, G.D., McHargue, J.S., and Hodgkiss, W.S. (1942). Determination of active manganese in soil. *Soil Science*. 54: 253–258.

110 Coker, E.G. (1971). Utilisation of sludge in agriculture. *Sludge Treatment and Disposal*. 87–104.

111 Stover, R.C., Sommers, L.E., and Silviera, D.J. (1976). Evaluation of metals in wastewater sludge. *Journal (Water Pollution Control Federation)* 48: 2165–2175.

112 Johnston, A.E. (1975). Woburn Market Garden Experiment, 1942–69. II. The effects of the treatments on soil pH soil carbon, nitrogen, phosphorus and potassium. *Rep Rothamsted Exp Stn*.

113 McLaren, R.G. and Crawford, D.V. (1973). Studies on soil copper I. *The fractionation of copper in soils. Journal of Soil Science*. 24: 172–181.

114 Lagerwerff, J.V., Biersdorf, G.T., and Brower, D.L. (1976). Retention of metals in sewage sludge I: Constituent heavy metals. *Journal of Environmental Quality*. 5: 19–23.

115 Jenkins, S.H. and Cooper, J.S. (1964). The Solubility of Heavy Metal Hydroxides in Water, Sewage and Sewage Sludge. 3. The Solubility of Heavy Metals Present in Digested Sewage Sludge. *Air and Water Pollution* 8: 695.

116 Leeper, G.W. (1972). Reactions of heavy metals with soils with special regard to their application in sewage wastes. *Report for Department of Army Corps of Engineers, USA, under contract No.* DACW 73-73-C-0026.

117 Bartlett, R.J. and Kimble, J.M. (1976). Behavior of chromium in soils: I. *Trivalent forms. Journal of Environmental Quality*. 5: 379–383.

118 Mercer, E.R. and Richmond, J.L. (1968). An investigation of the chemical form of copper in soil solutions. *Annu. Rep. Agric. Res. Counc. Letcombe Lab.* 19: 46–48.

119 Higgins, I.J. and Burns, R.G. (1975). *The Chemistry and Microbiology of Pollution*. London: Academic Press.

120 Doyle, P.J., Lester, J.N., and Perry, R. (1978) Survey of literature and experience on the disposal of sewage sludge on land, Final report to the UK Department of the Environment.

121 Albert, A. (1973). Arsenicals, antimonials and mercurials. In: *Selective Toxicity, 5th e, 392–397*. London: Chapman and Hall.

122 Paz-Ferreiro, J., Gascó, G., Gutiérrez, B., and Méndez, A. (2012). Soil biochemical activities and the geometric mean of enzyme activities after application of sewage sludge and sewage sludge biochar to soil. *Biology and Fertility of Soils*. 48: 511–517.

13

Microbial Fuel Cells for the Treatment of Wastewater

CHAPTER MENU

13.1 Introduction

Due to the overwhelming increase in energy demand and the need to explore eco-friendly means to tackle the challenge, non-conventional sources of energy such as solar power, wind, biomass and nuclear energy can be promising and reliable options. Keeping in view the global warming and climate change issues, to release the stored carbon cannot be a sustainable option.

Microbial fuel cells (MFCs) are advanced tools to derive the electric energy through biochemical reactions catalyzed by bacteria. A proton exchange membrane is used to diffuse the protons and electrons which get combined with oxygen to complete the

Aquatic Environmental Bioengineering: Monitoring and Remediation of Contamination, First Edition.
Rouf Ahmad Bhat, Mohammad Yaseen Mir, Gowhar Hamid Dar and Moonisa Aslam Dervash.

biochemical process. The difference between the anode potential and substrate redox potential is equivalent to the energy gained by bacteria through their metabolic activity. The application of MFCs in the treatment of wastewater has been successfully tested and showed very promising results like removal of COD of up to 50% and also efficient decontamination of carbon and nitrogen compounds. It has been found that the MFC system can optimally work at low COD/N ratios and doesn't rely on the carbon supply.

MFCs are the advanced and innovative source of energy with zero emission of pollutants and were first introduced in the previous century [1]. MFCs are wonderful biotools to convert the biochemical bound energy to electrical energy via a catalytic process carried out by bacteria in an anaerobic atmosphere [2]. MFC biotechnology is an advanced approach to derive electrical energy from biomass and organic wastes using potential microbes [3].

The positives of using MFCs are:

- MFCs have the ability to use biomass waste matter as fuel and microbes as catalysts to generate electrical energy.
- MFCs have a very simple distribution system and have no complexity involved like in hydrogen fuel cells.
- MFCs are more efficient in converting the energy derived from electrons (up to 90%) from electron transport system as compared to enzymatic fuel cells.

MFC technology is an amalgam of many basic and applied sciences. MFC technology utilizes the advanced and in depth knowledge of microbial physiology, electrochemistry, electrical and material engineering, environmental engineering and bioremediation to come up with promising sustainable, alternate, non-conventional source of energy.

13.2 Biochemical Sustenance of Microbes

Microbes are able to survive by utilizing the organic compounds derived from the substrate under different circumstances. Depending on the election accepter molecule, the electrons obtained from oxidation are transported to the terminal through an electron transport chain. Oxygen is an acceptor terminal in the case of aerobic processes and gets combined with electrons to form water. Chemiosmotic hypothesis states that translocation of protons is coupled with bacterial electron transfer chains which is in turn linked to the synthesis of ATP synthesis across the membrane [4]. The chemiosmotic principle governs the functioning of bacterial cell membranes and the movement of protons on the outer side of membranes develops an electrochemical gradient. This phenomenon is further amplified by pH gradient and generates the proton motive force. The ATP-synthetase enzyme allows the protons to re-enter and synthesize ATP again, which in turn is utilized by the microbes for their survival [5].

13.3 Functioning of MFCs

A typical MFC contains two fundamental components, viz., an anodic part and a cathodic part separated by a salt bridge which is electively permeable. An anolyte is normally available in an anode containing suspended anaerobic bacteria and a cathode possesses oxygen as an electron accepter. Shuttling agents and nanowires are used to transfer electrons from the anode to cathode by using an external circuit to complete the reaction and to provide a sustainable current [6–9].

13.3.1 Uses of MFCs

Currently MFCs have a very limited use because of their complex design, high cost and somewhat reduced performance. But research is ongoing and it is expected that MFCs will scale up in future and soon become promising non-conventional source of energy.

13.3.2 Wastewater Treatment

Microbes possess a unique ability to degrade pollutants and simultaneously derive energy from the same substrate. MFCs are currently under significant consideration as these devices have a capability to stabilize agricultural, municipal or industrial wastes, and also they generate electric energy from it. Microbes oxidize the organic matter in the sewage and simultaneously release electrons which yield a steady electric current. The focus in research is on increasing the power efficiency of MFCs so as to provide a very efficient and cost effective method of treating sewage, and to generate less sewage as compared to conventional methods both in developing and developed nations [10, 11].

13.3.3 Power Supply to Underwater Monitoring Devices

Sensors are used to acquire detailed data acquisition and information to study, understand and model the natural environment, but these sensors need a continuous and optimum supply of electricity for their operation. MFCs can emerge as possible solutions to such sensor devices which are used particularly in rivers and deep aquatic ecosystems where the battery replacement process is almost impossible. Sediment fuel cells are novel devices to monitor and assess the environmental condition of creeks, rivers, estuaries and oceans [12]. Currently, the power density of sediment fuel cells is a bit less due to lack of sufficient organic compounds in the system and the high internal resistance inside the MFCs [13].

13.3.4 Power Supply to Remote Sensors

Micro-electronics development has revolutionized the power requirement of most of the electronic devices. Most of the telemetry and chemical sensors use efficient batteries as the source of power, but repairing and replacing batteries regularly would be very costly, cumbersome and almost impossible practically. A practical possible solution to such a problem is to use MFCs which are self-renewable and can run on locally available resources [14].

13.3.5 BOD Sensing

MFCs may be efficiently applied for monitoring and control of pollutants and may be potentially used as pollutant sensors in cases of biological oxygen demand (BOD) in wastewater. The BOD value can be assessed by means of correlation assimilable organic matter and the coulumbic yield of MFCs. The beautiful part of using MFC in pollution sensing is that it can have a life span of five years or more than any other BOD sensor available in the literature [15].

13.3.6 Hydrogen Manufacture

Hydrogen generation by using MFCs is an interesting and efficient process. By using organic waste, anaerobic environments are regulated in the "cathode side" and an extra "voltage of around 0.25 V" is provided from the cathode side and protons get reduced to hydrogen. Such MFCs are known as "bioelectrochemically assisted microbial reactors" (BEAMR) [16].

13.4 Microbial Fuel Cells Treatment of Wastewater

The generation of biogas from sewage is a well-established process used by humans over the last few decades. Sewage and acid mine water treatment is an alarming problem in almost all settlements whether rural or urban. Normally, activated sludge, trickling filters and lagoons are used to stabilize wastewater, but due to large quantum of nutrient concentration in the effluent the discharged water proves to be disastrous for receiving water bodies. Wastewater treatment remains an economic burden to industries and the public. Domestic, agricultural, and industrial waste always contains a very large quantity of organic matter which contains almost 90% of energy which can be harvested [17]. Biological degradation can be a sustainable option to produce biogas and later the energy can be converted into electrical energy. The only problem with such an approach is the requirement of large digesters, a reduced efficiency factor and low thermal gas engine efficiency. The most feasible and widely used biological treatment used is the activated sludge process. But most of the energy is consumed in pumping and aeration, which makes it costly and less efficient.

Alternative approaches like MFCs are needed to utilize the treatment of industrial and municipal wastewater. This process is based on the anaerobic oxidation of organic compounds inside the cell. MFCs have the capability to convert organic matter into electrical energy with higher efficiency and involves a very simple process (Figure 13.1)

13.5 Microbial Fuel Cell Design

MFC is a biological electrochemical device that converts inorganic matter-based chemical energy into electrical energy using microbes through redox reactions. Dissolved organic matter present in the wastewater acts as the substrate to produce chemical energy. MFCs operate on the mechanism of cellular respiration. The overall process is exothermic and was first identified in 1911 by a botany professor at the University of Durham, M.C. Potter [18]. In a typical MFC, the anode and cathode are separated by a selective permeable membrane linked through an external conductor circuit. The bacteria oxidize the organic matter to produce ATP, and protons, electrons and carbon dioxide are derived as by-products in the electron transport system. Finally, oxygen available at the cathode again combines with hydrogen and electrons to produce water and the reaction gets completed.

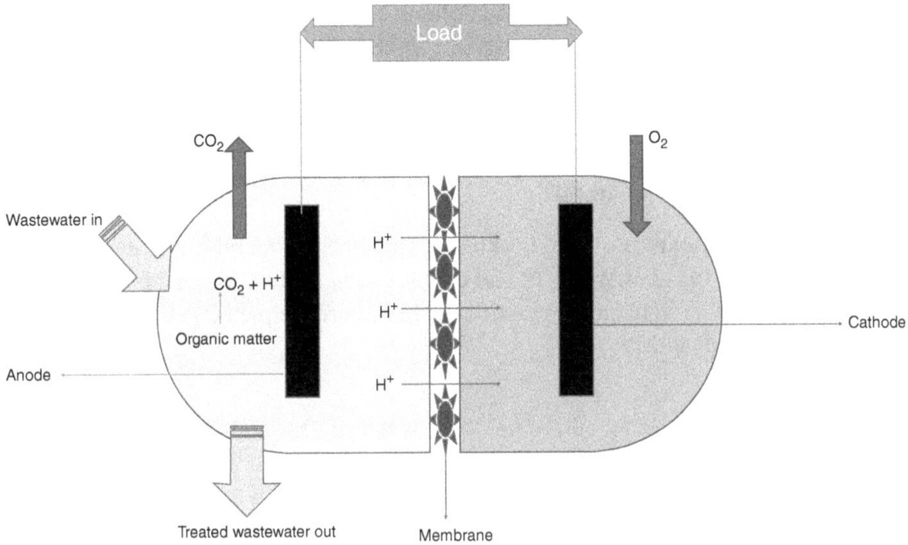

Figure 13.1 Two compartment microbial fuel cell.

The significant difference between the bacterial fuel cells and conventional batteries or supercapicitors is that the complexity involved in MFCs is greater than for conventional ones [19]. The remarkable differences between the MFCs and conventional low temperature fuel cells (direct methanol fuel cell or proton exchange membrane fuel cell) are described below.

- The biological electroactive bacteria or proteins are used at the anode.
- The optimum temperature range is between 15 and 45°C.
- The system working pH is neutral.
- Organic biomass is used as substrate at the anode.
- Life cycle environmental impact assessment is moderate.

13.6 Construction of MFCs

A typical MFC contains an anode and a cathode separated by a cation specific selectively permeable membrane. The organic substrate gets oxidized at the anode mediated with bacteria and generates the protons which cross the selectively permeable membrane and reach the cathode, then the electrons pass through the external circuit network to generate electric current. Electrons generated during the respiration of microbes lead to two types of MFCs: mediator and mediatorless. Mediatorless MFC is more efficient and promising than mediator and leads to two basic versions of the MFC: the two cell and the single cell.

13.6.1 Two Cell MFCs

Two cell MFCs are composed of two distinguished parts, anode and cathode separated by a selectively permeable membrane. The anode is the anaerobic part of the cell and contains substrate material which has an optimum organic matter quantity and acts as the substrate and is covered with microbial colonies. On the other hand, the cathode contains electrolytes and oxygen. The cathode cell contains the cathode and the electrolyte (Figure 13.1).

13.6.2 Single Compartment MFCs

The single compartment MFC uses an external air cathode which is separated from the inside of the cell by the membrane. The air cathode version gives a higher power density than the two chamber version. In practice the MFCs are coupled together in stacks to provide the required voltage.

MFC Anode

The anode is a very important part of the MFC as it forms the containment zone for the microbes and develops a coating over it. The activity of the MFC is decided by the quantum of bacteria inside the anode and its surface area. It is highly important to select the appropriate material for the electrode as it affects the electrochemical efficiency, microbial adhesion and electron transfer of the cell. In order to improve the working of MFCs, the costs of the material involved like carbon paper, carbon fibre, carbon nanotubes, etc., needs to be reduced drastically and the power density of material should be maximized. Furthermore, material like carbon nanostructures having a large surface area and less resistance should be used so as to improve the efficiency of MFCs and to obtain an optimum biofilm in the anodic component of the cell.

MFC Cathode

Cathodes are used for the redox process and oxygen gets reduced to produce water. Cathode materials are chosen to have catalytic properties for oxygen reduction. Platinized carbon electrodes are commonly employed as oxygen-reducing cathodes in MFCs. However, high cost and catalyst poisoning prevent the practical applications of Pt-based cathodes. The high-cost Pt is often replaced by other nonprecious electrodes like Mn_2O_3, and Fe_2O_3.

Air Cathodes

The single chamber version uses an air cathode, which allows air from the outside of the chamber to enter and react with the protons traveling from the anode. It has been found that air cathode cells have higher power density then immersed cathodes. A typical air MFC cathode has three layers: a diffusion layer, a conducting support material and a catalyst.

13.7 MFCs and Wastewater Remediation

Organic matter can be treated by this process, including domestic wastewater, brewery effluent and much else. Several plants have shown good results which are under operation. Use of MFCs for wastewater requires a design which permits the wastewater to flow through the cell over the anode surface. Various configurations have been adopted for this purpose, including the tubular MFC where the cathode is placed on the outside of the tube and the anode occupies the full internal space. Wastewater flows from one end to the other through the anode.

13.7.1 Microbial Fuel Cells for Wastewater Treatment and Energy Generation

Microbial fuel cell (MFC) systems employ the bioelectrochemical catalytic activity of microbes to produce electricity from the oxidation of organic, and in some cases inorganic, substrates. Theoretically, almost all the substrates are able to serve as the feedstock that can be metabolized by microorganisms in MFCs, such as simple chemical compounds, urban sewage, agricultural, brewery, food, and dairy farm and various industrial wastewaters. MFC technology represents a unique and novel platform to process waste and wastewater sources that allows for energy and resource recovery along with water sanitation in a single configuration. They have the potential to provide the paradigm shift for wastewater treatment from "environmental protection" to "resource recover". Energy and valuable resource content in the wastewater sources presents a promising potential for their recovery to develop sustainable technologies. Thus, into existing wastewater treatment systems MFCs can be integrated to enhance energy, water and nutrient resource recovery. This chapter provides an overview of wastewater treatment principles in MFCs, in comparison to other energy recovery alternatives from wastewater, substrate utilization in MFCs, process developments and organic, nutrient and metal removal capacities in MFCs.

13.7.2 Treatment of Sewage and Electricity Production by Microbial Fuel Cells

Among the newly developed concepts, microbial fuel cells (MFCs) integrated with constructed wetland appear to be very attractive because of direct electricity generation from organic compounds and treatment of wastewater. Constructed wetland coupled with microbial fuel cells (CW–MFC) is an emerging technology in recent years and has attracted a lot of attention from researchers in the fields of wastewater treatment and bioenergy production. In the fields of wastewater treatment and bioenergy, CW–MFC is a promising technology. However, at current power levels, the advantage of combining the two is mainly because of the enhancement of wastewater treatment in anaerobic zones within the wetland. To increase and utilize electricity output new operational strategies need to be explored.

13.7.3 Advanced MFCs for Wastewater Treatment

Due to the increase in the global population, existing conventional sources are expected to become exhausted within a century or so. Accordingly, the research and discovery of new alternate sources of energy is currently the need of the hour. MFCs are evolving as the most promising and eco-friendly means of energy to derive wealth from waste and have become a promising approach to generate cleaner and more sustainable electrical energy. To commercialize the MFC technology much research is going on and nanomaterials are studied to make this technology successful.

13.8 Wastewater Treatment by MFCs Coupled with Peroxicoagulation Process

In a very recent research study, it has been found that the electricity output increased remarkably when the MFCs were provided with a coagulation process using potassium permanganate. The development of biofilm along the anode has been efficiently increased by using peroxicoagulation techniques [20, 21]. Decolorization up to a remarkable limit of 98% has been achieved by this process. The overall process has been found to be very cost effective and energy efficient.

13.9 MFCs and Generation of Bioelectricity

Extensive scientific studies have found that application of MFCs can be remarkably improved by using a large variety of carbon sources and different microbial populations. Consequently, microbes are capable of biotransforming the wastes into energy and emerge as an environmentally friendly technique as well [22]. Therefore, in recent years, MFCs have emerged as the most potential technology to convert chemical energy into electric energy [8, 9, 23].

13.10 Electricigens in the MFCs

Electricigens are the microbes which are available in the anodic component of MFCs. Researchers have found that the efficiency of MFCs decreased slightly when cultures were prepared from pure colonies. The more astonishing results were shown by the mixed and miscellaneous consortia. It indicates that further research is needed on electricigens for their screening, domestication, modification and optimization in order to improve their electrochemical efficiency [8, 9].

13.11 Future Prospects

MFC technology is the most advanced and potential technology for converting different types of wastes into usable chemical energy [24]. However, further research is needed to study all the aspects of the biological roles and biochemical reactions in order to make this technology available for commercialization. Material usage and surface area parameters are to be studied intensively to improve the quality and electricity generation [25–28]. Recombinant DNA technology can be employed to study the genetic change possibility to derive some more potent microbes for this purpose [28–30]. Large quantities of wastewater can be easily treated and an intern optimum quantity of electricity can be generated once MFC technology is successfully commercialized [31–33].

13.12 Conclusion

MFCs have revolutionized the field of dealing with environmental pollutants. This is the only technique available to convert toxic wastes xenobiotics, chromium, agrowastes, nitrates and azo dyes into usable chemical energy with the final product being electrical energy. In MFCs, the wastes are utilized as substrates for the conversion of electricity or get biotransformed to some less toxic variants. However, the only drawback of MFCs is their lower power output and that is why the technology is not yet commercialized. Much advanced research is to be carried out to make this technology practically applicable and efficient. Being a non-conventional source of energy, this technique has a very promising future once it gets scaled up. Much research is going on in this direction to provide MFCs as an important alternative source of energy in the near future.

References

1 Potter, M.C. (1911). Electrical effects accompanying the decomposition of organic compounds. *Proc. R Soc. Lond. B* 84: 260–276. https://royalsocietypublishing.org/doi/10.1098/rspb.1911.0073.

2 Du, Z., Li, H., and Gu, T. (2007). A state of the art review on microbial fuel cells: A promising technology for wastewater treatment and bioenergy. *Biotechnol. Adv.* 25: 464–482.

3 Lovley, D.R. (2006). Bug juice: Harvesting electricity with microorganisms. *Nat. Rev. Microbiol.* 4: 497–508.

4 Nicholls, D.G. (1982). *Bioenergetics – An Introduction to the Chemiosmotic Theory*. London: Academic Press.

5 Moat, A.G., Foster, J.W., and Spector, M.P. (2002). *Microbial Physiology*, 4e. New York: John Wiley and Sons.

6 Rabaey, K., Boon, N., Siciliano, D., Verhaege, M., and Verstraete, W. (2004). Biofuel cells select for microbial consortia that self-mediate electron transfer. *Appl. Environ. Microbiol.* 70 (9): 5373–5382.

7 Bond, D.R. and Lovley, D.R. (2003). Electricity production by *Geobacter sulfurreducens* attached to electrodes. *Environ. Sci. Technol.* 69 (3): 1548–1555.

8 Reguera, G., Nevin, K.P., Nicoll, J.S., Covalla, S.F., Woodard, T.L., and Lovley, D.R. (2006). Biofilm and nanowire production leads to increased current in *Geobacter sulfurreducens* fuel cells. *Appl. Environ. Microbiol.* 72 (11): 7345–7348.

9 Logan, B.E. and Regan, J.M. (2006). Electricity-producing bacterial communities in microbial fuel cells. *Trends Microbiol.* 14 (12): 512–518.

10 Shukla, A., Suresh, P., Berchmans, S., and Rajendran, A. (2004). Biological fuel cells and their applications. *Curr. Sci.* 87: 455–468.

11 Kim, B.H., Chang, I.S., and Gadd, G.M. (2007). Challenges in microbial fuel cell development and operation. *Appl. Microbiol. Biotechnol.* 76: 485–494.

12 Bond, D.R., Holmes, D.E., Tender, L.M., and Lovley, D.R. (2002). Electrode-reducing microorganisms that harvest energy from marine sediments. *Science* 295: 483–485.

13 Logan, B.E., Hamelers, B., Rozendal, R., Schröder, U., Keller, J., Freguia, S., Aelterman, P., Verstraete, W., and Rabaey, K. (2006). Microbial fuel cells: Methodology and technology. *Environ. Sci. Technol.* 40 (17): 5181–5192.

14 Shantaram, A., Beyenal, H., Veluchamy, R., and Lewandowski, Z. (2005). Wireless sensors powered by microbial fuel cells. *Environ. Sci. Technol.* 39: 5037–5042.

15 Kim, J.D., Kim, D.J., Byun, H.G., Ham, Y.K., Jung, W.S., Han, D.W., Park, J.S., and Lee, H.L. (2007). The definition of basic TEDS of IEEE 1451.4 for sensors for an electronic tongue and the proposal of new template TEDS for electrochemical devices. *Talanta* 71: 1642–1651.

16 Chang, I.S., Moon, H., Jang, J.K., and Kim, B.H. (2005). Improvement of a microbial fuel cell performance as a BOD sensor using respiratory inhibitors. *Biosens. Bioelectron.* 20: 1856–1859.

17 Shizas, I. and Bagley, M. (2004). Experimental Determination of Energy Content of Unknown Organics in Municipal Wastewater Streams. *J. Energy Eng.* August. https://doi.org/10.1061/(ASCE)0733-9402(2004)130:2(45)

18 McAnulty, M., Poosarla, V.G., Kim, K.Y. et al. (2017). Electricity from methane by reversing methanogenesis. *Nat Commun* 8, 15419. https://doi.org/10.1038/ncomms15419.

19 Rahminijad, M. et al. (2015). Microbial fuel cell as new technology for bioelectricity generation: A review. *Alexandria Engineering Journal* 54: 745-756.

20 Logan, B.E. and Regan, J.M. (2006). Electricity-producing bacterial communities in microbial fuel cells. *Trends Microbiol.* 14: 512–518.

21 Davila 1, S., Wright, V.J., Khor, C.C., Sim, K.S., Binder, A., Breunis, W.B., Inwald, D., Nadel, S., Betts, H., Carrol, E.D., Ronald, de G., Hermans, P.W.M., Hazelzet, J., Emonts, M., Lim, C.C., Kuijpers, T.W., Martinon-Torres, F., Salas, A., Zenz, W., Levin, M., and Hibberd, M.L. International Meningococcal Genetics Consortium. (2010). Genome-wide association study identifies variants in the CFH region associated with host susceptibility to meningococcal disease. *Nat. Genet.* Sep 42

22 Tremblay, P.L., Aklujkar, M., Leang, C., Nevin, K.P., and Lovley, D. (2012). A genetic system for Geobacter metallireducens: Role of the flagellin and pilin in the reduction of Fe(III) oxide. *Environ. Microbiol. Environ Microbiol Rep.*, 4(1): 82-88.

23 Lovley, D.R. (2006). Bug juice: Harvesting electricity with microorganisms. *Nat. Rev. Microbiol.* 4: 497–508.

24 Lovley, D.R. and Nevin, K.P. (2011). A shift in the current: New applications and concepts for microbe-electrode electron exchange. *Curr. Opin. Biotechnol.* Jun 22 (3):441–448. doi:10.1016/j.copbio.2011.01.009. Epub 2011 Feb 16.

25 Pandey, B.K., Mishra, V., and Agrawal, S. (2011). Production of bioelectricity during wastewater treatment using a single chamber microbial fuel cell. International Journal of Engineering. *Sci. Technol.* 3 (4): 42–47.

26 Aelterman, P., Rabaey, K., The Pham, H., Boon, N., and Verstraete, W. (2006). Continuous electricity generation at high voltages and currents using stacked microbial fuel cells. *Environ. Sci. Technol.* May 15 40 (10): 3388–3394. 10.1021/es0525511.

27 Tront, J.M., Fortner, J.D., Plötze, M., and Michael, P. (2008). Microbial fuel cell technology for measurement of microbial respiration of lactate as an example of bioremediation amendment. September. *Biotechnol. Lett.* 30 (8): 1385–1390.

28 Heijne, A.T., Hamelers, H.V.M., and Buisman, C. (July 2007). Microbial Fuel Cell Operation with Continuous Biological Ferrous Iron Oxidation of the Catholyte. *Environ. Sci. Technol.* 41 (11): 4130–4134.

29 Cheng, S., Liu, H., and Logan, B.E. (2006b). Power generation in a continuous flow MFC with porous electrodes and reduced electrode spacing. *Environ. Sci. Biotechnology* 110–114.

30 Rabaey, K., Lissens, G., Siciliano, S.D., and Verstraete, W. (2003). A microbial fuel cell capable of converting glucose to electricity at high rate and efficiency. *Biotechnol. Lett.* 25: 1531–1535.

31 Cheng, S., Liu, H., and Logan, B.E. (2006a). Increased performance of single chambered MFCs using an improved cathode structure. *Electro. Chem. Biocommunity* 888–891.

32 Moehlenbrock, M.J. and Minteer, S.D. (2008). Extended lifetime biofuel cells. *Chem. Soc. Rev.* (6).

33 Korneel Rabaey, R.A., Rozendal, P.K., Dutta, Z.Y., and Yuan, Z. (2010). Bioelectrochemical Systems: From Extracellular Electron Transfer to Biotechnological Application. January.

14

Water Resources Planning and Management Paradigm Decision-Making

CHAPTER MENU

14.1 Introduction

There is an imperative international predicament of mounting freshwater paucity. As a direct upshot of unsafe drinking water and lack of sanitation, thousands of children die every year in poor countries of the world (FAO). Without proper management and governance, the problem will amplify in its magnitude under the influence of global population explosion and climate change. In mid latitudes, these catastrophic factors will boost the rate of incidence and relentlessness of droughts. As the volume of water per capita decreases, there will be an unprecedented increase in water conflicts within and between countries. Diversion of water from one catchment to another will emerge as a responsive factor in water-scarce areas, and this is a costly effort which can impose a detrimental impact on the ecology and hydrology of an area. In addition to industrial water consumption, agricultural industry (70% of total freshwater withdrawals) is one of the giant consumers of water resources. The strain on hydrological resources due to irrigational requirements for crop production will impact on food security. In order to judiciously curb water scarcity, *Water Resources Planning and Management* is strategically an efficient tool. In order to deal with any problem, a baseline analysis of the subject is required. The additional part of the predicament is re-treating water quality

Aquatic Environmental Bioengineering: Monitoring and Remediation of Contamination, First Edition.
Rouf Ahmad Bhat, Mohammad Yaseen Mir, Gowhar Hamid Dar and Moonisa Aslam Dervash.

on which billions of people rely. To take cognizance of failures, successes and delineating the benchmarks on the course of water conservation, planning and management of water through various approaches, goals, strategies, decisions and scenarios is mandatory to meet the conditions for *Integrated Water Resources Management*. An understanding of the agility and complexities of the hydrological cycle, globalization, disparity in supply and demand and its interconnectedness with varied ecosystems is quite important. Meticulously, an invariable theme is a need for policy-making to be supported by full-bodied support systems, not only in terms of data and modeling, institutional infrastructure and government policies but also through the involvement of key stakeholders in decision-making (local as well as global), including the common public, in planning processes.

14.2 Freshwater Stress

In the recent past the proportion of amplification in water exploitation on a global scale has surpassed twofold that of population growth. The ever-escalating populace and high per capita water usage inevitably changes the scenarios of water demands and invites water stress and scarcity in all the domains of society. Currently, the world population has reached the 7×10^9 mark and our world will need to supply food, energy and water for an additional 2 to 2.5×10^9 people by 2050 [1]. The human-dominated novel sphere of the Earth's *anthrosphere* is grabbing the spotlight on account of altered water consumption patterns, and many scientists render the present geological period as Anthropocene [1, 2].

14.3 Globalization

Globalization is a stimulating agent for the inception of novel and effective international trade protocols in terms of goods and services. The multinational companies are linked circuitously in water usage and transportation across the borders (virtual water) [1]. Virtual water across international borders is often the basis for critical arguments on account of expenditure, though the natural flow of any water course across the borders is not objectionable. But this international contract is steered by political agreements and negotiations [1].

14.4 Disparity in Supply and Demand

Conventionally, occurrence of any precipitation event is predicted on account of climatic factors, hydrological data, statistical analysis and predictive interpretations. Thereafter, water management and infrastructure like dams, reservoirs and rainwater harvesting strategies [3] are designed which require commission and implementation of planning strategies. But climate change possesses the power to change the entire probability of precipitation events (either a cause of floods or droughts) and water demand for production of energy and crops [1]. Climate and crop-water models are the technological interventions which are meant to remove the disparity between the supply and demand of water. Decision-makers and policy-makers rely upon these climate and crop-water models as far as sustainable agriculture is concerned [4].

14.5 Planning and Management Approaches

Broadly, planning and management possess two universal approaches. One is based on command and control and referred to as the top-down approach. And the other is based on grassroots strategy and known as the bottom-up approach. An integrated plan and management strategy will rely on the coordinated work of both approaches.

14.5.1 Top-Down Approach

Over the last five decades water resource experts have been analyzing strategies to prepare master plans for various river systems across the globe which must be multipurpose, flexible and integrated in manner. The master plan encompasses flowcharts of averments, published data and physico-chemical characteristics of water recourses and their utilization. While drafting a master plan alternative structural and nonstructural management strategies are suggested and assessed. On the account of these assessments, the more appealing plan is taken up for implementation. This method of drafting a master plan doesn't take suggestions from the common public; rather professionals dominate this approach. In the present arena of environmental consciousness, top-down approaches are becoming less acceptable as far as planning and management strategies are concerned.

14.5.2 Bottom-Up Approach

In the recent past, water resources planning and management strategies adopted active participation of all the stakeholders (concerned society, nongovernmental organizations and professionals in government institutions). The scenarios and plans were fabricated from the bottom-up rather than top-down through a process of consent, building for the establishment of adaptive comprehensive water management programs, policies and plans. The plans are primarily developed by professionals and active public participation in decision-making has a high threshold of success.

14.6 Integrated Water Resources Management

The concept of integrated water resources management (IWRM) has been evolved over the past decades. The IWRM is the kernel strategy to deal with growing water demands and scarcity due to overpopulation. Thus, reconstruction and reconsideration of development policies with respect to the management strategies is the neo-strategy in the present world to tackle water scarcity and water issues.

The concept of IWRM recedes from top-down "water master planning" that usually focuses on water accessibility, and preferably approaches the bottom-up "comprehensive water policy planning" that tackle the interface between diverse subsectors, seeks to ascertain precedence, considers institutional necessities and deals with the edifice of management facility.

14.7 Water Management and Planning: Goals, Strategies, Decisions, and Scenarios

In water management and planning, there is a defined set of protocols to be followed for project accomplishment. A goal in water management and planning defines what and how a target is to be accomplished. The goal most probably is to facilitate "user-oriented demand" management procedures in comparison to large-scale water delivery infrastructure improvement.

A strategy is defined as a coherent amalgamation of discrete measures or pronouncements [5] that accomplishes the intended objectives and persuades the limitations forced on the WRS. The erection of a reservoir and the broadening of the passage downstream and the amplification of the irrigation system is focused to trim down the menace to the agriculture sector in a drought-prone area. This is one strategy. Another strategy could be execution of that cropping orientation which consumes less water.

A decision is the execution of a specific approach or itinerary of accomplishments. A decision relies upon:

- Technical (structural) measures: alteration in canals, pumping stations, reservoirs and fish ladders. Technical measures often incorporate decision-making procedures in order to better utilize the infrastructure.
- Ecological (nonstructural) measures: better development in the performance of ecological activities, i.e; fish fry in spawning areas, or large herbivores.
- Economical measures: meant to persuade water consumers to modify their water consumption by altering the value of the resource use (through excise or subvention).
- Regulatory measures: Oriented to amend the water consumption through legal measures like permits, no pollution zone declaration, etc.
- Institutional measures: Meant to identify the specific roles to governmental agencies to meet requirements of the WRS, and promotion of the obligatory connections involving the civil and private sectors.

Finally, the scenario is depicted by the external environmental factors of an aquatic system under contemplation. Examples of scenario include environmental factors like climate, demographical trends, production functions (including crop water requirements) and precipitation including cost–benefit analysis.

14.8 Systems Approaches to Water Resource System Planning and Decision-Making

The systems approach to planning frequently accentuates the arithmetical practices. The use of statistical tools, nonetheless, is merely component of what composes a systems approach. The approach applied to multifaceted systems of many inter-reliant components involves [8]:

Production of prognostic models to elucidate system approach, creation of strategies that correlate interpretations with models outputs and informed findings, contrasting the optional strategies accessible to decision-makers, corresponding the outcomes to the decision-makers in consequential approach, proposing and producing decisions based on the information provided during these interactions between forecasters, planners, decision-makers and stakeholders, and examining and appraising the outcomes of the strategies applied.

System diagrams or conceptual models recognizing system components and their associations are significant tools in systems analysis. A system diagram represents cause–effect relations among the framework of the general system. Two problems may arise. First, the amount required may be bigger than the supply; second, they negatively influence the natural system (e.g., cause contamination or change the water level). The insight of these tribulations can stimulate investigation and planning actions, urging at management.

14.9 Analysis and Implementation Framework

A water resource planning study usually comprises five broad stages, as exemplified in Figure 14.1. The depiction of these phases including activities and the interactions among the activities is known as the analysis framework. A consistent set of models is

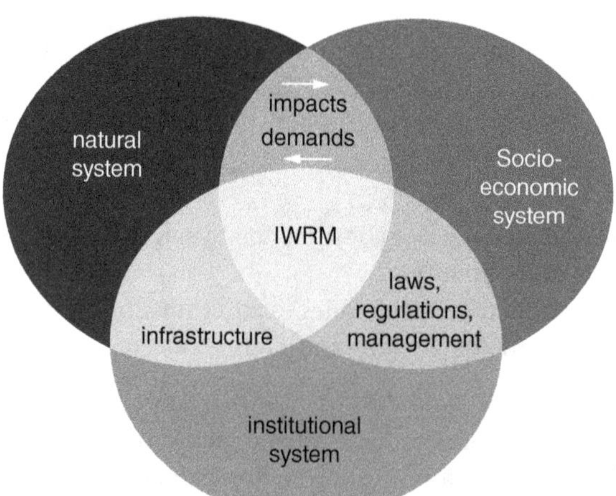

Figure 14.1 Interactions among the natural, institutional, and socioeconomic systems with respect to IWRM.

characteristically employed for the quantitative analyses aimed at recognizing and assessing substitute advantageous procedures and approaches.

A decision progression is not an uncomplicated linear chain of steps but engrosses responses to former steps. Component of the process is thus iterative. Feedback loops are desirable when:

- solutions do not convene existing conditions;
- novel insights modify the discernment of the predicament and its solutions;
- necessary system modules and associations are ignored;
- alteration in the goals or the scope of discourse.

Statement and interface with the decision-makers are indispensable right through a developmental venture and the execution of the preferred progress. To take no notice of this augments the hazard of producing tactics and guidelines which are no more significant to the customer. Customary coverage (commencement and intervening reports, etc.) helps in triumphant statement, but an unremitting dialogue is imperative throughout all stages of the investigation.

14.10 Decision-Making

Decision-makers and stakeholders must be implicated in each of the five stages of this scaffold. Stakeholder participation fetches both information and penchants to the planning progression – a procedure that characteristically will necessitate finding appropriate negotiations among all decision-makers and stakeholders if an accord is to be attained [6].

The scaffold engrosses a sequence of pronouncements at the conclusion of each phase. The "divergence–convergence" practice for concerning stakeholders in decision-making on the five investigation stages is demonstrated in the rhombus approach (Figure 14.2) [6].

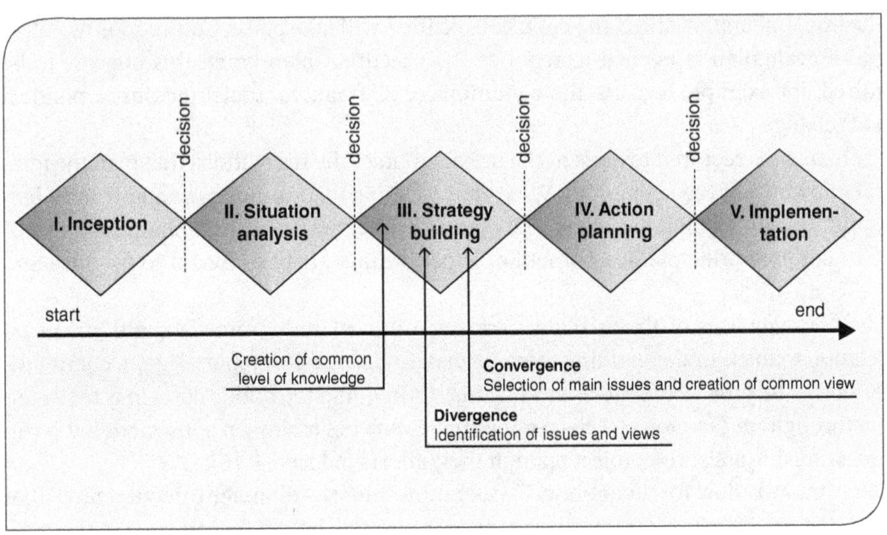

Figure 14.2 Divergence –convergence process in decision-making.

The first commencement stage of the route categorizes the question of the analysis (what is to be analyzed and under what circumstances), the objectives (the preferred outcomes of the analysis), and constraints (its restrictions). Based on this investigation, during which exhaustive communiqué with the decision-makers is important, an accord on the approach for the remainder of the investigation needs to be accomplished. The consequences of the inception stage can be accessible in an inception report, which comprises the work plan for the further phases of the analysis. In the situational analysis stage, the tools for the investigation of the water resource system are preferred or developed. Chief actions in this stage characteristically contain data compilation and modeling. The models will be exercised to enumerate the present and upcoming tribulations in the system. Scenarios will be devised to illustrate the future boundary circumstances for the system. Recognizing and screening optional judgements can happen at this stage. If feasible, no regret procedures will be recognized for instantaneous execution. A steady enhancement of the perceptive of diverse characteristics of the WRS is frequently obtained as the study advances from inadequate data sets and uncomplicated tools to more comprehensive data and models. Interface with the decision-makers will be significantly improved if their trustable stakeholders or advisors are implicated as a component of the investigation squad. More official communication can be planned throughout presentations of outcomes in conventions and intervening development reports.

In the strategy building stage, substitute approaches will be devised and discussed with the decision-makers/stakeholders. This will comprise adaptive management elements to ensure that the preferred strategy is adequately strong and flexible in case the prospect develops differently than expected. In the action planning stage, the preferred strategy will be arranged for execution. An implementation plan will be developed that portrays what will be done, by whom and how it will be funded, etc. This phase often requires supplementary labor on components of the strategy (such as feasibility and design studies), and environmental impact assessments (EIA). Endorsement of the chosen strategy is desirable to "sell" the planned actions to the community. Institutional arrangements will have to be prepared to make sure a smooth accomplishment. Finally, in the implementation stage, the definite execution will take place. Continuous monitoring and evaluation is needed to regulate the execution plan when this appears to be required, for example because the conditions (e.g., finances, social pressures, political mood) change.

Each phase is required to present the information to the institutions that plan the temporal knowhow of the activity [6]. What those leading institutions must know to be better, prior to making their decisions, will of course differ among diverse planning projects. The underlying principle of information is performing analyses meant to produce and correspond it.

The consequences of the analyses performed in a planning project should be of no revelation to those understanding them in an definitive project report. Again, communiqué among the project and the requesting institutions and the public domain is indispensable throughout the project. This communiqué may not assure an agreement but it can unquestionably assist the project team in their efforts to locate it [6].

The Organization for Economic Cooperation and Development documented that integrated water resources management as espoused by many countries cannot be appropriately implemented without considering a broader governance structure [7]. This would include not only sustainable water policies [6], but would also measure governing scientific, educational and technological issues, as well as communication and participation [1].

References

1 Cosgrove, W.J. and Loucks, D.P. (2015). Water management: Current and future challenges and research directions. *Water Resour. Res.* 51: 4823–4839. doi:10.1002/2014WR016869.

2 Williams, M., Zalasiewicz, J., Haywood, A., and Ellis, M. (2011). The Anthropocene: A new epoch of geological time, Philos. *Trans. R. Soc. A* 369: 835–1111.

3 Unganai, L. (2009). Adaptation to climate change among agropastoral systems: Case for Zimbabwe. *IOP Conf. Ser. Earth Environ. Sci.* 6: 412045. doi: 10.1088/1755-1307/6/41/412045.

4 IPCC. (2021). Summary for Policymakers. In: *Climate Change 2021: The Physical Science Basis. Contribution of Working Group I to the Sixth Assessment Report of the Intergovernmental Panel on Climate Change* (ed. V. Masson-Delmotte, P. Zhai, A. Pirani, S.L. Connors, C. Péan, S. Berger, N. Caud, Y. Chen, L. Goldfarb, M.I. Gomis, M. Huang, K. Leitzell, E. Lonnoy, J.B.R. Matthews, T.K. Maycock, T. Waterfield, O. Yelekçi, R. Yu, and B. Zhou). Cambridge University Press. In Press. Switzerland: IPCC.

5 King, J.M., Brown, C.A., and Sabet, H. (2003). A scenario-based holistic approach to environmental water flow assessments for rivers. *Rivers Res. Appl.* 19 (5–6): 619–639.

6 Loucks, D.P. and Beek, E. (2017). Water Resource Project Planning: Putting It All Together. In: *Systems Planning and Management- an Introduction to Methods, Models, and Applications*, (eds. D.P. Loucks and E. Beek), (ISBN 978-3-319-44232-7). 10.1007/978-3-319-44234-1.

7 Organisation for Economic Co-operation and Development (OECD). (2011). Water Governance in OECD Countries: A Multi-level Approach, OECD Studies on Water. 244. doi:10.1787/9789264119284-en.

8 Haasnoot, M., Van Deursen, W.P.A., Guillaume, J.H.A., Kwakkel, J.H., Van Beek, E., and Middelkoop, H. (2014). Fit for purpose? Building and evaluating a fast, integrated model for exploring water policy pathways. *Environ. Model. Softw.* 60: 99–120. ISSN 1364-8152.

Index

Aquatic Environmental Bioengineering: Monitoring and Remediation of Contamination, First Edition.
Rouf Ahmad Bhat, Mohammad Yaseen Mir, Gowhar Hamid Dar and Moonisa Aslam Dervash.

www.ingramcontent.com/pod-product-compliance
Lightning Source LLC
LaVergne TN
LVHW062209200926
848484LV00006B/1359